Hybrid Polymeric Nanocomposites from Agricultural Waste

Hybrid Polymeric Nanocomposites from Agricultural Waste examines the use of agricultural by-products for green production of new materials. It covers nanoparticle synthesis from agricultural wastes and nanocomposite development with a focus on polyethylene, polylactic acid, polymethylmethacrylate, and epoxy resins, and considers possible biomedical and engineering applications.

- Showcases agricultural waste as polymer reinforcements to replace expensive synthetic fibres that discourage wide polymeric nanocomposite applications
- Discusses green synthesis and characterisation of hybrid nanocomposites from polylactic acid, polymethylmethacrylate, recycled/new polyethylene, and epoxy resins
- Contrasts hybrid nanocomposites properties with standard nanocomposites, using automotive case studies

The book is aimed at researchers, advanced students, and industrial professionals in materials, polymer, and mechanical engineering and related areas interested in the development and application of sustainable materials.

Emerging Materials and Technologies

Series Editor: Boris I. Kharissov

Hybrid Polymeric Nanocomposites from Agricultural Waste

Edited by
Sefiu Adekunle Bello

CRC Press
Taylor & Francis Group
Boca Raton London New York

CRC Press is an imprint of the
Taylor & Francis Group, an **informa** business

First edition published 2023
by CRC Press
6000 Broken Sound Parkway NW, Suite 300, Boca Raton, FL 33487-2742

and by CRC Press
4 Park Square, Milton Park, Abingdon, Oxon, OX14 4RN

CRC Press is an imprint of Taylor & Francis Group, LLC

ISBN: 978-0-367-77269-7 (hbk)
ISBN: 978-0-367-77270-3 (pbk)
ISBN: 978-1-003-17054-9 (ebk)

DOI: 10.1201/9781003170549

Typeset in Times
by MPS Limited, Dehradun

Contents

Section I   Fundamentals

Section II Agricultural Waste–Based Polyethylene-Based Nanocomposites

Section III Agricultural and Domestic Waste–Based Nanocomposites of Other Polymers

Section IV Industrial Applications

Preface

Why use nanoparticles from agricultural wastes? Annual agriculture productions to meet food demands of increasing human populations are enormous. Once edible parts of the harvested crops are extracted, remnants (waste) are little or not tapped; meanwhile, they are potential sources of nanoparticles for green composite productions for fabricating artefacts to meet society's needs. This handbook features agricultural by-products for green production of new materials. It covers nanoparticle synthesis from agricultural wastes and nanocomposite development with focuses on polyethylene, polylactic acid, polymethylmethacrylate, and epoxy resins. This handbook establishes properties of a series of hybrid nanocomposites through rigorous experimentation and possible applications as emerging materials for biomedical and engineering applications. It is well explained, serving as a source of knowledge for undergraduate/graduate students at all levels and an important reference book for both academic and industrial professionals.

It aims to establish particulate technology with focuses on polymer reinforcement using nanoparticles sourced from agricultural productions. It comprises a review and proof of concepts achieved through rigorous experimental investigations establishing polylactic acids, polymethylmethacrylate, polyethylene, and epoxy resin nanocomposites as emerging materials for both engineering and biomedical applications. It covers needs for materials that are lightweight in transportation systems, potentials of agricultural products as sources of nanoparticles and fibres, properties of *Parquetina nigrescens* and *date seed* nanoparticles for biomedical applications, green synthesis, and characterisation of hybrid nanocomposites from polylactic acid, polymethylmethacrylate, recycled/new polyethylene and epoxy resins, and comparison of hybrid nanocomposite properties with a standard using a Toyota dashboard model as a case study.

Acknowledgement

On behalf of CRC, Taylor & Francis, the editor appreciates the support of all the contributors for making this project a success. Support of Dr Akeem Akinwekomi, Federal University of Technology, Akure; Dr Tunji Owoseni, Dr Kamoru Olufemi Oladosu, Dr Morakinyo Wasiu Kareem, Dr Taofeek Amoloye, Kwara State University, Malete and Dr Ibrahim Dauda Muhammed, University of Abuja are acknowledged for sparing their time and expertise in reviewing some chapters of this handbook. Efforts of Gabrielle Vernachio; Nick Mould; and Aimee Wragg, editorial assistant; and Allison Shatkin, senior publisher, materials, chemical, and petroleum engineering are acknowledged towards success of this project. Furthermore, special thanks are due to Professor Abdul Ganiyu Funsho Alabi, the Dean Faculty of Engineering and Technology and Professor Abdul Kabir Hussain Solihu, the Chairman of Library and Publication Committee, Kwara State University for their supports and Professor Boris I. Kharissov, series editor, for finding me worthy to be part of the Emerging Materials and Technology team. Finally, I appreciate CRC, Taylor & Francis for publishing this handbook.

Editor Biography

Dr Sefiu Adekunle Bello was born to late Mr Sunmola Akanji and Mrs Anifat Bello of Jigan Compound, Idioro, Elekuro, Ibadan, Oyo State, Nigeria. He attended Public Day School, Elekuro, Ibadan, and completed his senior school in 1992 at Methodist Secondary Grammar School, Elekuro, Ibadan. He graduated from the Department of Metallurgical and Materials Engineering, Federal University of Technology, Akure, in 2007. He completed his M. Sc. and Ph.D. studies in 2014 and 2017, respectively from the University of Lagos, Nigeria. He worked as a science teacher at St Isabel Comprehensive Model College, Benjamin, Eleyeile, Ibadan, Oyo State, Nigeria, from 2010 to 2012. From his undergraduate days till 2010, he engaged in home teachings and joined many tutorial centres in Ibadan for preparing youths for a Senior Secondary Certificate and Joint Admission Matriculation Board Examinations. In 2011, he set up G TEK Centre, where he trains youths on the use of Design Software (Auto CAD) and I SWEAR Classic Tutors as a youth training centre for advanced examinations. He joined the Department of Materials Science and Engineering, Kwara State University, Malete in 2014 as a member of the academic staff engaging in teaching, research, and community development. He served the Department of Materials Science and Engineering as an examination officer from 2017 to 2020 and HND Conversion/Top Up programme coordinator from 2018 to 2020. He served the Kwara State University as a member of the admission committee from 2018 to 2020. At the time of the writing of this biography, he is the project and postgraduate coordinator of the Materials Science and Engineering Department, associate editor of *Technoscience Journal for Community Development in Africa* published by Kwara State University Press, and deputy director of the Institute of Professional and Continuous Education, Kwara State University.

Dr Sefiu is an erudite scholar. His academic performances have demonstrated excellence over years. He completed his Ph.D. study in record time and has been leading authors in many projects with successful completion. He authored many publications including conference proceedings, journal articles, and book chapters. He serves as reviewers for many journal outlets including *Materials Letters, Tribology in Industry, The Journals of the Minerals, Metals & Materials Society (JOM), International of Journal of Engineering Research in Africa (JERA), Ceramica, World Journal of Civil Engineering and Construction Technology, Songklanakarin Journal of Science and Technology, Composite Communication, Multidiscipline Modelling in Materials and Structures,* and *Engineering Journal.* He served as the chair of the parallel session III at ICNMA 2019: XIII. International Conference on Novel Materials and Applications that took place at the Hilton Sharja Hotel Dubai, 19–20 December 2019.

Scholarly, Dr Sefiu received the Certificate of Honour from the Mathematics Association of Nigeria (MAN), Oyo State Branch for preparing a candidate to represent Oyo State at the 2012 National Mathematical Olympiad in Abuja, Nigeria. He won a Teaching Assistantship Award at the University of Lagos, Akoka Yaba Lagos, Nigeria in 2013; UNILAG Muslim Alumni (UMA) Scholarship Award, 1

Obafemi Awolowo Way, Ikeja in 2015; Pan African Material Institute Sponsorship for attending and presenting "Quasicrystal Al (1xxx)/Carbonized Coconut Shell Nanoparticles via Ball Milling: A Novel Synthesis and Characterization" at the eighth International Conference of the Africa Materials Research Society, Accra Ghana in 2015; second Best Researcher Award at the tenth UNILAG Research Conference and Fair, University of Lagos Nigeria in 2015; TETFUND Institution Based Research (Grant number: KWASUNGR/CSP/251116/VOL3/TETF/0037) in 2015; first Best Researcher Award at the 12th UNILAG Research Conference and Fair, University of Lagos Nigeria in 2017; TETFUND Institution Based Research (Grant number: KWASUIBR/CSP/250918/VOL5/TETF/0056) in 2018 to study structural modification of polyethylene using *Delonix regia* and eggshell particles: a composite development approach for automobile applications; TETFUND Institution Based Research (Grant number: KWASUIBR/CSP/090919/VOL6/TETF/0069) in 2018 to Design, Fabricate and Test Wear Tester for Laboratory Use, Nigeria Academy of Engineering-ARCO Petrochemical Company Postdoctoral Travel Grant, Lagos Nigeria in 2019 to research on polymeric hybrid composite for engineering applications and travelling to one university for a research collaboration; Best Paper Award at ICNMA 2019: XIII. International Conference on Novel Materials and Applications that took place at the Hilton Sharja Hotel Dubai, 19–20 December 2019; Islamic Development Bank (ISDB) Postdoctoral Award in 2020 and Islamic Development Bank (ISDB) Sponsorship to Attend 29th Engineering Assembly of the Council for Regulation of Engineering Practice in Nigeria (COREN), held 9–11 August 2021 at International Conference Centre, Abuja.

Dr Sefiu's research has been focusing on value additions to agricultural wastes. Success from his previous works motivated him to embark on this project. This handbook disseminates state-of-the-art wealth creations from agricultural waste and discusses properties of some agricultural wastes to ascertain their potentials as reinforcements in epoxy resin, polyethylene, polylactic acid, and polymethylmethacrylate to emerge new advanced nanocomposites for engineering and biomedical applications.

Contributors

Michael Moses Aba
Institute of Energy and Environment
University of Sao Paulo
Sao Paulo, Brazil

Lanre Ademola Adams
Department of Materials
  Science and Engineering
Kwara State University
Malete, Kwara State, Nigeria

Mohammed Kayode Adebayo
Department of Materials Science
  and Engineering
Kwara State University
Malete, Kwara State, Nigeria

Jeleel Adekunle Adebisi
Department of Metallurgical and
  Materials Engineering
University of Ilorin
Kwara State, Nigeria

Raphael Gboyega Adeyemo
Gateway (ICT) Polytechnic
Saapade, Ogun State, Nigeria

Timothy Adewale Adeyi
Department of Mechanical
  Engineering
Lead City University
Ibadan, Oyo State, Nigeria

Oluwole Daniel Adigun
Department of Materials and
  Metallurgical Engineering
Federal University Oye-Ekiti
Ekiti State, Nigeria

Ademola Agbeleye
Department of Metallurgical and
  Materials Engineering
University of Lagos
Lagos State, Nigeria

Johnson Olumuyiwa Agunsoye
Department of Metallurgical and
  Materials Engineering
University of Lagos
Lagos State, Nigeria

Winfred Emoshiogwe Aigbona
Department of Materials Science and
  Engineering
Kwara State University
Malete, Kwara State, Nigeria

Eniola Apena
Department of Metallurgical and
  Materials Engineering
University of Lagos
Lagos State, Nigeria

Emmanuel Kwesi Artur
Department of Materials Engineering
Kwame Nkrumah University of Science
  and Technology
Kumasi, Ghana

Kemi Audu
Department of Materials Science
  and Engineering
Kwara State University
Malete, Kwara State, Nigeria

Sunday Wilson Balogun
Department of Materials Science
  and Engineering
Kwara State University
Malete, Kwara State, Nigeria

Sefiu Adekunle Bello
Department of Materials Science
  and Engineering
Kwara State University
Malete, Kwara State, Nigeria

Babatunde Bolasodun
Department of Metallurgical and
    Materials Engineering
University of Lagos
Lagos State, Nigeria

Stephen Idowu Durowaye
Department of Metallurgical and
    Materials Engineering
University of Lagos
Lagos State, Nigeria

Luqman Babatunde Eleburuike
Department of Materials Science and
    Engineering
Kwara State University
Malete, Kwara State, Nigeria

Soliu Oladejo Abdul Ganiyu
Department of Civil and
    Environmental Engineering
University of Alberta
Edmonton AB, Canada

Suleiman Bolaji Hassan
Department of Metallurgical and
    Materials Engineering
University of Lagos
Lagos State, Nigeria

Farasat Iqbal
Interdisciplinary Research Centre in
    Biomedical Materials
COMSATS University
Islamabad, Lahore, Campus, Pakistan

Sikiru Oluwarotimi Ismail
Centre for Engineering Research
    Department of Engineering,
    School of Physics,
    Engineering and Computer Science
University of Hertfordshire, AL10 9AB
England, United Kingdom

Damilola Bukola Kolawole
Department of Materials Science and
    Engineering
Kwara State University
Malete, Kwara State, Nigeria

Ibiwumi Damaris Kolawole
Department of Food Science
University of Sao Paulo
Sao Paulo, Brazil

Funsho Olaitan Kolawole
Department of Metallurgical and
    Materials Engineering
University of Sao Paulo
Sao Paulo, Brazil

Shola Kolade Kolawole
National Agency for Science and
    Engineering Infrastructure
Abuja, Nigeria

Maruf Yinka Kolawole
Department of Mechanical Engineering
Kwara State University
Malete, Kwara State, Nigeria

Collieus Lebudi
Department of Chemical, Materials and
    Metallurgical Engineering
Botswana International University of
    Science and Technology
Palapye, Botswana

Chioma Ifeyinwa Madueke
Department of Materials and
    Metallurgical Engineering
Federal University Oye-Ekiti
Ekiti State, Nigeria

Henry Ekene Mgbemere
Department of Metallurgical and
    Materials Engineering
University of Lagos
Lagos State, Nigeria

Babatunde Abiodun Obadele
Department of Chemical, Materials
  and Metallurgical Engineering
Botswana International
  University of Science and
  Technology
Palapye, Botswana

Eugenia Obiageli Obidiegwu
Department of Metallurgical and
  Materials Engineering
University of Lagos
Lagos State, Nigeria

Enoch Nifise Ogunmuyiwa
Department of Chemical, Materials
  and Metallurgical Engineering
Botswana International
  University of Science
  and Technology
Palapye, Botswana

Saheed Olalekan Ojo
Bernal Institute, School of
  Engineering
University of Limerick, V94 T9PX
Castletroy, Ireland

Kehinde Adekunle Okunola
Department of Materials Science
  and Engineering
Kwara State University
Malete, Kwara State, Nigeria

Oluseyi Philip Oladijo
Department of Chemical, Materials and
  Metallurgical Engineering
Botswana International University of
  Science and Technology
Palapye, Botswana

Kazeem Koledoye Olatoye
Department of Food Science and
  Technology
Kwara State University
Malete, Kwara State, Nigeria

Boluwatife Olukunle
Department of Materials Science and
  Engineering
Kwara State University, Malete
Kwara State, Nigeria

Felix Adebayo Owa
Department of Materials and
  Metallurgical Engineering
Federal University Oye-Ekiti,
Ekiti State, Nigeria

Patience Bello Shamaki
Department of Chemical Engineering
University of Sao Paulo,
Sao Paulo, Brazil

Okhiria Dickson Udebhulu
Department of Mining and Petroleum
  Engineering
University of Sao Paulo
Sao Paulo, Brazil

Section I

Fundamentals

1 Emerging Materials in Polymer Reinforcement

Sefiu Adekunle Bello,
Luqman Babatunde Eleburuike,
Lanre Ademola Adams, and
Damilola Bukola Kolawole
Department of Materials Science and Engineering, Kwara
State University, Malete, Nigeria

Maruf Yinka Kolawole
Department of Mechanical Engineering, Kwara State
University, Malete, Nigeria

Emmanuel Kwesi Artur
Department of Materials Engineering, Kwame Nkrumah
University of Science and Technology, Kumasi, Ghana

Farasat Iqbal
Interdisciplinary Research Centre in Biomedical Materials,
COMSATS University, Islamabad, Lahore, Campus

CONTENTS

DOI: 10.1201/9781003170549-2

1.1   INTRODUCTION

Materials play a key role in the technological advancement of the world because of their wide range of engineering applications. This therefore has a direct impact on human development and progress. Every day, scientists do research to aid in the development of new and novel materials, as improved materials are beneficial to the world's economy, industrial growth, and domestic uses. There are four groups of materials, which are polymers, metals, ceramics, and composite materials with each of these materials having their own unique properties that are prerequisites to their applications. Due to the astounding rate of the development of engineering industries and needs for engineering materials for critical technological applications, there is a need to find ways of improving the physical, chemical, mechanical, and other properties of materials [1]. One of these materials with such improved properties is composite material. Composite materials are materials that have been modified by combining two or more insoluble materials to form a material with unique and preferred properties. A composite system is made up of two phases: a continuous phase that contains much of the solution, known as the matrix phase, and a discontinuous phase that contains the minority of the solution, known as the reinforcing phase.

Polymers and composite materials have a wide range of engineering applications; they are practically employed in every aspect of modern engineering production. These include the production of sporting goods, biomedical machinery and devices, protective clothing such as bullet-proof vests, and automotive and aerospace applications [1,2].

Polymers are usually reinforced with different materials to improve their properties to suit certain engineering applications. Some common forms of reinforcement in polymer composites are fibre, particles, fillers, and flakes. Moreover, this chapter focuses on overviews on composites containing more than one reinforcement or filler to discuss possible solutions to problems facing polymeric composite containing one reinforcement such as high strength-low toughness behaviour.

1.2   FIBRE-REINFORCED POLYMERS

Fibre-reinforced polymers are composite materials that contain polymer matrix and fibre reinforcement. Examples of the polymer matrix commonly used in the composite system include the epoxies, phenol resins, polyester plastics, and polyamides. Also, glass, carbon and aramid fibres are the most common examples of fibre reinforcements. Fibre-reinforced composites have become increasingly important in recent years due to their significant improvement in materials properties [3]. Their application ranges are broad, and they are utilised in almost every modern engineering industry. They are used in the fabrication of spacecrafts, construction of buildings and bridges, boats, vehicles, and other things. Scientists have developed innovative methods for reinforcing polymers with fibres such as cylindrical carbon nanotubes and nanoparticles with sizes ranging from 1 to 100 nanometers over the years [3]. Fibre reinforcement in polymer materials helps in improving the tensile strength, wear properties, thermal stability, and resistance to corrosion of materials [3].

1.3 PARTICLE-REINFORCED POLYMERS

Particle-reinforced polymer composites are made up of a polymer matrix with reinforcing particles such as aluminium, glass, chucky balls, and nanotubes (tiny particles less than 0.25 microns). When particles are added to a material system, the properties they possess include high tensile strength, high strength to weight ratio, oxidation resistance, and high operating temperature. They are used in the fabrication of helmets, device housings and casings, toys, and some other items that require a high level of wear resistance. It has been reported that when the carbon fibre content in the material system increases, there is an improvement in the Young's modulus of the composite material [4]. For instance, an increase in Young's modulus was disclosed on polypropylene reinforced with carbon fibre with an increase in the carbon contents [5]. It has also been observed by researchers that hybridising Kevlar fibres with glass or carbon fibres in the production of a composite improves the thermal resistance, impact, and tensile strength of the material produced [6].

1.4 FILLER-REINFORCED POLYMERS

Filler-reinforced polymers are composite materials having polymers as their matrix and filler as the reinforcement. Common filler examples include carbon particles, calcium carbonate, talc, asbestos, cellulosic silica, glass, and silicates. When fillers are added to a material system, the properties they possess include high abrasion resistance, improved rate of creep, increased strength, impact energy, and stiffness. The properties that are improved in a filler-reinforced polymer are highly dependent on the type of filler added to the composite material system. They are an important material in the production of automobile and aircraft parts, in construction of buildings and bridges, production of sporting goods, artificial bones, computer accessories, and high-impact shoe sole and vehicle tyres (Figure 1.1).

1.5 LIGHTWEIGHT MATERIALS, PROPERTIES, AND APPLICATIONS

Lightweight materials have become an important part of modern technology evolvements. When looking for ways to innovate a device, one of the most important things to be considered is the weight reduction of that device. Lightweight materials are easier to handle, more flexible, have less power consumption, and are more efficient [7]. One of the primary reasons why the automobile and automotive companies are actively researching ways to reduce the weight of their products is to reduce fuel consumption and improve vehicle fuel usage efficiency [8]. Examples of lightweight materials are aluminium alloys, polymeric composites, and magnesium.

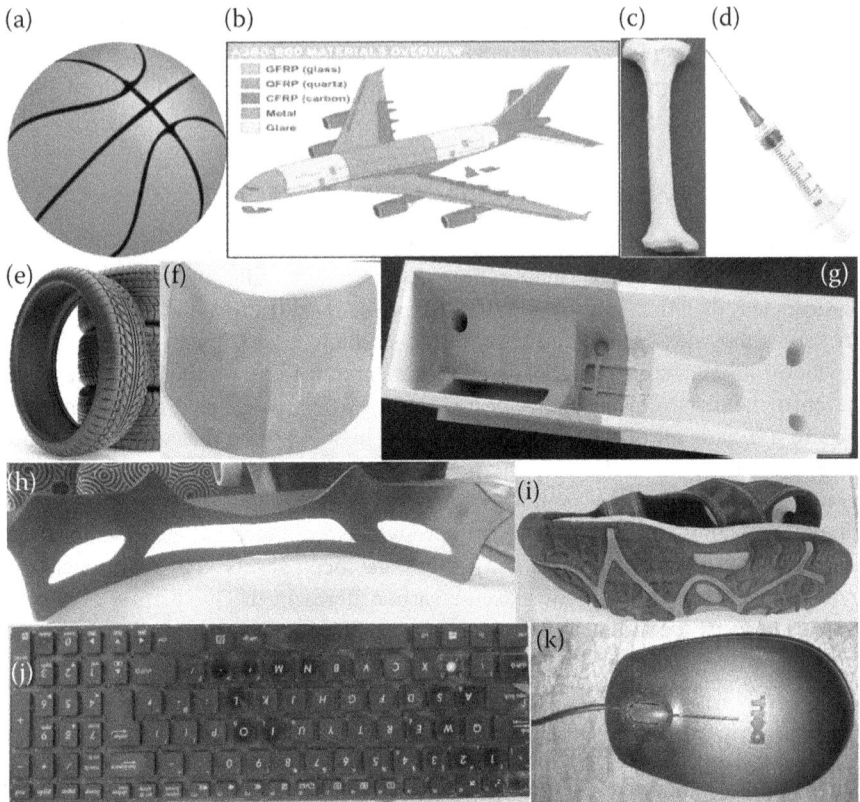

FIGURE 1.1 Industrial applications of emerging polymeric composites (a) sporting material, (b) aerospace materials, (c) artificial implant, (d) medical device, (e) vehicle tyre, (f) particle-reinforced polylactic acid vehicle bonnet, (g) particle-reinforced polylactic acid vehicle dashboard cover, (h) eggshell nanoparticle-reinforced recycled low density polyethylene bumper cover prototype, (i) high-impact carbon particle reinforced polymer shoe sole (j-k) computer accessories.

They are being applied in the production of gadgets, cars, airplanes, sporting goods, and all other related technologies (Figure 1.1).

Aluminium is a very important metal in the industrial world. Its application range is very wide, and it is a material that is found in our immediate environment, our home appliances, food cans and foils, buildings, bridge structures, automotives, airplanes, and offshore. A large variety of aluminium alloys exhibits outstanding corrosion resistance in many atmospheric and chemical conditions due to a naturally occurring tenacious surface oxide coating (Al_2O_3). In architectural and transportation applications, its corrosion and oxidation resistance are very essential. Aluminium is a superior electrical conductor than copper when weight and cost are equal. It is also applied in the production of kitchen appliances and radiators because of its strong thermal conductivity. It is used in manufacturing of hand tools, in transportation industries, especially in the aviation industries because of properties

TABLE 1.1

Properties of Lightweight Materials

Materials (Alloys)	Density (g/cm^3)	Tensile Strength (MPa)	Elastic Modulus (GPa)	Elongation (%)
Aluminium (Al)	2.64–2.81	90–690	70	3–35
Magnesium (Mg)	1.75–1.85	260	45	1.0–30.5
Beryllium (Be)	1.85	370	287	3
Titanium (Ti)	4.50	1170	105–120	54
Titanium aluminide (TiAl)	4.00	448–827	145–172	0.2–2.1
Polymer materials	0.83–2.15	40–150	1.08–4.83	1–600
Polymeric Composite Materials				
Cellulose fibres	0.7–1.5	20–1627	3.2–33.8	1.0–20.0
Glass fibres	2.5	2000–4570	70–90	2.3–3.0

like low density that it possesses. Plastic deformation technique formed wrought aluminium alloys have good properties of strength and ductility. One of the main reasons why aluminium alloys are more used in the industries is because of their nontoxic nature, ease of production, machining, and recycling [9].

Carbon fibre–reinforced polymers are composite materials with remarkable properties (Table 1.1) that are widely used in industries [10]. However, due to the carbonisation method used in manufacturing of carbon fibres, production costs of carbon fibre–reinforced polymers are extremely high. Their impacts in car and aircraft production cannot be overemphasised because of their lightweight leading to reductions in overall weight of the car or aircraft with more than 60% reduction of its initial weight of the transportation equipment made from heavy alloys/materials. This has therefore necessitated scientists to actively research on ways to reduce its production cost or research on agricultural wastes that can replace the carbon fibres either partially to develop hybrid polymeric nanocomposites or totally with the aim of developing ecofriendly advanced nanocomposites to replace the carbon fibre–reinforced polymer [10].

Magnesium alloys are materials with great properties (Table 1.1). They are light and have the lowest density of all alloys [11]. However, the industrial usage of magnesium is quite low because of limitations like high-cost production, they are not easily handled and formed at low temperature, poor ductility, and most of the other alloys are devoid of those limitations. Furthermore, the use of magnesium in multi-material systems raises difficulties of joining, corrosion, repairs, and recycling. Researchers [10] are actively working on ways to overcome these barriers and improve the efficiency of magnesium alloys. For example, Pacific Northwest National Laboratory researchers have found a new process of extruding magnesium alloys as well as a new method that can accurately predict their analysis of experimental measured attributes [10]. More so, some researchers at Pacific Northwest National Laboratory, Ford Engineering laboratory and the University of Michigan have found a modelling device made of microstructures that can be used to forecast the ductility in diecasting of magnesium, and they are still researching on ways to improve the model by validating the model and adding new features to it [10]. Other researchers from Oak Ridge National Laboratory (ORNL) are also researching ways to understand and determine the corrosion process of magnesium alloys [10].

1.6 POLYMERIC NANOCOMPOSITE

The size and structure of nanoscale materials vary from 1 to 100 nanomaterials, with diameters directly proportional to the type of nanomaterial. The properties of nanocomposite materials are determined by the size of nanomaterials, particularly nanoparticles, and the surface of the material gains momentum as the nanoparticles become smaller [12]. There are different classifications of nanomaterials that include nanoparticles, nanofibres, and nanoplates. Nanoparticles have three dimensions in their nanoscale, nanofibres have two dimensions, and nanoplates have just one dimension in their nanoscale. Due to their nanoscale dimensions, they display different physical and chemical properties; therefore making their areas of application different also [13]. Nanopolymers are environmentally friendly, and they possess excellent properties. Due to the tiny sizes of nanoparticles, it has a large surface area and high number of atoms attached to those surfaces [13]. Nanofibres also have a large surface area because of their tiny size, they conduct electricity well, and have good mechanical properties [14]. Nanofibres are divided into nanotube and nanorods, in which one is hollow and the other is solid, respectively. Nanoplates also have excellent properties; they are used as reinforcement in polymers and other light materials like aluminium and magnesium to build advanced materials with improved thermal conductivity, mechanical properties [15], and outstanding catalytic properties, such as photo- and thermo catalytic activity [16].

Nanotechnology is a science of forming and arranging atoms to produce innovative industrial materials. The field is currently at the forehead of the world's technological development, and it shows promising potentials to bring more development to the world in coming times [17]. The science of this field is being greatly utilized in electronic industries, boats and ships production, automobile, and aircraft industries. Polymer nanocomposites have piqued the interest of industry and academics in nanotechnology because of their wide range of applications [18]. Polymer nanocomposites are materials with polymeric matrix of thermoplastics, thermosets, or elastomers reinforced with nanoparticles [19]. Polymer nanocomposites are a newer type of polymer that can be used instead of normally filled polymers. Some of the advantages are high elastic modulus, strength, heat and fire resistance, and low weight. They are utilised in biotechnology to make optical and magnetic devices, as well as medication packaging, adhesives, flame retardants, coatings, and final goods.

Promising modern materials such as polymeric materials when joined directly or doped with coordination complexes give hybrid materials exciting and excellent properties. Unique modern materials such as polyacetylene, polythiophene, and polypyrole are all examples of doped or n-conjugated materials. These hybridised n-conjugated polymers are good electric conductors with good optical properties, and they are utilised to make batteries, electronic gadgets, electrochromic devices, LED screens, and other modern technological devices [18].

To overcome all the difficulties associated with the usage of nanoparticles, hybrid nanocomposites must be created by impregnating tiny particles over large, solid particles [20]. The relevance of nanotechnology and polymeric nanocomposites, as well as their continual development, is due to their unique features. As previously stated, polymer nanocomposite materials contain matrices with incorporated nanoparticles, nanofibres, or nanoplates featuring their single or multi-dimension phases. These materials offer excellent mechanical qualities, which are not available in natural monolithic polymeric materials [12]. Polymer nanocomposites have combined properties of nanoparticles and polymers, therefore giving the produced materials remarkable mechanical properties and compatibility, as well as unique physical and chemical properties. The decision taken by scientists to use polymeric supports, on the other hand, is usually influenced by their thermal and mechanical properties. Therefore, the properties of the materials used in the production of nanocomposites should be considered to produce materials with the right and intended properties [21].

1.7 PARTICULATE REINFORCEMENTS

Scientists and engineers have created a new generation of exceptional materials by producing materials with a two-phase material system of combined matrix, continuous phase surrounding a dispersed discontinuous phase. The created materials have properties of both the matrix and dispersed phase components, resulting in materials with unique and needed qualities [22]. Natural materials such as metal, ceramic, and polymer form the matrix phase, which is reinforced with a dispersed phase of materials that have been formed into particles, fibres, or structures.

The small particle reinforcement also known as dispersion strengthened reinforcement and the large particle reinforcement are the main types of particles used as reinforcement in composite materials. The small particle reinforcements are particles at their atomic forms with size ranging from 1 nm to 100 nm. Due to the small sizes of these particles, their interactions with the matrix phases are always excellent, therefore producing composite materials with improved properties (tensile, flexural, and compressive). The small size of these particles also helps to reduce crack propagation and plastic deformation. In the case of the large particle reinforcement, due to their large surface areas, they cannot be studied at atomic level, but rather using mathematical expressions to determine their mechanical properties. To get the peak of their properties, particles should be evenly distributed and the more the reinforcing particles, the better the mechanical properties once the matrix saturation has not been reached. Large particle reinforcement is a cheaper method of producing composite materials because in most cases cheap fillers are used as the reinforcement. They are used in the production of tungsten carbide, cemented carbide, cutting tools, and cermet generally.

1.8   RULE OF MIXTURE

The composition of matrix and dispersed phase in the production of a composite material has great effect on the properties of the produced material. For example, when making a particle-reinforced composite, the more the particle content, the better the mechanical property of the composite. Also, how well the matrix and dispersed phase in a composite system are mixed and arranged has a great effect on the strength of composite materials. The rule of mixture is a mathematical expression used to calculate the mechanical properties of a composite material. It has an upper and lower bound of the elastic modulus in a material. The rule of mixture has two models which are the Voigt model and the Reuss model. They are used in calculating the Young's modulus of a composite material. The Voigt model uses axial loading to calculate the Young's modulus of the composite materials, where the strain of the materials is assumed to be uniform [23]. The Reuss model uses transverse loading to calculate the Young's modulus of the composite materials, where the stress of the materials is assumed to be uniform [23]. Equations (1.1) and (1.2) show the theories in calculating materials elastic modulus for the upper and lower boundary [23,24].

$$E_c(u) = E_m V_m + E_p V_p \qquad (1.1)$$

$$E_c(l) = \frac{E_m E_p}{V_m E_p + V_p E_m} \qquad (1.2)$$

where E and V are the elastic modulus and volume fraction, respectively; subscripts c stands for composite; subscript m stands for matrix; and subscript p stands for particulate.

Metal, polymer, and ceramics in particulate or fibrous form are used as reinforcement in polymers to emerge new materials with special properties for advanced applications. An example of an advanced composite material is a cermet, a composite material containing ceramics and metal. Their carbide matrix is being reinforced with metals like tungsten, titanium, copper, and nickel. They are mainly used in the production of cutting tools because of their high hardness, strength, good thermal shock properties, and ability to withstand high temperatures. A typical example of large particle reinforcement is concrete, where matrix cement is reinforced with sand and gravel. They are used in the construction of buildings and bridges, water tanks, slabs, nuclear power plants, etc.

1.9   NEEDS OF HYBRIDISATION OF REINFORCING PARTICLES

Scientists do not stop researching and that is what keeps the world evolving. Materials that have been actively impacting the technological advancement of the world are lightweight materials. They have a big impact on recent aviation and automotive industries. Fibre-reinforced lightweight materials keep getting attention from industries and scientists because of their exceptional resistance to deformation, strength, and low density. Unfortunately, with their excellent property of strength and stiffness, their limitation of brittleness cannot be overlooked. In most cases, materials with high strength are always lacking toughness because of their low resistance to fracture. Therefore, to overcome this challenge of high strength, low toughness behaviour, researchers are finding ways to produce materials with less brittleness while still maintaining their good strength and consequently, a variety of ways have been developed.

One of the proposed solutions by the researchers is toughening the polymer fibre-reinforced composite with reinforcers like nanoparticles and rubbers [25,26]. Another potential way researchers came up with is the consideration of the properties and the structure of materials that are going to be used to produce the improved tough lightweight composite material [27–29]. For instance, if you consider using polymer fibres to balance the properties of these lightweight materials because of their good thermal resistance and good deformation resistance, their defect properties such as their low ductility and density make them less of a favourite. In addition, the use of metal fibre as reinforcement results in materials with good stiffness and strain resistance. However, the relatively high density of metals remains a challenge, which makes them less suitable for some engineering applications [27–29].

These solid reasons inspire ways to improve the toughness and general properties of lightweight materials; therefore, methods of hybridising materials must be introduced. Hybridisation is more of an advanced way of producing composite materials. Hybridised materials are materials that involve the use of multiple materials for composite matrix or reinforcement, meaning more than one type of material is used for a composite matrix or dispersed phase [30]. The main purpose for the hybridisation of materials is to have optimum combinations of material properties for required industrial applications. This therefore helps in the production of materials with more unique and precise properties for specific applications. It also helps bridge the barriers of lightweight materials with low toughness. An example is the substitution of carbon fibre with cheaper glass fibre in a laminate composite material to help reduce the amount that will be used to produce the ecofriendly composite material while retaining their properties [31].

In the production of hybrid materials, the two main types of reinforcement used are the low elongation and high elongation reinforcements. The low elongation reinforcement breaks easily when load is applied; they cannot stretch much more than 5%–9% of their length [32]. The high elongation reinforcement has higher strain value as they tend to resist breakage when force is applied. It should be known that how the reinforcement in a hybrid material behaves is dependent on the material system [33].

Matrices and reinforcements in hybrid systems can be arranged in three forms, which are:

- Inter-layer: This is the arrangement method of putting different material reinforcement per layer. Figure 1.2(a) shows an example of an inter-layered hybrid material.
- Intra-layer: The reinforcements that are to be used in the hybrid system are mixed first, then they are arranged in layers of the matrix. Figure 1.2(b-c) shows an example of an intra-layered hybrid material for both fibre- and particle-reinforced polymeric composites.

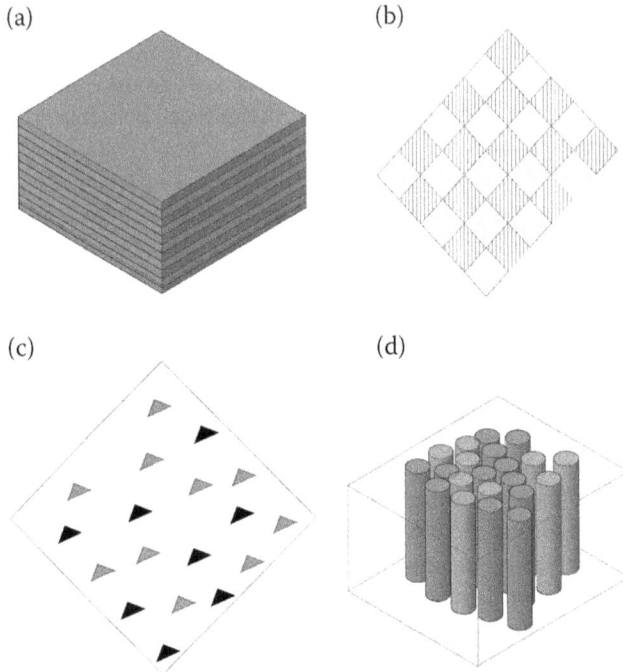

FIGURE 1.2 Hybrid materials pattern of arrangement: (a) Inter-layer arrangement, intra-layer arrangement for (b) fibre, (c) particles, and (c) intra-yarn arrangement.

- Intra-yarn: This method is the best of them all; it involves the combination of the reinforcements in a tow. Figure 1.2(d) shows an example of an intra-yarned hybrid material.

1.10 HYBRID EFFECT OF REINFORCEMENT IN COMPOSITE

The reinforcements in a hybrid composite must be properly arranged or mixed to get optimal properties. The reciprocal of the smallest repeated length is used to determine how well the reinforcements are mixed [34,35]. Different ways in which reinforcements can be mixed in a hybrid material system are shown at Figure 1.3. Figure 1.3(a) shows a poorly dispersed material system, the reinforcements are being arranged separately, and the properties of the material can be improved if additional layers of reinforcements are added to the material system as shown in Figure 1.3(b). Hybridising at the fibre bundle level, as shown in Figure 1.3(c), is another technique to increase dispersion. The dispersion that will produce hybrid material with optimal properties is the Figure 1.3(d) because of the dispersedly arranged reinforcements [31].

Due to the active research on ways to improve hybrid materials over the years, several hypotheses have been proven explaining the effects of hybrid materials as thus.

(a)

(b)

(c)

(d)

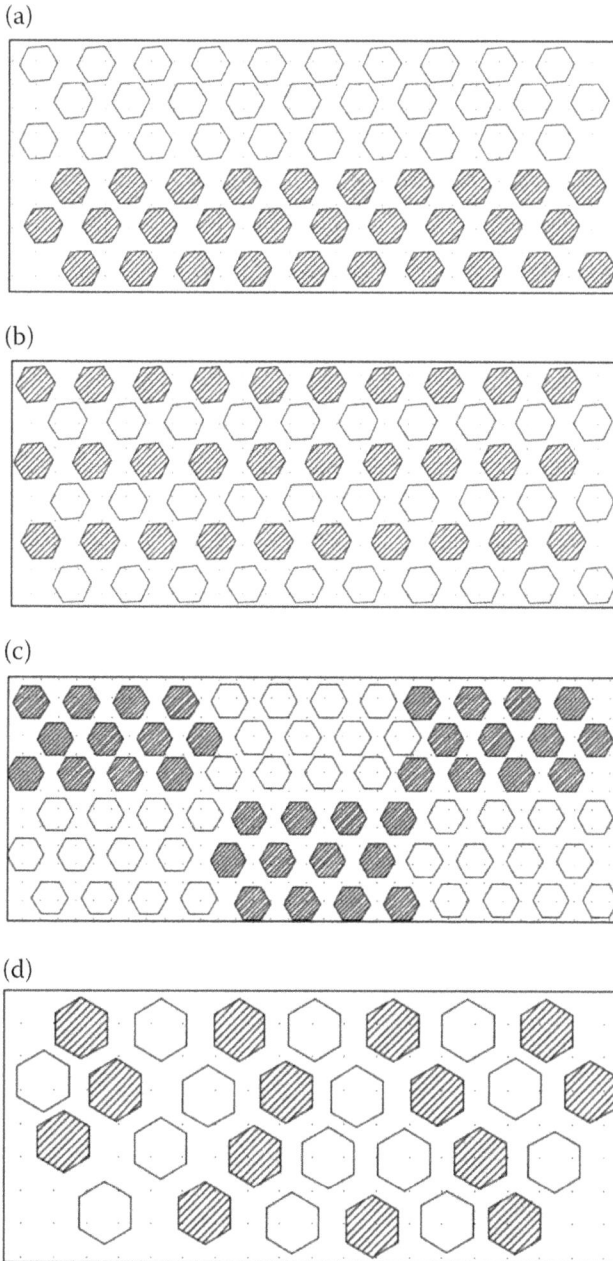

FIGURE 1.3 Arrangement of hybrid reinforcement (a) a poorly dispersed hybrid material system, (b) alternate layers dispersion of hybrid reinforcements, (c) hybridising at the fibre bundle level, and (d) hybrid reinforcements with even dispersion for both particulate and fibrous reinforcement.

a. Residual stresses induced by temperature

One of the major effects in the production of hybrid materials is the difference in thermal properties of the materials used in the production. Carbon fibres, for example, have thermal expansion of -1 to $+1 \times 10^{-6}k^{-1}$, whereas glass fibres have thermal expansion of 5 to $10 \times 10^{-6}K^{-1}$ [36]. Due to the difference in their thermal expansion, when the two materials are heated at the same temperature and cooled, the glass fibre shrinks while the carbon fibre retains almost the same size [36]. However, there is a restricted movement in a hybrid material system because of the multiple reinforcements or matrix present in the system. Due to this cause, residual stresses develop in the hybrid materials system when exposed to high temperature. For example, in the production of a hybrid material reinforced with glass and carbon fibre, residual stresses of tensile and compression are applied to the two fibres, respectively, because of their different thermal properties. Even with the importance of residual stress in solving hybrid effects, it only solves 10% of the effects because there are other structures in a hybrid material system that affects it [36,37].

b. Damage development

Due to the multiple types of reinforcements used in a hybrid system, the re-inforcements have materials of different properties. Therefore, their strengths are classified in distinct distribution, which vary the tensile loadings of the material significantly [33]. Hybridisation influences ineffective lengths as well as stress concentration factors for different types of fibres which have distinct elastic char-acteristics, radius, and interface quality [37]. Variations in stress redistribution has an impact on damage propagation. The hybrid materials delay breakage when a load is applied to the material in the direction of the reinforcement because of the presence of high elongation materials in the hybrid system. Another aspect that can influence the development of damage in hybrid materials is size effects, the better the properties of the hybrid, the smaller content of the reinforcements. Small amounts of reinforcements will only add a modest amount of volume to the system, allowing for the addition of more reinforcements, therefore, increasing the strength and other significant properties of the hybrid material [31].

c. Dynamic stress concentrations

The way a hybrid material fails is progressive and it constantly changes. Hedgepeth [38] discovered that when a force is applied to a fibre and it fails, it generates a wave in the axial direction, which was then validated by several other researchers (the wave formed is always dependent on the fibres elongation property) [39,40]. Hedgepeth focused mainly on stress concentration of fractured plane of materials; he used the shear lag theory for his research, but the theory has drawbacks when studying dynamic processes. Xing et al. [40] later improved the efficiency of Hedgepeth's theory. Some other research was made by Ruiz [41] on stress con-centration on glass and carbon hybrid materials; they noticed that glass hybrid materials had a 20% greater concentration [41]. Hybridised materials will always have a good dynamic stress concentration because reduced stress concentration will occur when load is applied [31]. The improved properties of hybrid materials will always make them relevant in industrial world for their applications in special areas. Although the properties a hybrid material after being developed cannot really be predicted because of the unpredictable interactions of materials in the hybrid system [33]. This is due to the limited dynamic models for predicting the stress concentration of hybrid materials, improved dynamic theories should be utilised to back up these findings.

1.11   CURRENT STATE OF HYBRID PARTICLE COMPOSITES

Hybrid particle polymer composites have diversified optimum combination of properties because of the addition of a filler to another that imparts some special properties to the resulting hybrid composites; therefore, they are applied industrially for production of artefacts of special applications. A brake pad with better performance than that of the Nissan Jeep Cherokee brake pad was reported in [42] to be produced from particulate steel slag (refractory filler), gum arabic (binder), aluminium chips (an abrasive material), graphite (for performance), gum Arabic (binder), and epoxy (binder). Moreover, production of automobile bumper fascia from epoxy (matrix) and aluminium powder containing coconut shell-based carbon particles was disclosed in [43]. In addition, Bello et al., 2020 [44] reveals productions of automobile bumper fascia from recycled low-density polyethylene containing mainly $CaCO_3$ and other minerals like K_2CO_3 and $CuCO_3$ sourced from eggshells. Besides the application of hybrid polymeric composites in the disclosed automobile applications, poly lactic-co-glycolic acid and polymethylmethacrylate-based hybrid composites have been reportedly used in biomedical engineering for bone cements, bone repair, and drug delivery system as reported in [45–52].

1.12   ADVANTAGES AND DISADVANTAGES OF HYBRID COMPOSITE

1.12.1   Advantages of Hybrid Composites

The advantages of hybrid composite are as follows:

   i. Hybrid materials can help at achieving excellent materials with greater strength and stiffness with less weight.
   ii. Laminate patterns in the part's ply buildup can also be developed to provide the necessary and required mechanical qualities in multiple directions.
   iii. Hybrid materials have exceptional corrosion, chemical, and outside weathering resistance.
   iv. It has a low cost of production. Hybrids can be created via a variety of procedures, with naturally occurring reinforcements being used.

v. There is a lot of design flexibility. Production of materials with desired and required properties can easily be fabricated.

1.12.2 Disadvantages of Hybrid Composites

The disadvantages of hybrid composites are as follows:

i. Sometimes the prediction of the behaviour of hybrid materials are flawed.
ii. They are susceptible to crack because of their increasing brittleness.

1.13 PRIOR STUDIES IN HYBRIDISATION OF MATERIALS

Table 1.2 provides a summary of some of the recent works of researchers on hybridised reinforced polymeric matrix.

TABLE 1.2
Related Research on Hybridisation in Polymeric Composites

Authors (Year)	Approaches	Remarks
Ravishankar et al. (2019) [53]	Review on hybrid polymer-based composites	Emphasis was made on the hybrid polymeric-based composites for automobile applications
El-Wasery (2017) [54]	Review on hybrid composites	This review was focused on review of previous studies on natural fibres/glass hybrid polymeric composites disclosing properties of the hybrid composites and their applications
Agunsoye et al. (2016) [42]	Modification of epoxy with graphite and glass particles	Improvement in mechanical, wear resistance, and thermal stability properties were reported
Kumar et al. (2020) [55]	Reinforcement of cashew nut cell oil with treated and untreated banana and sisal fibres	Treated fibre-reinforced hybrid composite was reported as having better mechanical properties than the untreated fibre counterpart
Komal et al. (2020) [56]	Modification of polylactic acid with banana fibres (having different phases of reinforcements) developed by direct injection moulding, extrusion injection moulding and direct compression moulding	An enhancement in the tensile and flexural properties; storage modulus, loss modulus, tan delta; and crystallinity of the PLA-based composite produced by extrusion injection moulding was revealed
Owuamanam and Cree (2020) [57]	Review on bio calcium carbonate, waste eggshell, and seashell fillers in polymer reinforcements	The review identifies varieties of eggshell, snail shells, their challenges, and summarises previous studies on calcium carbonate, waste eggshell, and seashell particles as fillers in polymers

1.14 CONCLUSIONS/SUMMARY

This chapter discusses the emerging materials in polymer reinforcement, covering nanocomposites. Their properties, importance in the world's advancement, applications, and limitations were discussed. The increase in the application of these new materials has accelerated the research and production of new concept materials and implementation of materials with multi structures. Composite materials are made using a variety of processes, each of which is appropriate for a particular material, because each material has various physical qualities. The effectiveness of the manufacturing procedure is determined by the way the matrix and reinforcement are joined and the way the phases react with each other. As a result, manufacturing procedures are determined by the material chosen. Research on emerging materials in polymer reinforcement is gaining a lot of research interest due to their excellent properties and because it holds a lot of potential for enabling new uses and applications in industries. They broaden the range of applications for a specific polymer in the most basic way. New studies will focus on ways to improve the already existing materials and ways to create new materials according to industrial demands.

REFERENCES

[1] Jose, J.P., and Joseph, K. (2012). Advances in Polymer Composites: Macro- and Microcomposites – State of the Art, New Challenges, and Opportunities, 10.1002/9783527645213
[2] Wang, R.M., Zheng, S.R., and Zheng,Y.P. (2011). Polymer Matrix Composites and Technology.
[3] Masuelli, M.A. (2013). Introduction of Fibre-Reinforced Polymers – Polymers and Composites: Concepts, Properties and Processes, Fibre Reinforced Polymers, https://www.intechopen.com/chapters/41941
[4] Rajak, D.K., Pagar, D.D., Menezes, P.L., and Linul E. (2019). Fibre-Reinforced Polymer Composites: Manufacturing, Properties, and Applications, https://www.mdpi.com/2073–4360/11/10/1667/pdf
[5] Nobe, R., Qui, J., Kudo, M., Ito, K., and Kaneko, M. (2019). Effects of SCF content, injection speed, and CF content on the morphology and tensile properties of microcellular injection-molded CF/PP composites. pp. 1371–1380, 10.1002/pen.25120
[6] Singh, T.J. and Samanta, S. (2015). Characterization of Kevlar Fibre and its composites: A review. pp. 1381–1387, 10.1016/j.matpr.2015.07.057
[7] Automakers have been experimenting for decades with lightweighting, as the practice is known, but the effort is gaining urgency with the adoption of tougher gas mileage standards, (2015). URL, https://www.mercurynews.com/2021/12/29/holmes-jury-resumes-deliberations-after-closed-door-meeting/
[8] Lightweight Materials – Leading the way in materials manufacturing for lightweight, efficient vehicles, (2021). Pacific Northwest National Laboratory. URL, https://www.pnnl.gov/lightweight-materials
[9] Campbell, F. (2012). *Lightweight Materials: Understanding the Basics*. ASM International.
[10] Vehicle Technology office (2013). *Lightweight Materials- R&D program*. U.S. Department of Energy.
[11] Niinomi, M. (2010). *Metals for Biomedical Devices*. Woodhead Publishing Series in Biomaterials. pp. 405–420.

[12] Agboola, O., Fayomi, O.S.I., Ayodeji, A., Ayeni, A.O., E. Alagbe, E., Sanni, S.E., Okoro, E.E., Moropeng, L., Sadiku, R., Kupolati, K.W., and Babalola Aisosa Oni, B.A. (2021). A Review on Polymer Nanocomposites and Their Effective Applications in Membranes and Adsorbents for Water Treatment and Gas Separation. 10.3390/membranes11020139

[13] Jeevanandam, J., Barhoum, A., Chan, Y.S., Dufresne, A., and Danquah, M.K. (2018). Review on nanoparticles and nanostructured materials: history, sources, toxicity and regulations. pp. 1050–1074, 10.3762/bjnano.9.98

[14] Saifuddin, N., Raziah, A.Z., and Junizah, R.A. (2013). Carbon Nanotubes: A Review on Structure and Their Interaction with Proteins. pp. 1–18, 10.1155/2013/676815

[15] Jian, S., Zhu, J., Jiang S., Chen, S., Fang, H., Song, Y., Duan, G., Zhang, Y., and Hou, H. (2018). Nanofibres with diameter below one nanometer from electrospinning. pp. 4794–4802, 10.1039/C7RA13444D

[16] Zheng, Y., Wang, W., Jiang, D., Zhang, L., Lia, X., and Wanga, Z. (2015). Ultrathin mesoporous Co3O4 nanosheets with excellent photo-/thermo-catalytic activity. *Journal of Materials Chemistry A*. pp. 105–112, 10.1039/C5TA07617J

[17] Gacitua, W.E., Ballerini A.A., and Zhang, J. (2005). Polymer Nanocomposites: Synthetic And Natural Fillers A Review. **7**(4): pp. 159–178, 10.4067/S0718-221X2005000300002

[18] Lagashetty, A. and Venkataraman, A. (2005). Polymer nanocomposites. pp. 49–57, 10.1007/BF02867106

[19] Denault, J. and Labrecque, B. (2004). *Technology Group on Polymer Nanocomposites.* PNC-Tech. Industrial Materials Institute.

[20] Zhao, X., Lv, L., Pan, B., Zhang, W., Zhang, S., and Zhang, Q. (2011). Polymer-supported nanocomposites for environmental application: A review. pp. 381–394, 10.1016/j.cej.2011.02.071

[21] Anuchai, S., Kelly, L. A., Richard, A.V., and Farmer, B.L. (2003). Influence of polymer matrix composition and architecture on polymer nanocomposite formation: Coarse-grained molecular dynamics simulation. *Joural of Polymer Science*, **41**(24), 3272–3284, 10.1002/polb.10696

[22] Askeland, D.R. (1996). The Science and Engineering of Materials: Composite Materials. pp. 549–594, 10.1007/978-1-4899-2895-5_16

[23] Callister, D.W. (2007). *Materials Science and Engineering: An Introduction.* New York: John Wiley & Sons.

[24] Physics:Rule of mixtures (2021). URL, https://handwiki.org/wiki/Physics:Rule_of_mixtures

[25] Qian, H., Greenhalgh, S.E., Shaffer, M., and Bismarck, A. (2010). Carbon nanotube-based hierarchical composites: a review. *Journal of Materials Chemistry*. pp. 4751–4762, 10.1039/C000041H

[26] Greef, D.N., Gortbatikh, L., Godara, A., Mezzo, L., Lomov, V.S., and Verpoest, I. (2011). The effect of carbon nanotubes on the damage development in carbon fibre/epoxy composites. pp. 4650–4664, 10.1016/j.carbon.2011.06.047

[27] Fratz, P. and Weinkamer, R. (2007). Nature's hierarchical materials. pp. 1263–1334, 10.1016/j.pmatsci.2007.06.001

[28] Gorbatikh, L., Lomov, V.S., and Verpoest, I. (2010). Original mechanism of failure initiation revealed through modelling of naturally occurring microstructures. *Journal of the Mechanics and Physics of Solids*, **58**(5), 735–750, 10.1016/j.jmps.2010.02.007

[29] Studart, R.A. (2012). Towards High-Performance Bioinspired Composites. **24**(37): pp. 5024–5044, 10.1002/adma.201201471

[30] Kaw, K.A. (2005). Mechanics of Composite Materials. 10.1201/9781420058291

[31] Swolfs, Y., Gorbatikh, L., and Verpoest, I. (2014). Fibre hybridisation in polymer composites: A review. pp. 181–200, 10.1016/j.compositesa.2014.08.027

[32] Mathangadeera, W.R., Hequeta, F.E., Kelly, B., Dever, K.J., and Kelly, M.C. (2020). Importance of cotton fibre elongation in fibre processing. 10.1016/j.indcrop.2020.112217

[33] Tavares, P.R., Melro, R.A., Bessa, A.M., Turon, A., Liu, K.W., and Camanho, P.P. (2016). Mechanics of hybrid polymer composites: analytical and computational study. pp. 405–421.

[34] Kretsis, G. (1987). A review of the tensile, compressive, flexural and shear properties of hybrid fibre-reinforced plastics. **18**(1): pp. 13–23. 10.1016/0010-4361(87)90003-6

[35] Manders, M.B. (1981). The strength of hybrid glass/carbon fibre composites. *Journal of Materials Science,* **16**(8): 2333–2245. 10.1007/BF00542386

[36] Manders, W.P. and Bader, G.M. (1981). The strength of hybrid glass/carbon fibre composites. *Journal of Materials Science*, **8**(16): 2246–2256.

[37] Zweben, C. (1977). Tensile strength of hybrid composites. *Journal of Materials Science*, **12**: 1325–1337. 10.1007/BF00540846

[38] Hedgepeth, J.M. (1961). *Stress Concentrations in Filamentary Structures*. Washington, D.C.: National Aeronautics and Space Administration. https://www.worldcat.org/title/stress-concentrations-in-filamentary-structures/oclc/61446694

[39] Ji, X., Liu, X., and Chou T. (1985). Dynamic Stress Concentration Factors in Unidirectional Composites. 10.1177/002199838501900305

[40] Xing, J., Hsiao, C.G., and Chou, T. (1981). A Dynamic Explanation of The Hybrid Effect. pp. 443–461, 10.1177/002199838101500504

[41] Xia, Y., and Ruiz, C. (1991). Analysis of damage in stress wave loaded unidirectional composites. *Computers & Structures*, **38**(3): 251–258. 10.1016/0045-7949(91)90103-S

[42] Agunsoye, J.O., Bello, S.A., Bamigbaiye, A., Odunmosu, K.A., and Akinboye, I. (2018). Recycled Ceramic Tile Composite for Automobile Applications, a Comparative Study with Nissan Jeep Cherokee Brake Pad. *Engineering and Applied Science Research*, **45**(3): 180–187. 10.14456/easr.2018.30

[43] Bello, S.A., Agunsoye, J.O., Adebisi, J.A., Adeyemo, R.G., and Hassan, S.B. (2020). Optimization of Tensile Properties of Epoxy Aluminum Particulate Composites Using Regression Models. *Journal of King Saud University – Science*, **32**(1): 402–411. 10.1016/j.jksus.2018.06.002

[44] Bello, S.A., Raji, N.K., Kolawole, M.Y., Adebayo, M.K., Adebisi, J.A., Okunola, K.A., and AbdulSalaam, M.O. (2021). Eggshell Nanoparticle Reinforced Recycled Low-Density Polyethylene: A New Material for Automobile Application. *Journal of King Saud University – Engineering Sciences*. 10.1016/j.jksues.2021.04.008

[45] Zhao, D., Zhu, T., Li, J., Cui, L., Zhang, Z., Zhuang, X., and Ding, J. (2021). Poly (Lactic-Co-Glycolic Acid)-Based Composite Bone-Substitute Materials. *Bioact Mater*, **6**(2), 346–360. 10.1016/j.bioactmat.2020.08.016

[46] Cui, X., Huang, C., Zhang, M., Ruan, C., Peng, S., Li, L., Liu, W., Wang, T., Li, B., Huang, W., Rahaman, M.N., Lu, W.W., and Pan, H. (2017). Enhanced Osteointegration of Poly(Methylmethacrylate) Bone Cements by Incorporating Strontium-Containing Borate Bioactive Glass. *J R Soc Interface*, **14**(131). 10.1098/rsif.2016.1057

[47] Chen, L., Tang, Y., Zhao, K., Zha, X., Liu, J., Bai, H., and Wu, Z. (2019). Fabrication of the Antibiotic-Releasing Gelatin/Pmma Bone Cement. *Colloids Surf B Biointerfaces*, **183**, 110448. 10.1016/j.colsurfb.2019.110448

[48] Khandaker, M. and Meng, Z. (2015). The Effect of Nanoparticles and Alternative Monomer on the Exothermic Temperature of Pmma Bone Cement. *Procedia Eng*, **105**, 946–952. 10.1016/j.proeng.2015.05.120

[49] Sa, Y., Yang, F., Wang, Y., Wolke, J.G.C., and Jansen, J.A. (2018). Modifications of Poly(Methyl Methacrylate) Cement for Application in Orthopedic Surgery. *Adv Exp Med Biol*, **1078**, 119–134. 10.1007/978-981-13-0950-2_7

[50] Wang, X., Kou, J.M., Yue, Y., Weng, X.S., Qiu, Z.Y., and Zhang, X.F. (2018). Clinical Outcome Comparison of Polymethylmethacrylate Bone Cement with and without Mineralized Collagen Modification for Osteoporotic Vertebral Compression Fractures. *Medicine (Baltimore)*, **97**(37), e12204. 10.1097/MD.0000000000012204

[51] Xia, X., Shi, R., Huang, J., Li, Y., Zuo, Y., and Li, J. (2020). Development of a Phase Change Microcapsule to Reduce the Setting Temperature of PMMA Bone Cement. *J Appl Biomater Funct Mater*, **18**, 2280800020940279. 10.1177/2280800020940279

[52] Ahmad, H., Arya, A., Agrawal, S., and Dwivedi, A.K. (2019). Plga Scaffolds: Building Blocks for New Age Therapeutics. pp. 155–201. 10.1016/b978-0-12-816901-8.00006-7

[53] Ravishankar, B., Nayak, S.K., and Kader, M.A. (2019). Hybrid Composites for Automotive Applications – a Review. *Journal of Reinforced Plastics and Composites*, **38**(18), 835–845. 10.1177/0731684419849708

[54] El-wazery, M.S. (2017). Mechanical Characteristics and Novel Applications of Hybrid Polymer Composites- a Review. *Journal of Materials and Environmental Sciences*, **8**(2), 666–675.

[55] Kumar, T.N., Muralidharan, K., Pradhan, R., and Suresh, R. (2020). Experimental Investigation on Equally Treated Banana/Sisal Fibers Based Hybrid Composite. **2283**: p. 020083. 10.1063/5.0025089

[56] Komal, U.K., Lila, M.K., and Singh, I. (2020). PLA/Banana Fibre Based Sustainable Biocomposites: A Manufacturing Perspective. *Composites Part B: Engineering*, **180**, 107535. 10.1016/j.compositesb.2019.107535

[57] Owuamanam, S. and Cree, D. (2020). Progress of Bio-Calcium Carbonate Waste Eggshell and Seashell Fillers in Polymer Composites: A Review. *Journal of Composites Science*, **4**(2), 70. 10.3390/jcs4020070

2 An Overview of the Sources, Structure, Applications, and Biodegradability of Agricultural Wastes

Funsho Olaitan Kolawole
Department of Metallurgical and Materials Engineering, University of Sao Paulo, Sao Paulo, Brazil

Department of Materials and Metallurgical Engineering, Federal University Oye-Ekiti, Ekiti State, Nigeria

Ibiwumi Damaris Kolawole
Department of Food Science, University of Sao Paulo, Sao Paulo, Brazil

Okhiria Dickson Udebhulu
Department of Mining and Petroleum Engineering, University of Sao Paulo, Sao Paulo, Brazil

Chemical and Petroleum Engineering Department, Alex-Ekwueme Federal University Ndufu-Alike, IkwoAbakaliki, Ebonyi State, Nigeria

Shola Kolade Kolawole
National Agency for Science and Engineering Infrastructure, Abuja, Nigeria

Michael Moses Aba
Institute of Energy and Environment, University of Sao Paulo, Sao Paulo, Brazil

Enermics Consulting Limited, Ikeja, Lagos, Nigeria

Patience Bello Shamaki
Department of Chemical Engineering, University of Sao Paulo, Sao Paulo, Brazil

DOI: 10.1201/9781003170549-3

CONTENTS

2.1   INTRODUCTION

Agricultural residues are gaining global attention due to the valuable products which can be obtained from them. These agricultural wastes from different agricultural products, if not properly utilised can lead to pollution causing nuisance in the environment and affecting global warming [1]. Most often agriculture residues are discarded or burnt and release enormous volumes of air pollutants and gasses to the environment such as carbon dioxide, sulphur dioxide, non-methane hydrocarbon, semi-volatile, and volatile organic compounds [2]. Agricultural waste can be obtained from maize stalk, cassava stem and leaves, banana stem and peels, rice husk, coconut shells, eggshell, moringa leaves and pods, *Daniella oliveri* plants, and several others. It is very important to study and categorize agricultural waste based on their sources, their structure, biodegradability, and possible applications, especially in hybrid polymeric nanocomposites. The source and structure of agricultural waste can determine the processing that is needed to convert them to other useful products while biodegradability on the other hand would determine the field they can be applied to.

Bioproducts are biobased materials that can be obtained from agricultural waste that is much cheaper and can be used to replace the conventional products produced mostly from fossil fuel or metals. The bioproducts from agricultural waste can be used as household utilities, building materials, packaging materials, office supplies, automobile parts, and clothing [3]. Nanoparticles (silver, bismuth, zinc oxide, gold, iron oxide) have been synthesised from *Moringa oleifera* pods/seeds [4]. Adeyanju and Olatoyinbo (2018) [5] reported large-scale biosynthesis of nanoparticles from *Daniella oliveri* seeds. Banana stems can be used to produce sacks, ropes, textiles, tents, screens, backpack, covers, blankets, carpets (with wool), pop fibre, hair extensions, sanitary pads, and renewable energy [6]. Eggshells are important agricultural waste that can be used for nutritional supplementation, chicken feed supplementation, compost enrichment, pest control, and beauty treatments [7]. Lignin, corncob, sawdust, wool, durian peel, sludge, rice husk, palm oil ash, cocoa pod, palm shell, leaves, tree bark, sugarcane bagasse, orange peel, banana peel, and several major agricultural biomass wastes are used as adsorbents for colour-adsorbing and filtering materials [8].

Agricultural waste can be converted into consumable energy by using techniques such as biochemical conversion and thermochemical processes. Bioenergy derived from agricultural waste is often considered sustainable because of the ability to be renewed from feedstock. Although, it is well known that energy derived from waste is much cheaper than conventional energies, it might be very complex to make in comparison because of the low efficiency of the energies obtained from agricultural waste [9]. In this chapter, the sources, structure, applications, and biodegradability of agricultural waste; challenges of burning of agricultural waste; bioproducts; nanoparticles from *Moringa oleifera* pods and *Daniella oliveri* seeds; importance of banana stem and eggshells; and colour adsorbent and filtering materials will be discussed. Lastly, a discussion on techniques, impacts, sustainability, and cost of energy derived from agricultural waste will be presented.

2.2 BIOPRODUCTS

Bioproducts are consumer and industrial goods produced fully or in part from renewable biomass. The present era is marked by wealth expansion, notwithstanding still faced by great levels of pollution that consequently deteriorate public health due to an increase in fossil fuel usage for production development in the industry. However, the world has been trying to find alternative routes for energy production. As a result of this, some institutions in several countries have tried to promote bioproducts as replacements for conventional energy sources [10]. The use of plants and their residues for production of a variety of bioproducts, including materials and items manufactured for automobiles, construction industries, and homes, is becoming an interesting part of research [11]. Presently, the chemical manufacturing plant is solely petroleum-based, but a shift in growth of biocatalytic processes is anticipated to have at least 30% renewable sources of the entire chemical industry in about 30 years from now [12]. A considerable number of bioplant products such as succinic, lactic, itaconic, and bioethanol are obtained from corn syrup and other sources of sugar [13].

The most known plant bioproducts are made from corn and soyabeans. However, bioproducts can also be made from other plant resources such as canola, sunflower, rapeseed, castor beans, potatoes, wheat, switchgrass, sugarcane, flax, palm oil, coconut oil, and forest-derived materials [14]. Bioproducts are as diverse as any conventionally produced product. Items that can be biobased include personal care items, household cleaners, building materials, packaging materials, office supplies, clothing, and lubricants. Many bioproducts are derived from forestry, biological, and biological waste. An example of agricultural waste that is used in producing bioproducts is wheat straw, leftover after wheat is harvested and chaff removed, can be used for making baskets, animal bedding, and has the capacity to be used as biofuel or creating composite lumber [3].

The most sustainable replacement for solving the global reduction in supply of fossil fuels is vegetable oils, which were the earliest biofuels and serve as precursor material. Biodiesel aids the reduction of greenhouse gas emissions as compared with convectional diesel as follows: 48% carbon monoxide, 100% sulphur dioxide, and 47% particles [15]. However, due to the controversy of food versus fuel, lignocellulosic biomass is the most common source for production of bioproducts and is known as the most abundantly available renewable source for replacement of petrol derivatives [16]. Plant biomass is generally composed of cellulose, lignin, and hemicellulose and has the ability for energy storage through the process of conversion of sunlight into chemical energy.

Bioproducts are gaining increased scientific attention and public acceptance due to the necessity for a rise in energy dependability and concern with greenhouse gas emissions generated when fossil fuels are used [17]. Several chemicals are synthesised from bioproducts by direct fermentation of ethanol, acetone, and butanol from corn stover [18]. Aracil *et al.* (2012) [19] evaluated the prospective benefit of bioproduct generation from agricultural residues and their findings disclosed that production of bioproducts would lead to clear immediate climate benefits in countries with abandoned landfills.

2.3 AGRICULTURAL WASTES, BURNINGS, AND THEIR CHALLENGES

Most often, the agricultural residues are dumped into rivers and lakes, and burnt on farmlands, causing serious environmental hazards. An environmentally friendly solution and alternative economic use for this agricultural residue can be used to solve these problems [20].

2.3.1 AGRICULTURAL WASTES

The increasing world population is responsible for the rising demand for food, leading to an increase in food production worldwide. However, ventures that are agriculturally based represent profitable businesses in developed and developing nations. The large number of farming activities is strongly increasing the production volume of agricultural products, leading to an overall increase in agricultural residue production [1]. Municipal solid waste management in many developing countries focuses on waste collection with inappropriate consideration for waste disposal or treatment [21]. While collection would remove waste from the generators, wastes collected are in most times disposed of in an open landfill with no consideration for environmental degradation and human impacts [22]. The number of agricultural wastes that are usually produced depends on factors like culture and geography of a nation. Agricultural wastes are residues often generated from agricultural products. These wastes include manure, spoiled food waste, abattoir pesticides that enter soil, silt drained from fields, poultry house waste, and solid waste obtained from harvest [1].

Agricultural wastes are utilised differently depending on the source where they are obtained. Sometimes this waste can be used in raw or processed form, which is determined by the end product, and may be used in composting, animal feed, bioenergy production and cultivation (mushroom and others) [23]. Nowadays, bioenergy and compost can be produced from crop residues; several countries like Indonesia, China, Thailand, Nepal, Japan, Nigeria and Malaysia, utilise this technique [24].

2.3.2 BURNING AND THEIR CHALLENGES

Burning is a commonly used method to get rid of agricultural waste such as straw from farms. Crop straws are made up of majorly crop stubble and stalk. Burning is used to remove crop stubble from the field [25]. Ma (2009) [26] reported that burning was the conventional method used to get rid of approximately 40% of wheat stubble. Cutting on the other hand was mainly used to remove stalk for feeding livestock, selling, and biogas production. Burning cleans the land more rapidly compared with cutting and other methods. A country like China generates the largest farm waste in the world and is faced with huge challenges to remove these agricultural wastes after harvests due to limited time given to farmers to remove the residues to ensure timely planting of crops for another season [25]. This limited time for removal of agricultural waste could be due to rain or other inclement weather. Burning became a popular method for removing agricultural waste because of urbanisation and rapid development during the last three decades [27].

The burning of agricultural waste generates many environment problems. It can cause damages to both human and natural systems. Air pollution is the most adverse effect of waste burning, because the process of burning releases enormous volumes of air pollutants and gasses to the environment such as carbon dioxide, sulphur dioxide, non-methane hydrocarbon, and semi-volatile and volatile organic compounds [2]. Other damages caused include deterioration of soil fertility, loss of biodiversity of agricultural lands, greenhouse gas emissions that contribute to global warming, and increased levels of particulate matter and smog leading to health hazards [1]. Burning can affect human health in several ways such as causing lung infection, difficulty in breathing, and increased susceptibility to respiratory diseases [25]. There are also ecological impacts on straw crop burning in soil nutrient and conservation. Even though burning reverts some portions of the mineral elements into the soil, almost all nutritional contents and organic materials from the straw are adversely affected during burning [28]. Additionally, frequent burning on a farm site can cause soil surface layers to be impaired.

2.4 BENEFITS OF *MORINGA OLEIFERA* TREE: PODS AS SOURCES OF REINFORCING NANOPARTICLES

2.4.1 *MORINGA OLEIFERA* PARTS

Moringa oleifera belongs to the *Moringaceae* family and is a sacred plant, which for many decades has been cultivated around the world. *Moringa oleifera* parts (leaf, pod, and stem) are well known for their effective and interesting health benefits [29]. They are also referred to as miracle trees and have been used to cure over 300 diseases in mostly Africa and India [30]. It is important to know that *Moringa oleifera* offers nutritional values to the human body as they can be eaten as food, too [29]. Vital quantity of amino acids and carotenoids have been found in the leaves of *Moringa oleifera*. Seeds from *Moringa oleifera* have been used as water purifiers to remove elements such as lead, nickel, and cadmium from drinking water [31]. Fresh *Moringa oleifera* fruits and open dried fruits containing pods and seeds are shown in Figure 2.1.

2.4.2 SYNTHESIS OF NANOPARTICLES (NPS) FROM *MORINGA OLEIFERA*

Apart from the medicinal, nutritional, and water purification abilities of *Moringa oleifera*, nanoparticles for several applications can be extracted from both the leaves and pods, as reported by several authors. Silver [4], bismuth [32], zinc oxide [33],

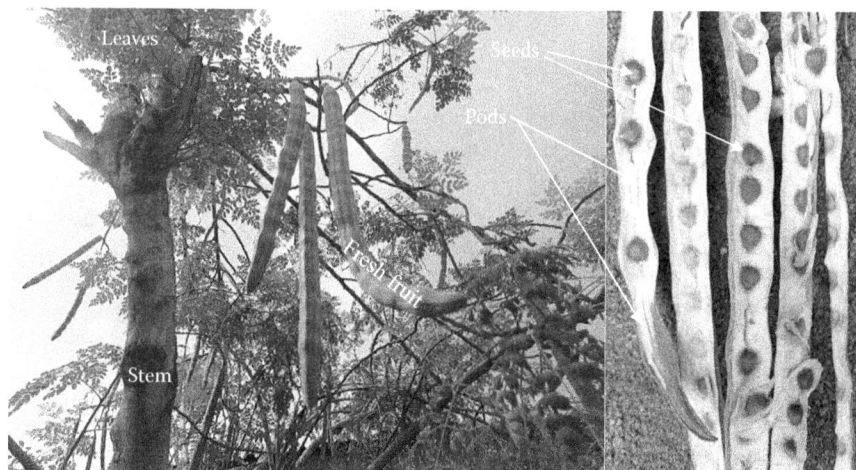

FIGURE 2.1   Parts of *Moringa oleifera* tree.

gold [34], and iron oxide [35] nanoparticles (NPs) have been synthesised from *Moringa oleifera* pods/seeds.

Nanocomposites are fundamentally multiphase solid material, as well as porous media, colloids, gels, and copolymers. Silver nanoparticles (AgNPs) from *Moringa oleifera* seed composites have enhanced graphene structures by aiding in pollutant removal efficiency from liquid industrial waste, especially for water treatment [36]. In 2019, some researchers published the manufacturing of *Moringa oleifera* oil/PVC-Ag based polymeric bio-nanocomposite [37]. Silver nanoparticles (AgNPs) by green synthesis make use of aqueous *Moringa* extract and are reinforced in polyvinyl alcohol (PVA) as a food packaging material was developed by Abdelhamid et al. (2020) [38]. Wood polymer nanocomposites (WPNCs) using nano-ZnO and nano-clay were produced during the impregnation of melamine formaldehyde–furfuryl alcohol copolymer, 1,3-dimethylol-4,5-dihydroxyethyleneurea (DMDHEU) from the *Moringa oleifera* plant under vacuum condition [39].

2.5   SEED STRUCTURE AND APPLICATIONS OF *DANIELLA OLIVERI* (ROLFE) TREE

2.5.1   IMPORTANCE OF *DANIELLA OLIVERI*

The plant *Daniella oliveri* is found in places like the Amazon region in Brazil, Nigeria, and other countries within South America and Africa [40]. The *Daniella oliveri* tree bark has several uses, which include herbal medicine, firewood, paper making, soap production, and fermentation process. Gum can also be obtained from the tree, while the leaves can be used as an edible soup and cattle feed and the seeds and fruit are eaten [41].

FIGURE 2.2   *Daniella oliveri* tree and seeds obtained from dried fruits.

2.5.2   SEED STRUCTURE OF *DANIELLA OLIVERI*

The seeds (Figure 2.2) of the *Daniella oliveri* are enclosed in a flat, oblong pod, and are mostly green when fresh and turn brown when dried. The seeds usually possess extremely high protein and relatively high carbohydrates, approximately ~33.4% and ~44.6%, respectively. Detailed compositions of the *Daniella oliveri* are 2.85 ± 0.23% moisture, 8.32 ± 0.42% crude fat, 33.40 ± 0.4% crude protein, 4.30 ± 0.35% ash, 6.35 ± 0.53% crude fibre, and 44.60 ± 0.30% carbohydrate. Their bulk density, fatty acids, and energy are 0.41 ± 0.52 gcm^{-3}, 6.66 g/100 g, and 1634 kJ/100 g, respectively [42,43].

Ca is the most abundant mineral in the *Daniella oliveri* seed, while Zn, Cu, and Mn are the smallest minerals. Some compounds such as cyanide, polyphenols, phytate, and oxalate in *Daniella oliveri* seeds are anti-nutritional and are present in low quantities [43]. Mineral contents of the *Daniella oliveri* seed (in mg/100 g of the seed) as presented in [43] are 263 Ca, 168 Mg, 680 K, 7.4 Na, 26.4 Fe, 6.47 Mn, 1.52 Zn, 2.23 Cu, and 384 P while 1.22 tannic acid, 0.49 polyphenol, 0.42 cyanide, 7.13 phytate, 0.19 oxalate, 7.48 alkaloid, 3.27 saponin, and 0.86 flavonoid are the antinutrients in the *Daniella oliveri* seed.

2.5.3   PRESENT AND FUTURE APPLICATIONS OF BIOPRODUCTS FROM THE *DANIELLA OLIVERI* SEED

The physicochemical properties and mineral contents in the *Daniella oliveri* seed show that it is possible to extract chemicals and minerals, further applying them in areas where they are needed. For instance, *Daniella oliveri* seed can be used as an antioxidant in medicine. It contains biomolecules such as proteins, amino acids, and polysaccharides, which are used for large-scale biosynthesis of nanoparticles [5]. The seeds also serve as feed ingredients for livestock and can be processed based on the seed structure into useful engineering materials. Ash content in the *Daniella oliveri* seed can serve as a carbonisation agent if reinforced in metals. Seed extracts can also be used as green corrosion inhibitors, due to the presence of tannic and alkaloid, which assist in inhibiting the corrosive elements in the environment. Nanoparticles that can be synthesised from the seeds of the *Daniella oliveri* can be used as reinforcements in polymers, clay, or metallic alloys to form suitable composites that are relatively economically available. Studies on the seeds of the *Daniella oliveri* should be carefully carried out for further application in other areas of engineering materials.

2.6 STRUCTURE AND APPLICATIONS OF BANANA STEMS

Banana production worldwide has been estimated to be approximately 48.9 million tonnes, India contributing 10.4 million tonnes, which is the highest in the world, followed by Brazil, Ecuador, Philippines, and China. About 220 tonnes of residue can be obtained per hectare. In addition, banana stems can be allowed to rot in the soil with the intention of replenishing the nutrients for growth of the next banana season; nevertheless, a huge loss of biomass and a lot of carbon dioxide is generated, producing unpleasant odour that contributes to environmental hazards [44]. Bananas are regarded as an important fruit in the subtropical and tropical region, and it is possible to harvest about 100 t/ha of banana stems [45].

2.6.1 BANANA STEMS

The stem is the largest part of the banana plant, which is very fleshy, with tons of water and possesses low lignin content [46]. Banana stems are an abundantly available resource in sub-tropical and tropical regions and, if not converted to useful products, will end up as environmental nuisance and be dumped. Figure 2.3 shows the outer layer, middle layer, inner layer, and core of the banana stem.

2.6.2 CHEMICAL, PHYSICAL, AND MECHANICAL PROPERTIES OF BANANA STEMS

A report on the chemical composition of banana stems in [46] shows that the stem has 39.12% cellulose, 72.71% holocellulose, 8.88% Klason lignin, 1.9% acid soluble lignin, 8.2% ash contents, 3.05% extracts, and 0.27% pectin [46].

The banana stem biomass possesses a good modulus of elasticity, tensile strength, stiffness, including a porous structure, moisture content, and hydrophilic nature of cellulose [6,47]. These properties make them suitable for papers, fibres, and other related applications.

2.6.3 APPLICATIONS OF BANANA STEMS

Over the past decades, several products have been developed from banana stems, such as sacks, ropes, textiles, tents, screens, backpacks, covers, blankets, carpets (with wool), pop fibre, hair extensions, and sanitary pads [6]. Renewable energy production

FIGURE 2.3 Picture of banana stem showing different layers stacking over one another.

from both fresh and dried banana stems have been successful using anaerobic digestion in recent times [48]. Car mats were made for the Ford Lincoln using banana stem fibre composites [49]. Construction materials are also beginning to benefit from the use of banana stem fibre in building, with excellent properties [50].

2.6.4 ECONOMIC SIGNIFICANCE OF BANANA STEM WASTE

Every year during the banana harvest season, the stems, leaves, and other parts being cut off are dumped on farm sites or along the roads, which end up polluting the environment and occupying large land areas. However, based on the physiochemical properties of the banana stem, it is necessary to develop useful materials for different applications, which will help in providing solutions to the environmental issues caused by the dumping of banana stem biomass. An improved waste management system will utilise the harvested waste from banana stems to create high-demand products that can be used as packaging, padding, and waterproof disposable materials. This can assist in driving the Gross Domestic Product (GDP) of many developing nations. In general, this brings about the development of alternative materials from biomasses that are durable, economical, and environmentally friendly.

2.7 IMPORTANCE AND CHALLENGES OF EGGSHELLS

Eggshells are obtained from the eggs of birds and most reptiles. In most climes of the globe, eggshells are perceived as agricultural wastes due to the lack of awareness of their benefits. Lots of benefits are lost by disposing eggshells from homes and farms. Besides calcium and protein, strontium, fluoride, magnesium, and

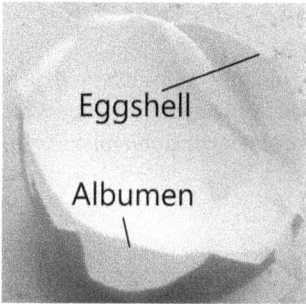

FIGURE 2.4   Picture of eggshell.

selenium can be found in eggshells [51]. Eggshells comprise three distinct layers: hard, chalky layer covered by the bloom cuticle to prevent dust and bacteria from passing through the pores, the second layer is the inner membrane, and lastly is the outer membrane, both composed of strong and flexible proteins [7].

2.7.1   THE IMPORTANCE OF EGGSHELLS IN AGRICULTURE

1. Nutritional Supplementation

Eggshells (Figure 2.4) are edible when thoroughly cleaned and prepared appropriately. Eggshells are rich in calcium as well as strontium, fluoride, magnesium, and selenium. Eggshells' nutritional benefits include growth and development of new bones; regulation of heart rhythm; muscular activity dynamics; control levels of magnesium, phosphorous, and potassium in the blood; treatment of osteoporosis by bone strengthening; promotion of bone metabolism; and stimulation of cartilage growth, protection, and remineralisation of teeth enamel.

2. Chicken feed supplementation

Allowing chickens to peck and eat the eggshells helps to promote healthier eggs from improved calcium intake.

3. Compost enrichment

For compost enrichment like nitrogen and phosphorus, plants require calcium for optimal growth and development. Calcium deficiency could cause stunted growth, twisted leaves, and dark spots. Homemade compost can be enriched using eggshells. Crushing the eggshells into the soil provides calcium and other minerals for compost.

4. Pest control

Eggshells can be used to deter garden pests like cats, snails, and slugs. Spreading crushed eggshells around your prized plants helps deter slugs, grubs, and snails from snacking on your garden. The smell of eggshells also keeps the deer away.

5. Beauty treatment

Eggshells are used for beauty treatments such as exfoliation and increased face glow; improved hair growth and gloss; and as a teeth whitening agent. Eggshells are also used as part of the ingredients for skin treatments against blemishes, inflammations, rashes, and some skin infections.

2.7.2 CHALLENGES OF EGGSHELLS

1. The separation of the eggshell from their membrane is a very difficult task, and a lot of time and energy is expended in the process of separation [52]. The value of the eggshells is limited in cases where the membrane is not properly separated from the eggshell.
2. Cleaning of the outer surface of eggshells is another difficult task. Some eggshells are either stained with excrete or blood.
3. Eggshells may harbour germs and infectious diseases, and if not handled hygienically can transmit diseases to humans.

2.8 AGRICULTURAL WASTES AS COLOUR ADSORBING AND FILTERING MATERIALS

Wastewater discharged into streams and river channels after their usage from textile, cosmetic, leather, paper, dye, distillery, printing, and carpet industries, gives rise to dangerous effects to aquatic and human life [53]. The textile industry alone discharges approximately 1.5×10^8 m^3 of coloured effluents annually according to Ip *et al.* (2010) [54]. A reduction in photosynthetic activity occurs due to the transmission of sunlight through wastewaters [53]. Hence, it is a best practice to ensure the removal of synthetic dyes, preventing them from entering downstream bodies of water. Biological oxidation and chemical precipitation are the most common methods used for colour removal from effluents. However, due to their expensive cost and only applicable where solute concentrations are relatively high, thus, alternative methods are required [55].

2.8.1 CONVENTIONAL METHODS OF ABSORBING COLOUR

Conventional methods utilised in removing dye from wastewater are categorised into three: biological, chemical, and physical methods [56]. However, all of them have benefits and pitfalls. Ditching cost and complications are the major challenges of conventional methods used in treating wastewater generated from dye industries and limiting their applications at a substantial level in the fabric and paper industries [57]. Notwithstanding, the availability of several techniques for removal of pollutants from wastewaters is a legal requirement. Methods such as coagulation, chemical oxidation, membrane separation process, electrochemical, and aerobic and anaerobic microbial degradation are partially successful because of numerous limitations [58].

2.8.2 Agricultural Wastes as Adsorbents

Low-priced adsorbents from agricultural waste possess excellent colour-removal capabilities [55]. Adsorption using agricultural waste is a cheap and pragmatic method in removing different pollutants has proven efficient for the removal of various pollutant types, like heavy metals, chemical oxygen demand phenol, gasses, and dyes [8,55]. Adsorbents from lignin, corncob, sawdust, wool, durian peel, sludge, rice husk, palm oil ash, cocoa pod (*Theobroma cacao*), palm shell, leaves, tree bark, sugarcane bagasse, orange peel, banana peel, and several major agricultural biomass wastes are utilised in dyes sequestration [8,59]. They are non-conventional, low cost, and effective. Agricultural waste materials are abundantly available, cheap, high carbon, and low inorganics, thereby serving as precursors to produce activated carbon (AC), which aids in colour removal from wastewater [59].

Activated carbon (AC) is amorphous in nature and possesses a high level of porosity that is used for removing pollutants from wastewaters. Porosity is the driving force for the performance of AC. The filtering function and extremely sizeable surface area create numerous sites where adsorption of impure molecules can occur. AC preparation requires two basic stages: carbonaceous raw materials are carbonised at < 1000 °C, in an inert atmosphere, and carbonised products are activated and can be regarded as either physical or chemical.

2.8.2.1 Physical Activation

This occurs when raw material is grown into activated carbon using gasses, performed by carbonisation; forming non-porous char, when pyrolysis is carried out on the precursor at a temperature range of 400°C–850°C, and could reach 1000°C, in an inert atmosphere (mostly nitrogen). Activation is the next step, which necessitates contacting the char with oxidising gas, like carbon dioxide, steam, air or their mixtures, at a temperature range of 600°C and 900°C. At this stage is the elimination of disordered carbon and creation of a well-refined micropore structure. Mostly CO_2 is used as an activation gas due to its clean nature, easy handling, and expedited control of the activation process due to the slow reaction rate at around 800°C [60].

2.8.2.2 Chemical Activation

Chemical activation involves impregnation with chemicals such as $ZnCl_2$, KOH, NaOH, H_3PO_4, or K_2CO_3 after which they are heated under a nitrogen flow between 450°C–900°C. In the chemical activation process, carbonisation and activation are performed concurrently. Chemical activation has some advantages because it can be performed at lower temperatures in a single step; thus, creating a better porous structure. The main concern is the environmental issue of using chemical agents for activation that could be developed, although part of the added chemicals can be recovered easily [61].

However, a two-step process can be used as an alternative. Steam-pyrolysis is an additional one-step treatment, and is the heating of raw agricultural residue usually performed at a temperature range (500°C–700°C) under a flow of pure steam or could be heated between 700°C–800°C under a flow of steam. Several agricultural residues from straw, bagasse, apricot stones, cherry stones, grape seeds, nutshells, almond shells, oat hulls, corn stover, and peanut hulls have been studied using this method [60,61].

2.9 TECHNIQUES AND IMPACTS OF AGRICULTURAL WASTE–DERIVED ENERGY

Agricultural wastes contain a high carbon content, which makes it a suitable fuel for industrial processes, energy generation, service sectors, and household energy source. Agricultural waste–derived energy includes all forms of energy produced from the conversion of wastes, such as biofuels, biogas, heat, and electricity. Conversion techniques involve biochemical conversion and thermochemical processes depending on the desired energy product and state of matter. The following subsections discuss the details of the various conversion techniques.

2.9.1 BIOCHEMICAL CONVERSION TECHNIQUES

Agricultural wastes are composed of cellulose, hemicellulose, and lignin. The processes of biochemical conversion involve the use of cellulose and hemicellulose breakdown to cellulose, to produce fuels. These techniques involve the chemical conversion of biomass through routes such as anaerobic digestion, fermentation, and transesterification applied to convert a "wet" biomass to fuels [62,63]. These conversions routes are described further below.

2.9.1.1 Anaerobic Digestion

This involves the microbial conversion of biomass with 85%–90% moisture content in the absence of oxygen to generate gasses comprised of carbon dioxide (CO_2), (biogas), and other trace gasses such as hydrogen sulphide (H_2S) with byproducts that could be used as agricultural fertilizers. Anaerobic digestion takes places in three major phases: hydrolysis (breakdown of complex biomolecules to simple biomolecules), fermentation (conversion of cellulose to alcohols, fatty acid, acetic acids, and gasses such as hydrogen and carbon dioxide), and methanogenesis (metabolysis of the gasses to biogas). Biogas generated from this process possesses approximately 20%–40% of the lower heating value (LHV) of the energy content in the biomass feedstock and can be utilised for electricity generation in combined heat and power (CHP) plants [9,62]. One of the ways to maximise nutrient extraction from a biomass is via anaerobic digestion, and it is a flexible technology with several advantages for production of energy from agricultural and other biodegradable wastes both for domestic and industrial processes.

2.9.1.2 Fermentation

The third step in the production of alcohols after a biomass pre-treatment is to increase its surface and hydrolysis of cellulose and hemicellulose to simple sugars is fermentation. It is the enzymatic conversion of simple sugars to alcohols in the presence of microbes such as bacteria. It converts fermentable substrates into products like organic acids and alcohols using microorganisms and/or enzymes.

Crude alcohol produced from the process is distilled to concentrate it. Presently, the most desirable product from fermentation is ethanol; however, the production of ethanol is associated with other high-value co-products such as hydrogen, methanol, and succinic acid [62].

2.9.1.3 Transesterification

Transeterification is the conversion reaction of fats and oils to form esters and glycerol in the presence of catalysts. The process produces biodiesel and glycerol. A two-step esterification-transesterification technique is used to produce biodiesel. In comparison to conventional fossil diesel fuel, biodiesel offers certain advantages such as sulphur free fuel, lower carbon monoxide emission, less smoke, and particulate substances hydrocarbons free compound. It also has higher free oxygen content, which results in complete combustion and reduced emissions. Other advantages include its biodegradability, higher flashpoint, and characteristic lubricity over the conventional Petro diesel [9,64].

2.9.2 THERMOCHEMICAL CONVERSION TECHNIQUES

Thermochemical routes utilise heat to break down "dry" organic matter via chemical reactions. It is a high-temperature chemical process that involves breaking chemical bonds and organic matter reformation into biochar (solid), synthesis gas, and highly oxygenated bio-oil (liquid). These techniques include gasification, pyrolysis, and liquefaction. Several factors determine the choice of conversion technique applied, such as the type and volume of biomass feedstock, energy end use, project specifications, etc. [9,65]. The techniques are discussed in detail in the following sections.

2.9.2.1 Gasification

Gasification involves the conversion of chemical energy present in a solid fuel biomass into a gas product with a capacity to generate work. It is a process in which organic matter undergoes thermal decomposition in the presence of stoichiometric oxidizing agents in reaction process. This process takes place in the absence of oxygen at pressures up to 33 bar and high temperatures (500°C–1400°C). A mixture of flammable gas is comprised of carbon oxide (CO_x), hydrogen (H_2), hydrocarbon, steam, and tars. The low-calorific value gas produced can be directly combusted or utilised as fuel for gas engines and gas turbines in electricity generation [9,62,65].

2.9.2.2 Combustion

Combustion accounts for 97% of bioenergy use and involves the partial or complete oxidation of solid biomass to convert chemical energy to heat. The efficiency of the combustion process is dependent on factors like the characteristics of the fuel, operating conditions, types of furnaces, etc. Application of this technique spans from drying to concurrent production of thermal and electrical energy through steam turbines [62,65].

2.9.2.3   Pyrolysis

Pyrolysis is an endothermic process that involves thermal degradation of a biomass at moderate temperatures (300°C–800°C) in comparison to high gasification temperatures (800°C–1200°C). The process requires a set of chemical mechanisms that produces radical elements, bio-oil, biochar, and syngas. Generally, the characteristics of the products vary according to the techniques applied. These include three main groups: slow, intermediate, and fast. Each technique applies different rates of heating, temperatures, and vapor/solid feedstock residence times. Liquid products from this process could be utilised as energy carriers for various industrial applications or in turbines, engines, and processing plants. Pyrolysis oil can also be co-processed in conventional refineries, after its upgrade into transportation fuels using refinery hydrogen and pyrolysis off-gas can be used in the refinery [62,66].

2.9.2.4   Liquefaction

Liquefaction is a process like the pyrolysis technique that produces bio-oil or bio-crude from biomass. It involves two conversion techniques: thermochemical and hydrothermal (HTL). The thermochemical process produces bio-oil at low temperatures and high pressure in the presence of hydrogen with or without a catalyst. The HTL process, on the other hand, uses subcritical water at temperatures ranging from 250°C–340°C and operates at pressures ranging from 40–220 bars to convert biomass to bio-oil. This technique uses biomass with high moisture content, which potentially minimises the cost of drying or dewatering the biomass. The process is expensive and produces a tarry lump, which is difficult to handle [9,67].

2.10   SUSTAINABILITY AND EVALUATION OF ENERGY GENERATED FROM AGRICULTURAL WASTES

Bioenergy systems are often considered sustainable because of the renewability of their feedstock; however, they may also present potential negative consequences. This connotes a contest on their opinion about their sustainability, which is because such systems are inherently complex, normative, subjective, and ambiguous. Therefore, there is a need for proper sustainability evaluation of each project or technology [68,69]. As discussed in section 2.8, there are several techniques to convert agricultural waste to energy; however, each technique significantly differs in efficiency, scale, and impact. This makes the sustainability assessment project or technology specific and makes it impossible to grow a quantitative procedure encompassing a wide range of various projects under the bioenergy category [70].

Sustainability evaluation is important to achieve the needed government and community support to enable such projects attain political and economic feasibility. Consequently, several factors such as economic, environmental, and social aspects must be considered. The following sections provides detailed discussion on the factors.

2.10.1 ECONOMIC CONSIDERATIONS

Economic considerations are generally associated with the assessment of the competitiveness of bioenergy projects in comparison to their fossil fuel counterparts. The methods adopted are usually techno-economic analysis, net energy balance, market analysis, and cost-benefit analysis of projects. The fundamental of techno-economic analysis is the calculation of the internal rate of return (IRR), net present value (NPV), and discounted payback period, commonly adopted by researchers because of its relatively low threshold and universality [70,71].

Common economic considerations and their indicators include feasibility of the business (net present value, minimisation of costs, adequate funding), long-term perspective (long-term commitments, contracts, and management plans), reliability of resources (security of supply), reliability of the used technology (adequately proven technology), risk minimisation strength and diversification of local economy, reliability of energy, and impact on other desirable developments [68].

2.10.2 ENVIRONMENTAL CONSIDERATIONS

This principally investigates the expected environmental impacts of bioenergy systems. These involve the emission of greenhouse gasses and poor air quality; however, often touch on tenable land and water management. The common technique applied is the Lifecycle Assessment (LCA) methodology. The LCA is a "cradle-to-grave" method, which accounts for the environmental impact across the life cycle stages of the energy product. LCA comprises of a set of tools that represent material and energy flow of in and out of a product's life cycle. When high-quality data is available, LCA can be used for a comprehensive simulation analysis to evaluate environmental and ecological impacts of product systems [69–72].

Common environmental considerations comprise the quality of environmental resources (air, water, soil), land use change, use of ground and surface water, biodiversity and preservation of sensitive ecosystems, deforestation, desertification, and drought landscape view, including landscape variation, conservation of typical elements, use of agrochemicals, waste management, use of energy, and use of fossil resources [68].

2.10.3 SOCIAL CONSIDERATIONS

This is usually the least considered in studies and often considers issues around the impact on rural regions and their incomes from agricultural commodities. The most-used approach is the multi-index analysis, which consists of several indicators; for instance, job creation and income change. One social benefit related to bioenergy development is the amount of job opportunities created by bioenergy-related industries [69–71,73]. Common social considerations and their indicators include labor conditions (freedom of association, collective bargaining, minimum wages, and discrimination), protection of human safety and health (protection of human health, safe and healthy work environment), protection of human rights (rights of children, women, indigenous people, and discrimination), etc. [68].

2.11 COST OF ENERGY GENERATED FROM WASTES COMPARED WITH ENERGIES FROM OTHER SOURCES

Comparing the cost of energy from agricultural wastes with energies generated from other sources is quite complex because each of the various energy sources available tends to have numerous contrasting factors affecting their generation and deployment, and several different real-world market conditions, which is further convoluted by the necessity to consider the time variations, by means of a uniform discount rate when evaluating the cost. To enable "an apple for an apple" kind of comparison, it's imperative to compare the unit cost of energy generated for each of the energy resources considered. This can be represented as $/kWh (dollar per kilowatt-hour) or $/MWh (dollar per megawatt-hour). Typically, at par, the kind of comparison in cost per unit energy is also known as the "Levelized Cost of Energy (LCOE)", as presented in Figure 2.5.

The LCOE is calculated as the ratio between all the discounted costs over the lifetime of an electricity generating plant divided by a discounted sum of the actual energy amounts delivered. The LCOE is used to compare different methods of electricity generation on a consistent basis. The LCOE represents the average revenue per unit of electricity generated that would be required to recover the costs of building and operate a generating plant during an assumed financial life and duty cycle. Inputs to LCOE are chosen by the estimator. They can include cost of capital, decommissioning, fuel costs, fixed and variable operations and maintenance costs, financing costs, and an assumed utilisation rate [74].

Several firms provided their LCOEs for various energy sources in the open literature. Some such firms include Lazard, BNEF (Bloomberg New Energy Finance), IRENA (International Renewable Energy Agency), IPCC (Intergovernmental Panel

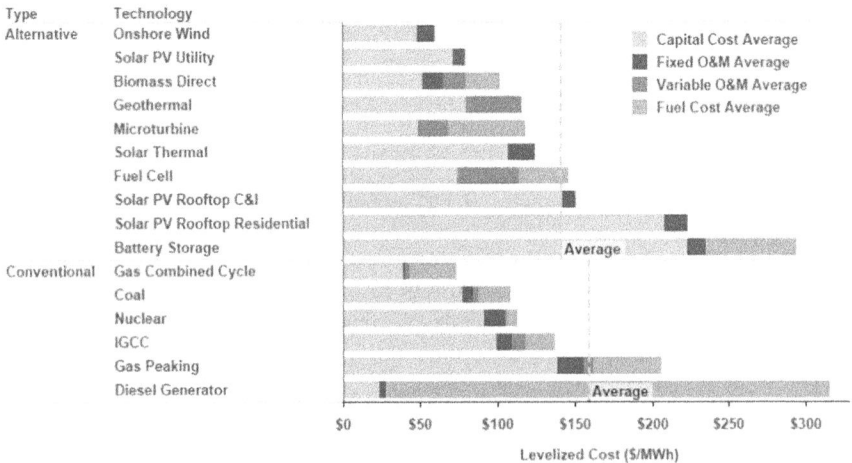

FIGURE 2.5   Components of levelized cost of energy [76].

Source: LAZARD, (2014).

on Climate Change), OECD (Organization for Economic Cooperation and Development), among others. The following section discusses the Lazard's Levelized Cost of Energy, Version 8.0, 2014 [75].

Lazard is an international financial advisory and asset management firm. They provided several versions of LCOEs analysis for various energy sources in the open literature. Though more recent versions with interesting modifications and updates are currently available, with the most recent been the "Lazard's Levelized Cost of Energy Analysis – Version 14.0, October, 2020", their LCOE, Version 8.0, 2014 is selected in this study because of the simplicity of its graphical illustration, which shows and compares the energy generation cost of 16 different energy sources, 6 of which are conventional energy sources, while the remaining 10 are alternative energy sources. The alternative sources are mainly low-carbon, renewable technologies while the conventional sources are fossil fuels and nuclear-based [76].

The energy cost for each of the energy sources analysed is shown in Figure 2.5 in the Lazard's LCOE-Version 8.0 is a summation of all the cost of the components involved, the components include capital, fuel, and operations and maintenance (O&M). The figure reveals that as of 2014, energy generation from the onshore wind resources provide the cheapest renewable alternative energy source and was seconded by the solar PV, while the biomass direct (including agricultural wastes) being the third was also interestingly cost-effective, costing approximately $101/MWh. The overall cheapest energy source was the conventional gas combined cycle ($74/MWh).

Lazard's LCOE shows that as the cost of renewable energy continues to decline, certain alternative sources like the onshore wind, the solar PV, and the biomass direct, which became cost-competitive with conventional energy generation sources several years ago, would maintain competitiveness with lower marginal cost expectations in the future as technologies and innovations evolve [76]. Although alternative, cleaner, renewable energy sources are preferred to minimise dependence on fossil fuels, these alternatives still possess significant differences between, according to viaspace.com. However, biomass (agricultural wastes inclusive), the leading energy alternative to conventional fossil-based sources, offers the most realistic and sustainable alternative to fossil fuels. Biomass is interestingly versatile; it can be used to produce biogas to substitute natural gas and biofuels to substitute gasoline and diesel fuel. It can also be used to produce biochemicals and bioplastics [77].

2.12   CONCLUSIONS

This chapter discusses agricultural wastes from several sources, their structure, applications, and biodegradability. Presently, a lot of bioproducts from agricultural wastes are commercially available and are replacing the conventional products that are more expensive and difficult to dispose of after been used. Instead of burning agricultural wastes, which causes environmental hazards, its preferable to convert them into useful products. In this chapter, it is important to note that agricultural wastes such as *Moringa* pods, *Daniella oliveri* seeds, banana stems, and eggshells can be used to produce nanoparticles for several applications. Lastly, agricultural wastes can be used as filtering materials and alternative energy. This chapter contributes to the understanding and applications of agricultural waste as hybrid polymeric nanocomposites.

ACKNOWLEDGEMENT

The authors are grateful to the Petroleum and Technology Development Fund (PTDF), Nigeria for funding this research work and University of Sao Paulo, Sao Paulo (USP), Brazil for their support for this research work.

REFERENCES

[1] Bhuvaneshwari, S., Hettiarachchi, H., and Meegoda, J. (2019). Crop Residue Burning in India: Policy Challenges and Potential Solutions. *International Journal of Environmental Research and Public Health*, **16**(5), 832.

[2] Guerrero, L.A., Maas, G., and Hogland, W. (2013). Solid Waste Management Challenges for Cities in Developing Countries. *Waste Management*, **33**(1), 220–232.

[3] Alvarez, C., Reyes-Sosa, F.M., and Diez, B. (2016). Enzymatic Hydrolysis of Biomass From Wood. Microbial *Biotechnology*, **9**, 149–156.

[4] Bindhua, M.R., Umadevib, M., Esmailc, G.A., Al-Dhabic, N.A., Arasu, M.V. (2020). Green Synthesis and Characterization of Silver Nanoparticles from Moringa Oleifera Flower and Assessment of Antimicrobial and Sensing Properties. *Journal of Photochemistry & Photobiology, B: Biology*, **205**(111836), 1–7. 10.1016/j.jphotobiol. 2020.111836

[5] Adeyanju, O. and Olatoyinbo, F.A. (2018). Physicochemical Characterization of Daniella oliveri Exudate Gum. A Potential Biomaterial for Green Chemistry Approaches to Nanotechnology. *Journal of Pharmaceutical and Applied Chemistry An International Journal*, **4**(3), 183–187. 10.18576/jpac/040304

[6] Subagyo, A. and Chafidz, A. (2018). Banana Pseudo-Stem Fiber: Preparation, Characteristics, and Applications. In Jideani, A.I.O. Banana Nutrition – Function and Processing Kinetics. *IntechOpen*, 1–19. DOI: 10.5772/intechopen.82204

[7] Adeyeye, E.I. (2009). Comparative Study on the Characteristics of Egg Shells of Some Bird Species. *Bulletin of the Chemical Society of Ethiopia*, **23**(2), 159–166.

[8] Kharat, D.S. (2015). Preparing Agricultural Residue Based Adsorbents for Removal of Dyes from Effluents – A Review. *Brazilian Journal of Chemical Engineering*, **32**(1), 1–12. 10.1590/0104-6632.20150321s00003020

[9] Lee, S.Y., Sankaran, R., Chew, K.W., Tan, C.H., Krishnamoorthy, R., Chu, D.-T., Show, P.-L. (2019). Waste to Bioenergy: A Review On The Recent Conversion Technologies. *BMC Energy*, **1**(1), 1–22. 10.1186/s42500-019-0004-7

[10] Antonio, T., (2009). A Review on Biomass: Importance, Chemistry, Classification, and Conversion. *Biofuel Research Journal*, **22**, 962–979.

[11] Ramos, J.L., García-Lorente, F., Valdivia, M., and Duque, E. (2017). Green Biofuels and Bioproducts: Bases for Sustainability Analysis. *Microbial Biotechnology*, **10**(5), 1111–1113.

[12] Philp, J.C., Ritchie, R.J., and Allan, J.E.M. (2013) Biobased Chemicals: The Convergence of Green Chemistry with Industrial Biotechnology. *Trends in Biotechnology*, **31**, 219–222.

[13] Ramos, J.L., Udaondo, Z., Fernandez, B., Molina, C., Daddaouda, A., Segura, A., and Duque, E. (2016a). First and Second Generation Biochemicals from Sugars. Microbial *Biotechnology*, **9**, 8–10.

[14] Okullo, A., Temu, A., Ogwok, P., and Ntalikwa, W. (2012). Physico-chemical Properties of Biodiesel from Jatropha and Castor oils. *International Journal of Renewable Energy Resources*, **1**, 47–52.

[15] Caye, M.D., Nhuan, P., and Terry, H.W. (2008). *Biofuel's Engineering Process Technology*. McGraw-Hill, USA.

[16] Ramos, J.L., Valdivia, M., Garcıa-Lorente, F., and Segura, A. (2016b) Benefits and Perspectives on the use of Biofuels. *Microbial Biotechnology*, **9**, 436–440.

[17] Valdivia, M., Galan, J.L., Laffarga, J., Ramos, J.L. (2016.). Biofuels 2020: Biorefineries Based on Lignocellulosic Materials. *Microbial Biotechnology*, **9**, 585–594.

[18] Wang, L., and S Chen, H. (2011). Increased Fermentability of Enzymatically Hydrolyzed Stream-Exploded Corn-stover for Butanol Production by Removal of Fermentation Inhibitors. *Process Biochemistry*, **46**, 604–607.

[19] Aracil, C., Haro, P., Giuntoli, J., and Ollero, P. (2012) Proving the Climate Benefit in the Production of Biofuels from Municipal Solid Waste Refuse in Europe. *Journal of Cleaner Product*, **142**, 2887–2900.

[20] Ingale, S., Joshi, S.J., and Gupte, A. (2014). Production of Bioethanol Using Agricultural Waste: Banana Pseudo Stem. *Brazilian Journal of Microbiology*, **45**(3), 885–892.

[21] Vidanaarachchi, C.K., Yuen, S.T.S., and Pilapitiya, S. (2006). Municipal Solid Waste Management in the Southern Province of Sri Lanka: Problems, Issues, and Challenges. *Waste Management*, **26**(8), 920–930.

[22] Slagstad, H. and Brattebø, H. (2013). Influence of Assumptions about Household Waste Composition in Waste Management LCAs. *Waste Management*, **33**(1), 212–219.

[23] Hayashi, K., Ono, K., Kajiura, M., Sudo, S., Yonemura, S., Fushimi, A., Saitoh, K., Fujitani, Y., and Tanabe, K. (2014). Trace Gas and Particle Emissions from Open Burning of Three Cereal Crop Residues: Increase in Residue Moistness Enhances Emissions of Carbon Monoxide, Methane, and Particulate Organic Carbon. *Atmospheric Environment*, **95**, 36–44.

[24] Lohan, S.K., Jat, H.S., Yadav, A.K., Sidhu, H.S., Jat, M.L., Choudhary, M., Peter, J.K., and Sharma, P.C. (2018). Burning Issues of Paddy Residue Management in North-West States of India. *Renewable and Sustainable Energy Reviews*, **81**, 693–706.

[25] Shi, T., Liu, Y., Zhang, L., Hao, L., and Gao, Z. (2014). Burning in Agricultural Landscapes: An Emerging Natural and Human Issue in China. *Landscape Ecology*, **29**(10), 1785–1798.

[26] Ma, J. (2009). The Reasons for Crop Straw Burning in China: Profit Comparisons and Constrain Condition Analyses. *Agricultural Economy Technology*, **2**:77–84.

[27] Tian, H., Deng, W., Shen, Z., Chen, H., and Du, Z. (2010) Advances in the Study on the Environmental Impacts of Crop Straw Burning. Proceedings of Atmospheric Physics and Environment, the 27th Annual Meeting of China Meteorological Society.

[28] Li, X., Wang, S., Duan, L., Hao, J., Li, C., Chen, Y., and Yang, L. (2007). Particulate and Trace Gas Emissions from Open Burning of Wheat Straw and Corn Stover in China. *Environmental Science & Technology*, **41**(17), 6052–6058.

[29] Ayerza (h), R. (2012). Seed and Oil Yields of Moringa Oleifera Variety Periyakalum-1 Introduced for Oil Production in Four Ecosystems of South America. *Industrial Crops and Products*, **36**, 70–73. DOI: 10.1016/j.indcrop.2011.08.008

[30] Daba, M. (2016). Miracle Tree: A Review on Multi-purposes of Moringa oleifera and Its Implication for Climate Change Mitigation. *Journal of Earth Science & Climate Change*, **7**(8), 366, 1–5. DOI: 10.4172/2157-7617.1000366

[31] Aziz, N.A.A., Jayasuriya, N., and Fan, L. (2016). Adsorption Study on Moringa Oleifera Seeds and Musa Cavendish as Natural Water Purification Agents for Removal of Lead, Nickel and Cadmium from Drinking Water. *IOP Conference Series: Materials Science and Engineering*, **136**(012044), 1–9. doi:10.1088/1757-899X/136/1/012044

[32] Das, P.E., Majdalawieh, A.F., Abu-Yousef, I.A., Narasimhan, S., and Poltronieri, P. (2020). Use of A Hydroalcoholic Extract of Moringa oleifera Leaves for the Green Synthesis of Bismuth Nanoparticles and Evaluation of Their Anti-Microbial and Antioxidant Activities. *Materials*, **13**(876), 1–19. doi:10.3390/ma13040876

[33] Rajeswari, M., Pushpa, A., Roopa, G.S., Akshay, J.A., and Praveen, K.G. (2018). Green Synthesis and Characterization of Multifunctional Zinc Oxide Nanomaterials using Extract of Moringa Oleifera Seed. *Materials Today: Proceedings*, **5**, 20996–21002.

[34] Belliraj, T.S., Nanda, A., and Ragunathan, R. (2015). In-vitro Hepatoprotective Activity of Moringa Oleifera Mediated Synthesis of Gold Nanoparticles. *Journal of Chemical and Pharmaceutical Research*, **7**(2), 781–788.

[35] Pradhipa, S. and Priya, R.I. (2018). Green Synthesis of Iron Nanoparticles from Moringa Oleifera Seeds. *International Journal of Science and Research*, **8**(8), 632–640. DOI: 10.21275/ART20199923

[36] Beyene H.D. and Ambaye, T.G. (2019). Application of Sustainable Nanocomposites for Water Purification Process, in *Sustainable Polymer Composites and Nanocomposites*. I.S. Thomas, R.K. Mishra, and A.M. Asiri, Editors. Springer Nature, Switzerland AG, pp. 387–392. DOI: 10.1007/978-3-030-05399-4

[37] Amina, M., Al Musayeib, N.M., Alarfaj, N.A., El-Tohamy, M.F., Orabi, H.E., Bukhari, S.I., and Mahmoud, A.Z. (2019). Exploiting the Potential of Moringa oleifera Oil/Polyvinyl Chloride Polymeric Bionanocomposite Film Enriched with Silver Nanoparticles for Antimicrobial Activity. *International Journal of Polymer Science*, **5678149**, 1–11. DOI: 10.1155/2019/5678149

[38] Abdelhamid, A.E., Yousif, E.A.A., El-Saidi, M.M.T., El-Sayed, A.A. (2020). Polyvinyl Alcohol Food Packaging System Comprising Green Synthesized Silver Nanoparticles. *Indonesian Journal of Chemistry*, **21**(2), 1–11. DOI: 10.22146/ijc.55483

[39] Hazarika, A., and Maji, T.K. (2017). Ultraviolet Resistance and Other Physical Properties of Softwood Polymer Nanocomposites Reinforced with Zno Nanoparticles and Nanoclay. *Wood Material Science & Engineering*, **12**(1), 24–39. DOI: 10.1080/17480272.2014.992471

[40] Langenhein, J.H. (1983). In: *Tropical Forest Ecosystems in Africa and South America: A Comparative Review* (Meggers, B.C., Ed.) Smithsonian Institution Press, Washington, DC, pp 30–35.

[41] Djoueche, C.M., Azebaze, A.B., and Dongmo, A.B. (2011). Investigation of Plants Used for the Ethnoveterinary Control of Gastrointestinal Parasites in Bénoué Region, Cameroon, *Tropicultura*, **29**(4), 205–211.

[42] Okunade, S.A., and Olafadehan, O.A. (2019). Rolfe (Daniellia oliveri) Seed Meal as a Protein Source in Locally Produced Concentrates for Lambs Fed Low Quality Basal Diet. *Journal of the Saudi Society of Agricultural Sciences*, **18**, 83–88.

[43] Adubiaro, H.O., Olaofe, O., and Akintayo, E.T. (2011). Chemical Composition, Calcium, Zinc and Phytate Interrelationships in Albizia lebbeck and Daniellia oliveri seeds. *Oriental Journal of Chemistry*, **27**(1), 33–40.

[44] Padam, B.S., Tin, H.S., Chye, F.Y., and Abdullah, M.I. (2012). Banana by-products: an under-utilised renewable food biomass with great potential. *Journal of Food Science and Technology*. 10.1007/s13197-012-0861-2

[45] Zhang, C., J. Li, C. Liu, X. Liu, J. Wang, S. Li, and L. Zhang. (2013). Alkaline pretreatment for enhancement of biogas production from banana stem and swine manure by anaerobic codigestion. *Bioresour. Technol.* **149** (Dec), 353–358. 10.1016/j.biortech.2013.09.070

[46] Li, K., Fu, S., Zhan, H., Zhan, Y., and Lucia, L.A. (2010). Analysis of the Chemical Composition and Morphological Structure of Banana Pseudo-Stem. *BioResources*, **5**(2), 576–585. DOI: 10.15376/BIORES.5.2.576-585

[47] Jirukkakul, N. (2018). Physical Properties of Banana Stem and Leaf Papers Laminated with Banana Film. *Walailak Journal Science & Technology*, **16**(10), 753–763.

[48] Li, C., Liu, G., Nges, I.A., Deng, L., Nistor, M., and Liu, J. (2016). Fresh Banana Pseudo-Stems As A Tropical Lignocellulosic Feedstock For Methane Production. *Energy, Sustainability and Society*, **6**(27), 1–9. DOI: 10.1186/s13705-016-0093-9

[49] Assisa, F.S., Margema, F.M., Cordeiroa, T.C., Figueiredob, A.B., Bragab, F.O., and Monteiro, S.N. (2015). Photoacoustic Thermal Characterization of Banana Fibers. *Materials Research*, **18**(2), 240–245. DOI: 10.1590/1516-1439.368914

[50] Prakash, C.T., Solanki, G.S., Sakale, R., and Singh, H.P. (2018). Application & Analysis of Banana Stem Fibre use as Construction Material. *International Journal of Trend in Scientific Research and Development*, **2**(2), 235–246.

[51] Hincke, M.T., Nys, Y., Gautron, J., and Mann, K. (2012). The Eggshell: Structure, Composition and Mineralization. *Frontiers in Bioscience*, **17**, 1266–1280. https://draxe.com/nutrition/eggshell

[52] Macneil, J.H., 2001. Method and Appatus for Separating a Protein Membrane and Shell Material in Waste Egg Shells. The Penn State Research Foundation, *US PATENTS*.

[53] Feng, Y., Yang, F., Wang, Y., Ma, L., Wu, Y., Kerr, P.G., and Yang, L. (2011). Basic Dye Adsorption onto an Agro-Based Waste Material – Sesame Hull (Sesamum indicum L.). *Bioresource Technology*, **102**, 10280–10285.

[54] Ip, A.W.M., Barford, J.P., and McKay, G. (2010). A Comparative Study on the Kinetics and Mechanisms of Removal of Reactive Black 5 by Adsorption onto Activated Carbons and Bone Char. *Chemical Engineering Journal*, **157**(2–3), 434–442.

[55] Adegoke, K.A. and Bello, O.S. (2015). Dye Sequestration using Agricultural Wastes as Adsorbents. *Water Resources and Industry*, **12**, 8–24.

[56] Young, L., Jain, Y.U. (1997). Ligninase-catalyzed Decolorization. *Water Resources*, **31**, 1187–1193.

[57] Devi, R., Singh, V., and Kumar, A. (2008). COD and BOD Reduction from Coffee Processing Watewater Using Avacado Peel Carbon. *Journal of Bioresource Technology*, **99**, 1853–1860.

[58] Sulak, M.T., Demirbas, E., and Kobya, M. (2007). Removal of Astrazon Yellow 7GL from Aqueous Solutions by Adsorption onto Wheat Bran. *Journal of Bioresource Technology*, **98**, 2590–2598.

[59] Singh, K.P., Mohan, D., Sinha, S., Tondon, G.S., and Gosh, D. (2003). Color Removal from Wastewater Using Low-Cost Activated Carbon Derived from Agricultural Waste Material. *Industrial & Engineering Chemistry Research*, **42**(9), 1965–1976. DOI: 10.1021/ie020800d

[60] Zhang, T., Walawender, W.P., Fan, L.T., Fan, M., Daugaard, D., and Brown, R.C. (2004). Preparation of Activated Carbon from Forest and Agricultural Residues through CO_2 Activation. *Chemical Engineering Journal*, **105**, 53–59.

[61] Okieimen, F.E., Okiemen, C.O., and Wuana, R.A. (2007). Preparation and Characterization of Activated Carbon from Rice Husks. *Journal of Chemical Society*, **32**, 126–136.

[62] Ben-Iwo, J., Manovic, V., and Longhurst, P. (2016). Biomass Resources and Biofuels Potential for the Production of Transportation Fuels in Nigeria. *Renewable and Sustainable Energy Reviews*, **63**, 172–192. 10.1016/j.rser.2016.05.050

[63] Sims, R.E.H., Mabee, W., Saddler, J.N., and Taylor, M. (2010). An Overview of Second Generation Biofuel Technologies. *In Bioresource Technology*, **101**(6), 1570–1580. 10.1016/j.biortech.2009.11.046

[64] Gebremariam, S.N., and Marchetti, J.M. (2018). Economics of Biodiesel Production: Review. *Energy Conversion and Management*, **168**, 74–84. 10.1016/j.enconman.2018. 05.002

[65] Thermie (1995). Combustion and Gasification of Agricultural Biomass-Technologies and Applications.

[66] Santos, J., Ouadi, M., Jahangiri, H., and Hornung, A. (2020). Thermochemical Conversion of Agricultural Wastes Applying Different Reforming Temperatures. *Fuel Processing Technology*, **203**, 106402. 10.1016/j.fuproc.2020.106402

[67] Goyal, H.B., Seal, D., and Saxena, R.C. (2008). Bio-fuels from Thermochemical Conversion of Renewable Resources: A review. *Renewable and Sustainable Energy Reviews*, **12**(2), 504–517. 10.1016/j.rser.2006.07.014

[68] Buytaert, V., Muys, B., Devriendt, N., Pelkmans, L., Kretzschmar, J.G., and Samson, R. (2011). Towards Integrated Sustainability Assessment for Energetic Use of Biomass: A State of the Art Evaluation of Assessment Tools. *In Renewable and Sustainable Energy Reviews*, **15**(8), 3918–3933. 10.1016/j.rser.2011.07.036

[69] Palmeros Parada, M., Osseweijer, P., and Posada Duque, J.A. (2017). Sustainable Biorefineries, an Analysis of Practices for Incorporating Sustainability in Biorefinery Design. *Industrial Crops and Products*, **106**, 105–123. 10.1016/j.indcrop.2016.08.052

[70] O'connell, D., Keating, B., and Glover, M. (2005). Sustainability Guide for Bioenergy: a Scoping Study. http://www.rirdc.gov.au

[71] Wang, J., Yang, Y., Bentley, Y., Geng, X., and Liu, X. (2018). Sustainability Assessment of Bioenergy from a Global Perspective: A review. *Sustainability (Switzerland)*, **10**(8), 1–19. 10.3390/su10082739

[72] Arodudu, O., Helming, K., Wiggering, H., and Voinov, A. (2017). Towards a more Holistic Sustainability Assessment Framework for Agro-Bioenergy Systems—A Review. *In Environmental Impact Assessment Review*, **62**, 61–75. 10.1016/j.eiar. 2016.07.008

[73] Lawal, K.O. (2015). Hydro-based, Renewable Hybrid Energy Sytem For Rural/ Remote Electrification in Nigeria. 2015 Clemson University Power Systems Conference, PSC 2015, 1–6. 10.1109/PSC.2015.7101691

[74] Lai, C.S. and McCulloch, M.D. (2017). Levelized Cost Of Electricity For Solar Photovoltaic And Electrical Energy Storage. *Applied Energy*. **190**, 191–203. doi: 10.1016/j.apenergy.2016.12.153

[75] Transparent Cost Database. (2021). Open Energy Information (en). Accessed April, 1st, 2021: https://openei.org/wiki/Transparent_Cost_Database"

[76] Lazard. (2014). Lazard's Levelized Cost Of Energy Analysis—version 8.0, (September), 0–19.

[77] Liu, W., Wang, J., Richard, T.L., Hartley, D.S., Spatari, S., and Volk, T.A. (2017). Economic and Life Cycle Assessments of Biomass Utilization for Bioenergy Products.

3 Nanocomposites Based on Nanoparticles from Agricultural Wastes

Henry Ekene Mgbemere,
Eugenia Obiageli Obidiegwu,
Johnson Olumuyiwa Agunsoye,
and Suleiman Bolaji Hassan
Department of Metallurgical and Materials Engineering,
University of Lagos, Lagos State, Nigeria

CONTENTS

DOI: 10.1201/9781003170549-4

45

3.1 INTRODUCTION

Agriculture is one of the financial bedrocks for many nations of the world, especially the developing countries, because it constitutes a major part of their Gross
Domestic Product (GDP). Agricultural wastes refer to either low-value or unwanted
materials that are generated due to agricultural practises like planting and harvesting
crops, livestock farming, and even processing of agricultural products. Agricultural
wastes are also known as biomass, which can exist in different forms like solid,
liquid, and even as lubricants [1].

Globally, it is estimated that about 140 billion metric tonnes of waste products are
generated yearly from agricultural operations [2]. According to the Food and
Agricultural Organisation (FAO), it is also estimated that 20%–30% of fruits and
vegetables produced during the harvest season are left to spoil, thereby producing a lot
of waste [3]. In India, it is estimated that between 18% and 40% of all fresh fruits and
vegetables end up as waste [4]. Some of these wastes are used for energy generation
purposes, but the majority are burnt openly, especially in developing countries [5].
Some of these wastes can be converted to raw materials as input, which can then serve
as feedstock for different applications as seen in Table 3.1. If these wastes were to be
converted to crude oil, for instance, it will be approximately equivalent to about
50 billion tonnes [6]. Some of the benefits that can be derived from these wastes
include a reduction in greenhouse gas emissions, provision of feedstock for renewable
energy which can power about 1.6 billion people around the world who do not have
access to stable electricity, and creation of jobs for the unemployed [1].

3.2 CATEGORIES OF AGRICULTURAL WASTES

Agricultural operations can be classified into livestock farming, crop farming, and
processing of primary agricultural products into secondary products either for
consumption or for other purposes. The waste products from all these operations
include manure, farm waste, slaughterhouses, postharvest waste, peels from orange,
grape pomace, pineapple, sugarcane bagasse, lemon, rice husks, palm oil fibres,
fruit seeds, coconut shells, wheat straw, potato peels, soy protein isolates, sugar
palm, sugarcane bagasse, sawdust, neem bark, corn starch, straw, animal husbandry
waste, roots, leaves, husk, nuts, seed shells, waste wood, etc. [7–12].

TABLE 3.1

Cost Comparison on Commonly Used Nanofillers

Types of Nanofillers	Cost
Graphite	$2 to $5 per kilogrammme
Clays	$2 to $5 per kilogramme
Carbon nanofibers	$95 to $1500 per kilogramme
Single-walled carbon nanotubes	$170,000 per kilogramme
Multiple-walled carbon nanotubes	$0.5–100 per gramme
Double-walled carbon nanotubes	$10 per gramme
Nanosilica	$8.50 per kilogramme
Delonix regia seed	$0.65 per kilogramme
Coconut shell	$0.52 per kilogramme

The above-mentioned agricultural wastes initially have very little or no economic value but through conversion to nanoparticles, their value can be increased. Plastic wastes can equally be generated from agricultural practises through the shading of nets from fishing operations, materials used for mulching, pesticide containers, etc. As the global demand for food increases, the generation of agricultural wastes will continue to increase [3]. Most of these agricultural operations, especially crop farming, led to the generation of more than one type of waste. Table 3.2 shows a list of some of the major crops used in agriculture and the types of wastes that they generate.

With the help of the right technologies, a lot of advances have been made in recent years to convert biomass in the form of agricultural wastes into useful raw

TABLE 3.2

Major Crops Usually Planted and the Type of Wastes Generated

Crop	Waste Generated from Crops
Coconut	Shell, front, husk
Coffee	Ground, hull, husk
Corn	Stover, leaves, cob, stalks, husk
Cotton	Stalks
Groundnut	Shells
Peanuts	Shell
Rice	Straw, hull/husk, stalks
Sugarcane	Bagasse
Pepper	Stalks

materials for different applications using the right technologies. This ranges from controlled burning of wastes like rice husks and sugarcane bagasse to the gasification of other residues [1]. The utilisation rate for most agricultural wastes is still very low which makes them either to rot away or be openly burnt in the fields. This occurs mostly in countries that do not have strong policies for such environmentally harmful practises. They are routinely openly burnt, leading to a lot of air pollution that poses respiratory risks to humans and the environment. The challenge, therefore, is to convert these agricultural wastes into a resource for energy and other productive uses [1]. Presently, most of the agricultural wastes are used in landfills, which generate different gases, thereby contributing to global warming. Using these wastes for commercial applications will certainly reduce the rate of global warming although some contribute to it. According to a report from researchers in Italy, recycling agricultural wastes through fertiliser production and composting increases the carbon emissions rate such that between 64 and 67 kg of CO_2 is generated per mg as well as anaerobic digester-based compost [3]. A piece of information that shows some of these wastes, technologies used for their conversion, and the resulting materials has been reported in literature [1]. Bio-reduction was used to convert pineapple leaves and sugarcane residue to industrial absorbents, animal feed and additives for making beverages. Fertiliser and wheat straw polypropylene pelletised raw materials are produced from cotton gin and wheat straw using bio-reduction. Lumber materials and panelboards are manufactured from wheat straw and sugarcane bagasse by bio-refinery system. and coconut coir is converted to coconut twines by twining.

It is clear that wastes like rice husks are being used to produce nanoparticles of silicon and other metallic nanoparticles. To have an understanding of the contents of these wastes, chemical analysis must be carried out. A chemical composition analysis of some waste particles that indicate their actual composition shows that organic components of the waste materials namely cellulose, hemicellulose, and lignin constitute the bulk [3] while the ash content is the source of the nanoparticles, although nanoparticles are also producable from the organic components. Chemical composition of sugarcane which contains 28.2% cellulose, 54.7% hemicellulose, 11.4% lignin, 1.9% ash, and 3.8% moisture buttresses to the fact that agricultural waste has mainly organic components.

The incorporation of nanoparticles inside a polymer matrix leads to the enhancement of its properties and the simplest process strategy is by adding the appropriate nanoparticles. To select the appropriate agricultural waste that will result in the required nanoparticle, certain factors have to be considered. They include the ability of the agricultural waste to biodegrade, the content of starch, hemicellulose, cellulose, and lignin present as well as the complexity of the route that will lead to the conversion from waste to nanoparticles.

3.2.1 Methods of Nanoparticle Synthesis from Agricultural Wastes

To convert these agricultural wastes into nanoparticles, certain processing steps have to be followed. Nanoparticles have been produced in the last few years

Agricultural waste e.g rice husk

↓

Disc grinding and ball milling (Optimisation of milling parameters are imperative)

↓

Alkali extraction

↓

Leaching using inorganic acids e.g HCl, H_2SO_4, HNO_3 or Organic aid e.g ethanoic acid

↓

Separation of sol from gel by filtration

↓

Nanoparticles

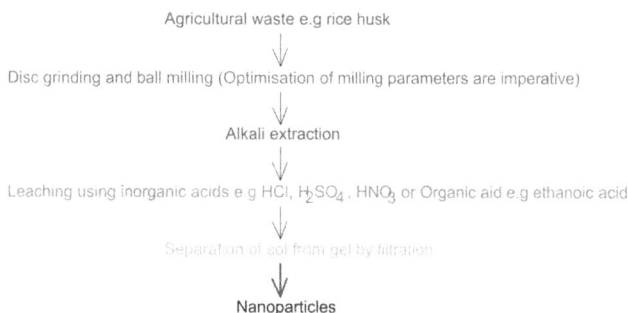

FIGURE 3.1 Flow chart of the synthesis method for nanoparticles from agricultural wastes using chemical treatment.

through a lot of different techniques. They usually have high aspect ratios and are known to serve as good reinforcing agents when dispersed in polymeric materials. Most of the plant-based agricultural wastes have similar compositions with the difference being the content of ash and lignin, except for nuts. Nanoparticles usually serve as reinforcement for polymers and both SiO_2 and clay are the most widely researched type of particles [13]. Wood and agricultural wastes usually serve as the raw material for the carbonisation process. One of the most commonly available inorganic nanoparticles is based on SiO_2. It is produced commercially using expensive chemicals like alkoxysilane, tetraethylorthosilicate (TEOS), etc. There is a need for cheaper, safer, readily available, and more environmentally friendly sources of silica [13]. There is also the awareness by materials scientists to ensure that sustainability in the use of materials is maintained. Methods of extraction are chemical treatment, thermal treatment (pyrolysis), and microbial treatment. A typical example of the chemical treatment of agricultural wastes to be converted into nanoparticles is shown in Figure 3.1.

A very common method to convert agricultural wastes e.g., coconut shells into nanoparticles is through thermal treatment at 600°C–1000°C using a pyrolyser or furnace and ball milling and subsequent incorporation of nanoparticles into a polymer to form nanocomposites, as illustrated using Figure 3.2.

Biological treatment is also known as biogenic synthesis and is used to produce nanoparticles. To obtain nanoparticles using this method, the agricultural waste is suspended in an inoculum containing fungi and/or bacteria which is then biotransformed. Another biogenic synthesis method involves the use of proteins for biomimetic silicification otherwise known as bio-silicification. The products from these two methods are calcined to produce nanoparticles. Ag nanoparticles have been produced when orange peel extract was used as a reducing agent [14]. Fe_3O_4 nanoparticles have also been synthesized using pawpaw leaves while coconut shell extract has been used to synthesize silver nanoparticles. The structure and properties of silica obtained from agricultural wastes depend on the extraction method, the temperature of calcination, as well as the nature of the chemicals.

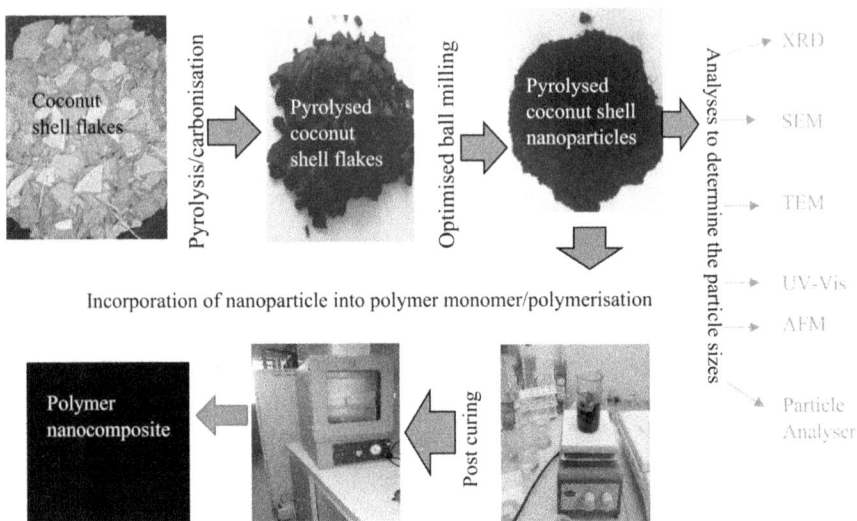

FIGURE 3.2 Thermal treatment of coconut shells into nanoparticles and subsequent incorporation into an epoxy matrix to form polymer nanocomposites.

3.2.2 COMPOSITES

A composite is a combination of two or more materials consisting of more than one chemically dissimilar phase with a distinct interface between them [15]. The continuous phase in a composite is known as the matrix while the dispersed phase(s) in this matrix is called the reinforcement. Nanocomposites on the other hand are defined as those composites where one or more of the reinforcements have dimensions in the nanometre size range. The properties of composites are usually determined by the size and surface chemistry of the reinforcing material [15].

The properties that are found in nanocomposites are usually not found in a conventional composite, ensuring that their application areas are wide-ranging from the packaging of materials to taking care of biomedical challenges. They are gradually being used to replace conventional and even monolithic materials in device manufacture and are termed "the materials of the 21st century" due to these characteristics [16]. This is made possible by the fact that at the nanoscale, the material features can be controlled, making them exceptional in their mechanical characteristics as well as other very important properties [17].

Metal matrix nanocomposites (MMNC), ceramic matrix nanocomposites (CMNC), and polymer matrix nanocomposites (PMNC) are the three main types of nanocomposites. Metal matrix nanocomposites usually have either a ductile metal or an alloy as the matrix and a nano-sized material as reinforcement. Their structures are usually ductile and tough translating to high modulus and high strength. Processing techniques like spray pyrolysis, electrodeposition, liquid metal infiltration, vapour-based techniques, etc. are used in the production of these types of nanocomposites. The ease of manufacture, their light weight, and low cost are some of the reasons which make polymer matrix nanocomposites to be used in a lot of industrial applications [18]. The emphasis of this chapter is the nanoparticle-reinforced polymer matrix nanocomposites.

The wide interest in polymer nanocomposites is in part because even at low contents of nanoparticles, significant improvements can be observed for the produced nanocomposite. One of the major challenges of applying nanoparticles is the fact that the dispersion and distribution of these particles are still a problem. They pose a challenge by reducing their potential for use in flame retardancy, mechanical properties [19].

Challenges do exist in preparing nanocomposites because it is difficult to control their stoichiometry and elemental compositions, especially at the nanometre range. Certain characteristic combinations and unique designs are some of the properties which they exhibit that can't be found in conventional composites. The choice of the appropriate reinforcement, the control of its morphology, and getting the right chemistry of the formulation are the necessary factors in the formulation of a nanocomposite.

According to [20], the merits of nanocomposites in comparison to normal composites include:

   i. Improvement in modulus
  ii. Gas barrier characteristics
 iii. Higher heat distortion temperature
  iv. Ability to resist small molecule permeation
   v. Improvement in ablative resistance
  vi. Increment in atomic oxygen resistance
 vii. Ability to retain high impact strength

3.2.3   FORMULATION OF A NANOCOMPOSITE

To process nanocomposites using nano-reinforcements for possible scale-up, the following conditions need to be met:

  1. The proposed reinforcement needs to be controllable and have reproducible morphologies with consistent properties.
  2. Dispersion of the reinforcement in the solvent and subsequent incorporation into the polymeric matrix should be as easy as possible.
  3. In modifying the surface of the nano-reinforcements for use, the process involved should be simple and reproducible as fast as possible with simple apparatus.

3.3   PROCESSING OF POLYMER NANOCOMPOSITES

To produce polymer nanocomposites, polymers like condensation polymers, polyolefins, vinyl polymers, specialty polymers, etc. are typically used. The reinforcements are usually in the nanosize range and are mixed with the polymers to produce nanocomposites. Synthetic and natural reinforcements are the two types of reinforcements used in nanocomposites. They are usually crystalline and include materials such as metal powders, metal oxides, clays, and other layered silicates [21,22].

The techniques applied in the preparation of a majority of these reinforcements include chemical methods, thermomechanical methods, and biological methods. To obtain a homogenous distribution of the reinforcements with a low volume of agglomerates as well as improvements in the interfacial bonding to the polymer matrix, surface modifications are usually carried out [18,23]. The choice of the particular method to be used is dependent on the type of polymer, the nanofiller, and the type of application being planned for the product [24].

Some of the already used methods to prepare polymer nanocomposites are:

   i. Intercalation of the polymer or pre-polymer from solution
   ii. A direct mixture of polymer and particulates
   iii. Melt intercalation/blending
   iv. Sol-gel process
   v. In-situ polymerisation
   vi. Template synthesis

To obtain enhanced performance of the polymer nanocomposites, a high degree of particle dispersion inside the polymer is necessary. Three possible ways to get proper dispersion of the nanoparticle in the polymer are:

   1. in-situ polymerisation in the presence of nanoparticles.
   2. mixing of the nanoparticles as discrete phases.
   3. combination of in-situ formation of nanoparticles and in-situ polymerisation.

3.3.1   IN-SITU POLYMERISATION

In-situ polymerisation requires that the nanoparticles be very well dispersed in the monomer solution before the commencement of the polymerisation process. This will ensure that the polymer will be formed between the nanoparticles. Here, the particles are initially dispersed inside the monomer and then the mixture is subsequently polymerised. A typical example is polyamide-silica nanocomposites prepared by first removing the moisture in the silica and blending with ε-caproamide as well as an initiator which is added simultaneously [25]. Polymerisation of the mixture at elevated temperature under a nitrogen atmosphere leads to well-dispersed nanocomposites that contain silica of ~50 nm. Agglomerates were also observed for nanoparticles with a size ~12 nm.

The polymerisation process is usually initiated with heat, using an appropriate initiator. When it is obvious that dispersion of the nanoparticles is difficult, modifiers may be added to improve the dispersion of the nanoparticles. The process is illustrated using the schematic diagram shown in Figure 3.3.

The in-situ method is divided into solution blending and melt blending. The solution blending method makes it possible to apply higher amounts of nanoparticles without the challenge of serious agglomeration. Nanocomposites from this

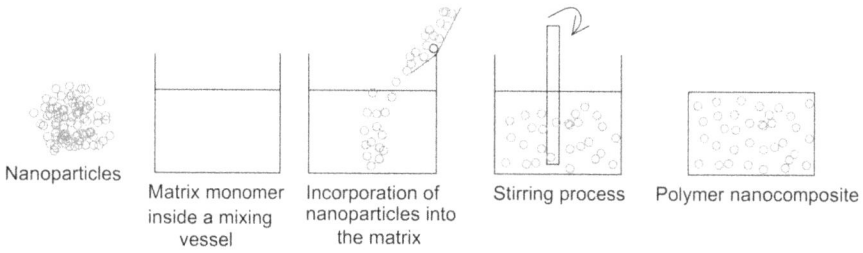

FIGURE 3.3   Schematic diagram of the steps involved in the in-situ polymerisation method.

method have enhanced properties and tend to form covalent bonds between the nanoparticle functional groups as well as polymeric chains. This process works for both thermoset and thermoplastic polymers. The challenge is in uniformly dispersing the nanoparticles inside the polymer matrix.

3.3.1.1   Solution Blending

With the aid of an appropriate solvent, solution blending can be used to easily disperse nanoparticles through magnetic stirring, ultrasonic irradiation, or shear mixing, as shown in Figure 3.4. The permeability of gases is reduced thereby making it easy to operate all types of nanoparticles for both thermoset and thermoplastic polymers. Some of the factors that affect solution blending are economic issues, environmental issues, and agglomeration.

Nanocomposites consisting of polymethyl methacrylate (PMMA) with even dispersion of alumina nanoparticles have been produced using this technique [21]. First, the nanoparticles were mixed with the methylmethacrylate (MMA) and, through sonication, uniformly dispersed in a low viscous solution. A chain transfer agent and an initiator is introduced and the mixture is polymerised under nitrogen atmosphere and dried in a vacuum. Conventional techniques are used to make samples for further processing.

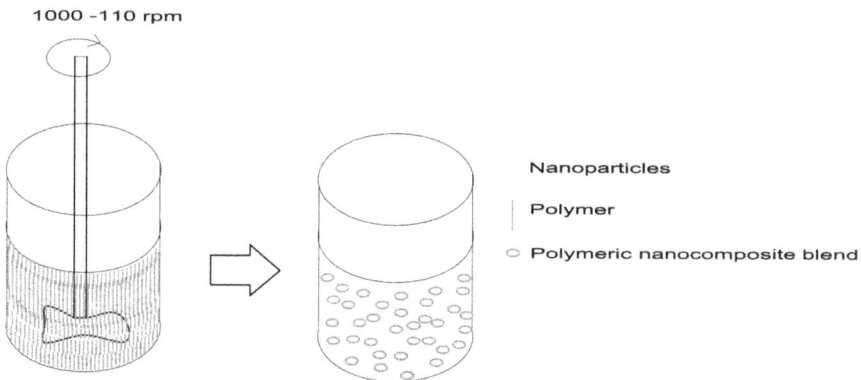

FIGURE 3.4   Schematic illustration of the solution blending method.

3.3.1.2   Melt Blending

In this method, nanoparticles are dispersed directly in the molten polymer at high levels of shear stress to minimise the size of the agglomerate. Extruders whether single screw or twin-screw extruders are used for melt blending but the temperature of the process can be high such that optimisation is reached, else it will be difficult to disperse the nanoparticles. This method has been used for polymer nano-composites with polycarbonate, poly(lactic acid), polypropylene, poly(methyl methacrylate), poly(vinyl chloride), or polyamide matrices [21]. It helps to disperse the nanoparticles, enhance the heat stability, improve the mechanical properties, and is both cost and environmentally friendly.

3.3.2   Direct Mixing

Direct mixing of known polymer processing techniques can be used to prepare polymer nanocomposites. Nanocomposites made from polypropylene and $CaCO_3$ particles (44 nm) have been produced using a direct combination of the various components with a Haake mixer [26]. When the filler volume fraction is low (4.8% and 9.2%), well-dispersed nanocomposites can be produced but at higher volume fractions, aggregation occurs. Polyurethane-silica nanocomposites have been prepared by combining silica with polyol and then curing the mixture with di-isocyanate at 100°C for 16 hours by adding 0.1% of the catalyst [25,27]. Particles that are spherical with a 12 nm average diameter with a narrow size distribution (10–20 nm) were obtained. Samples with very good quality were produced with up to 40 wt%, filler volume fraction. Grafting of styrene monomers to surround nanoparticles has also been reported to yield good dispersion [25,28].

3.3.3   Sol-Gel Method

The sol-gel process has two major reaction steps which include the formation of the sol and gel. Sol is a colloidal suspension of solid particles in a liquid phase, while the gel helps to form the interconnected network produced between phases. The reactions that produce the sol and gel are hydrolysis and condensation reactions, respectively. This method can be used to produce nanoparticles from wastes as well as to prepare the nanocomposites. It is compatible with both polymers and nano-particles where the polymerisation process allows the formation of nanoparticles in the presence of organic molecules. In nanocomposite production by the sol-gel technique, organic molecules and monomers are embedded on sol-gel matrices, introducing organic groups through formation of chemical bonds through in-situ formation of the sol-gel matrix within the polymer and/or simultaneous production of inorganic/organic networks.

3.3.4   Production of Polymer Nanocomposites

1. The mixing of the precursor particle with either a copolymer or polymer matrix followed by sol-gel synthesis is used to obtain a nanoparticle. After drying, the phases in the polymer blocks tend to separate while the silica coalesce.

2. Precursors of the nanoparticle materials like tetraethylorthosilicate (TEOS) are mixed with polymers like epoxy and polyimide. This method has been used to produce nanocomposites containing polymer and TiO_2. When the nanoparticle is mixed directly with polystyrene maleic anhydride (PSMA), there is a terrible aggregation. As a result, the polymer (PSMA) is usually dissolved in chemicals like tetra butyl titanate and tetrahydrofuran (THF) and before introduction under appropriate conditions, the structure of the nanoparticle is checked to obtain well-dispersed TiO_2 nanocomposites [25].

The reaction between maleic anhydride and the uncondensed TiOH ensures that the polymer coats the nanoparticles to minimise the formation of agglomerates. Polyimide-TiO_2 nanocomposite films are obtained with the sol-gel process. Another way of producing the polymer nanocomposite is by combining the direct mixing and melt blending and they have been used to synthesise polymer nanocomposites like polydimethylsiloxane (PDMS)-silica, as reported in [29]. The network swells with TEOS and the sol-gel reaction becomes catalysed.

3.3.5 STRATEGIES TO ACHIEVE IMPROVED DISPERSION QUALITY (CHEMICAL AND PHYSICAL MODIFICATION)

Strategies to obtain improved dispersion quality are surface modification, in-situ polymerisation, mechanical mixing, ultrasonic vibration, and electromagnetic fields.

1. Surface modification
   This helps to enhance the compatibility between the matrix and the re-inforcement/filler through the following procedures: grafting of organo-silanes, using long-chain alkyl ammonium clay intercalating ions.
2. In-situ polymerisation
   In highly viscous polymeric media, this method is used to enhance a good dispersion of the nanoparticles.
3. Mechanical mixing
   The use of high-speed mixers, extruders, etc. aid in the dispersion of the particles.
4. Ultrasonic vibration
   Polymers both in solution and melt polymers use ultrasonic vibration as a means to ensure that the agglomerates are effectively dispersed in the matrix.
5. Electromagnetic fields
   When electromagnetic fields, high shear/compression are applied in a particular direction, it helps to disperse the nanoparticles [19].

Methods used to process polymer-based nanocomposites are mixing of polymer or monomer with reinforcing materials, which is used in processing polymethyl methacrylate and nanoparticles; in-situ polymerisation which encompasses dispersion of nanoparticles into a precursor containing the polymer used for developing thermoset (epoxy, vinyl ester, etc.) based nanocomposites and polymerisation of a mixture of monomer, nanoparticles, and catalyst (hardener or accelerator) [18]; and the sol-gel technique already discussed in Section 3.2.3 of this chapter.

3.4 PROPERTIES AND APPLICATIONS OF AGRICULTURAL WASTE PARTICLE–REINFORCED POLYMERS

Polymer matrix nanocomposites (PMNCs) can be described as products of two or more materials where the matrix material is a polymer while the reinforcement phase has a minimum of one dimension in the nanometer range (< 100 nm) with adequate compatibilisation. They are also defined as materials that contain a mixture of two or more phase-separated materials, where one or more dispersed phase has dimensions in the nanoscale and a polymer as the main phase. A material is said to be in the nanoscale range when a minimum of one of its dimensions is between 1 nm to 100 nm [19].

Three broad types of dispersed phases are used in polymer nanocomposites as nano-reinforcement: layered materials (nano-clay, graphene), nanoparticles, and fibres [15]. The emphasis of this write-up is on the nanoparticle-reinforced polymers.

One of the features that makes polymer nanocomposites stand out is that the size of the nanofillers results in a very pronounced increase in the specific surface area. This large interface then enables a sizable volume fraction of the polymer to have properties that are different from the bulk even at low nanoparticle addition. The factors that determine the effect of the nanoparticle on the properties of the resulting nanocomposites include filler size, filler-matrix interactions, shape, and aspect ratio.

3.4.1 ADVANTAGES OF POLYMER NANOCOMPOSITES

Some of the advantages of using polymer nanocomposites include:

1. High levels of stiffness and strength even with less dense materials as a result of their light weight.
2. Significant improvements in their thermal barrier properties when compared to the neat polymer.
3. Superior mechanical and thermal properties behaviour.
4. Excellent flammability characteristics as well as higher levels of degradation in biodegradable polymers.

3.4.2 DISADVANTAGES OF POLYMER NANOCOMPOSITES

Some of the drawbacks are:

1. Nanocomposites usually have lower toughness and poor optical clarity because of the nanoparticles present.
2. The viscosities of the nanocomposites are much lower than the virgin polymers leading to difficulties during processing.
3. Due to the diverse nature of polymeric materials, it is very easy to obtain polymer nanocomposites with diverse properties. Nanoclay as a

reinforcement is the most used in the preparation of polymer nano-composite [30].

4. Minimal agglomeration and uniform distribution of the dispersant in the polymer is the main target in the formulation of polymer nanocomposites. When the volume fraction of the reinforcement increases considerably, agglomeration of the reinforcements starts to take place, which results in reduced mechanical and optical properties [15].

3.4.3 CHARACTERISATION OF POLYMER NANOCOMPOSITES

The following are some of the characterisation techniques for analysing the properties of polymer nanocomposites:

Transmission electron microscopy (TEM) is a tool that is similar to the light microscope. It is used to image thin samples by the passage of electrons that help to generate a projection image.

Scanning electron microscopy (SEM) uses a beam of focused electrons that have low energy to study the surfaces of solid objects. A probe is used to scan the surface of the specimen in a regular pattern.

Wide-angle X-ray diffraction (WAXD) is used to analyse diffraction peaks that are scattered to wide angles. The crystallinity, chemical composition, texture, crystallite size, stress, and structure of inorganic and organic polymers are investigated using this method.

Small-angle X-ray scattering (SAXS) is used to characterise formulated dendrimers and determine the radius of the gyration in the solution. It helps to determine the nano or microscale structure of particle systems for parameters such as average particle sizes, pore sizes, shapes, density differences, distribution, characteristic distances of partially ordered materials, and surface-to-volume ratio.

Thermogravimetric analysis (TGA) helps to determine the nanocomposite's thermal stability and the fraction of the volatile components. This is done by calculating the weight change that takes place on decomposition as the sample is being heated at a constant rate.

Dynamic mechanical thermal analyzer (DMTA) is used to determine the mechanical properties of polymer nanocomposites like brittleness, stiffness, impact resistance, or damping, etc.

Atomic force microscopy (AFM) is a powerful tool that helps to image surfaces of materials such as polymers, ceramics, composites, glass, and biological samples. The forces in the material such as adhesion strength, magnetic forces are measured in their localised state and used to determine the mechanical properties at the nanoscale level.

Solid-state nuclear magnetic resonance (NMR) is used to determine the molecular structure of solids especially when it is insoluble or exhibits polymorphism.

3.4.4 STRUCTURE OF POLYMER NANOCOMPOSITES

Structurally, nanocomposites consist of a material that contains the matrix as well as the nanosized reinforcements in the form of particles, fibres, whiskers, etc.

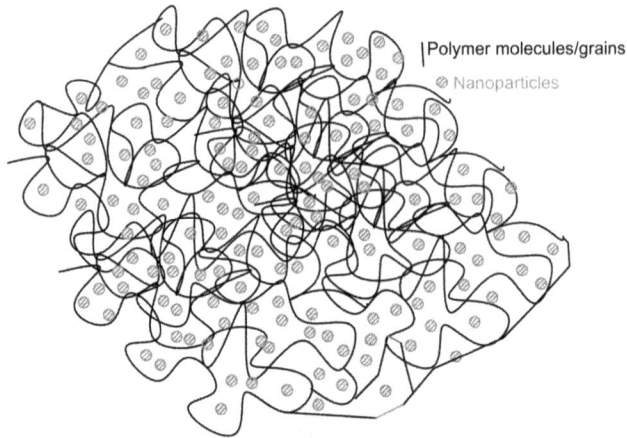

Polymer molecules/grains
Nanoparticles

FIGURE 3.5   A typical structure of particulate-reinforced polymer nanocomposites.

Understanding the relationship between the structure of a material and its properties as it relates to polymer nanocomposites depends on the surface area to volume ratio of the nano-reinforcements. The size at the nanoscale and the surface-area-to-volume ratio is typically about three times greater when nano-reinforcements are used compared to micro-reinforcements. The surface chemistry of the reinforcement therefore to a great extent determines the properties of the polymer nanocomposite [15].

The structure of polymer nanocomposites that have been reinforced with unidimensional particles resembles those of ceramic and metal nanocomposites. The reinforcement here is dispersed over the entire polymer matrix. The microstructure of some particle-reinforced polymer nanocomposites has been shown in Figure 3.5.

The structure of a material is directly related to its properties. In most cases, the properties that are observed for polymer reinforcement systems are proportionally related to their structure. A typical example is polystyrene-alt-maleic anhydride-CdSe nanocomposite used for photoluminescence applications. It exhibits an emission peak at 540 nm, which is close to the absorption edge of the nanoparticles.

3.4.5   Properties of Polymer Nanocomposites

Properties of polymers are known to be enhanced significantly through the introduction of very small amounts of nanoparticle fillers. The effect of nanoparticle size on the properties of polymer nanocomposites is not completely clear and a lot of research is ongoing to achieve that. The properties obtained from nanocomposites are usually markedly different from those of conventional composites. Some of the very critical properties include mechanical strength, thermal stability, optical transparency, barrier properties, novel electrical properties, and flexibility. For instance, nylon-6-layered silicates possess improved tensile modulus, tensile stress, storage modulus, and reduced flammability. Poly-lactide-layered silicates have enhanced bending strength and modulus, storage modulus, distortion at the break, gas barrier properties, and biodegradability. Epoxy-layered silicates have excellent tensile strength and modulus. SiC reinforced epoxy possesses improved micro-hardness, storage, and elastic moduli as reported in [18].

3.4.5.1 Mechanical Properties

The modulus and strength of polymers are usually low for applications as engineering materials. Inorganic additives like fibres, whiskers, and particles are often employed to improve their mechanical properties. The hardness, as well as scratch resistance of nanocomposites from nanoparticles added to a polymer matrix, was reported to increase significantly. Exfoliated polymer silicate materials exhibited better mechanical properties compared to conventional ones. At low silica loadings, it was discovered that Young's modulus improved by about 300% for 3 wt% silicate addition but the toughness decreased by 20% but the stress at the break did not change significantly [25].

Coconut shell–based carbon nanoparticle (CCSnp) reinforced epoxy demonstrates increases in tensile strength as percentage by weight of carbon nanoparticle increases up to 12 wt%. This is confirmed from both experimental investigation conducted and the mathematical model as expressed in Figure 3.6. Stiffness of carbon nanoparticles, which improves load bearing of the matrix, is linked with the increase recorded up to 12 wt% of the nanoparticle addition. The curves shows that binding capacity of the epoxy matrix has not reached maximum or saturation which implies that more nanoparticle could be added with possibly further enhancement in the tensile strength. Moreover, the tensile strength of the composite can be related to the role of matrix-particle interfacial adhesion. With an increasing amount of carbon nanoparticles in epoxy, a good adhesion at filler-matrix interface is proposed since properties of materials depends on their microstructures.

The nanocomposites showed good dispersion and better adhesion at the interface between the polymer and the carbon nanoparticles. The application of external stress to the nanocomposites led to better stress transfer at the interface which translates to enhanced strength. A similar result has been observed for the fracture

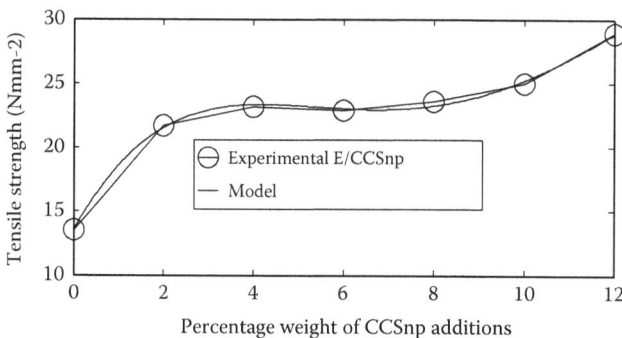

FIGURE 3.6 A graph showing the effect of carbon nanoparticle on the tensile strength of epoxy-based nanocomposites.

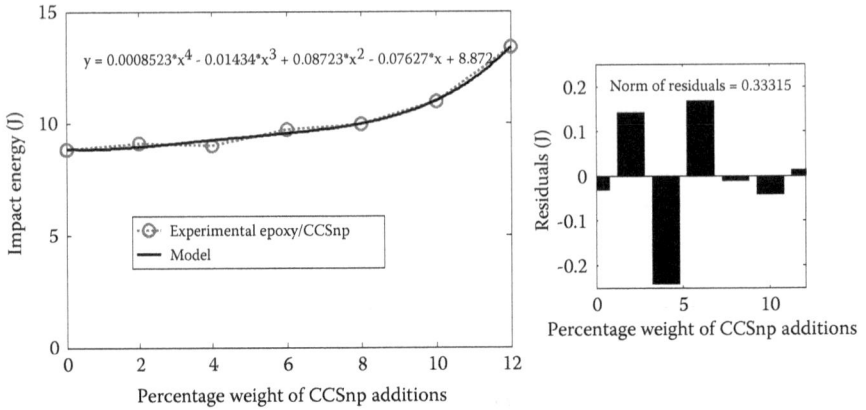

FIGURE 3.7   Impact energy of carbon nanoparticle-reinforced epoxy.

energy of silica-reinforced nylon 6 nanocomposites reported in [25]. The Young's modulus values for a nanoparticle-reinforced polymer has been significantly improved based on the report for silica-filled nylon 6 nanocomposites, with particle sizes between 50 and 110 nm [31].

The Young's modulus values for polyimide-silica (PI) nanocomposites were reported to increase almost linearly as the silica content increases [25]. A similar trend was observed when the composite was filled with constant-sized particles. The modulus value for a nanocomposite does not depend strongly on the particle size, unlike the strength which has a strong dependence, else the relationship that exists between Young's modulus and silica content will not be linear.

Moreover, result of experimental study on impact energy of carbon particle-reinforced epoxy reveals dependence of the impact energy of the epoxy based nanocomposites on increase in the carbon contents (Figure 3.7) like the tensile strength. This implies that degree of energy absorption by the nanocomposites increases with the carbon additions. The enhanced impact energy could make the materials a candidate material in an application demanding enhanced impact energy like materials for automobile bumper applications.

In addition, a report on storage modulus of polyvinylchloride-$CaCO_3$ nanocomposites at different amounts of the $CaCO_3$ nanoparticles shows dependence on temperature such that at temperatures below 60°C, the higher the content of the nanoparticle, the higher is the storage modulus. Between 60°C and 80°C, there is a steep drop in the storage modulus values for all the samples and above 80°C, the value drops to zero. It is seen that when the $CaCO_3$ nanoparticles are well-dispersed within the polyvinylchloride, the stiffness of the composite increases [25].

3.4.5.2   Barrier Properties

The barrier properties of polymer nanocomposites compared to the clean polymer were enhanced when there was a uniform dispersion of nano-sized inorganic layers within the polymer matrix. The gas transport characteristics of both nano- and micron-sized particles dispersed in natural rubber matrix were studied and increases

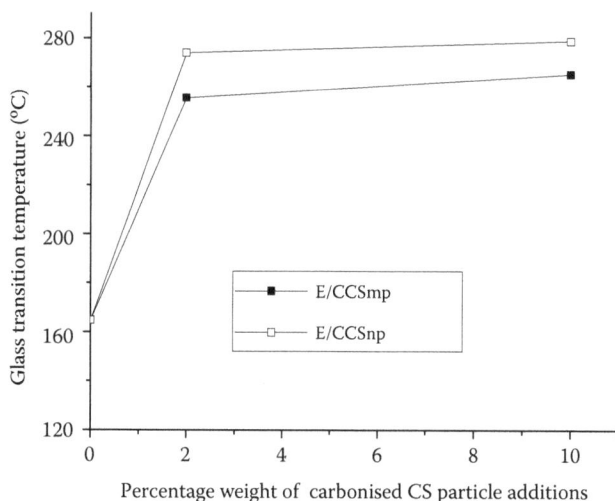

FIGURE 3.8 Thermal stability of carbon particle-reinforced epoxy nanocomposites.

in the gas barrier properties of the latex membranes were observed. This was attributed to the exfoliation of the silicates due to the uniform dispersion of the organic and inorganic phases.

3.4.5.3 Thermal Properties

Glass transition temperature (Tg) of coconut shell–based carbon particle-reinforced epoxy obtained from differential scanning calorimetry (DSC) is shown in Figure 3.8. It was observed that Tg of the epoxy (Diglicidyl ether of bisphenol A) increases due to carbon particle additions. Both size and amount of carbon particles affect the Tg of the composite. This is affirmed from greater Tg of the carbon particle-reinforced epoxy nanocomposite than those of its microcomposite. At 2 wt% of carbon particle additions, Tg of both composites are smaller than those at 10 wt% additions. Since carbon particles are refractory, their presence in epoxy delays thermal influence on the epoxy which is reflected from sluggish vibration of polymeric composite chains when compared with those of the pristine epoxy without carbon nanoparticles additions. This implies that greater thermal energy is needed to transform the composites from the hard to the rubbery phase. The refractoriness increases with the decrease in sizes but increases in the amount of nanoparticles. This explains a basis for greater values of Tg, as observed in Figure 3.8.

Moreover, the improvement in stability is possible as a result of the high thermal stability/refractoriness of the carbon particles. As the particle size decreases, the interfacial area of the nanoparticles increases.

3.4.5.4 Microstructural Properties

Transmission electron microscope (TEM), scanning electron microscope (SEM), optical microscope (OM), and atomic force microscope (AFM) techniques are used to

(a)   (b)   (c)   (d)

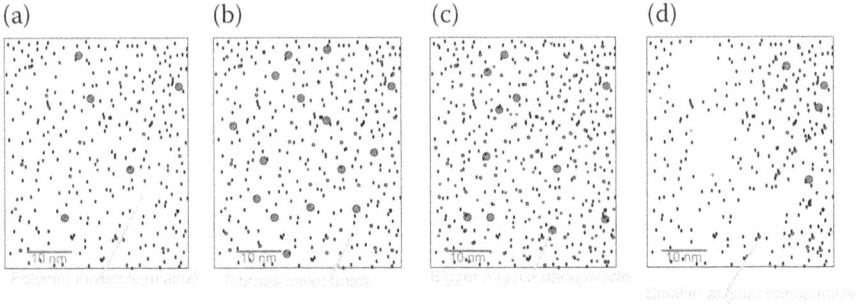

FIGURE 3.9 Two-dimensional geometric model representing TEM image, demonstrating dispersion of hybrid reinforcing particles in the polymeric matrix (a–c) even dispersion with increasing proportion of hybrid reinforcing particles (d) sparse or nonuniform dispersion of hybrid reinforcement leading to different properties at varied points within the nanocomposites.

characterise the degree of nanoparticle dispersion in polymers. They are important to determine degree of homogeneity or uniformity of the nanocomposites to ascertain if the nanoparticles/reinforcements are well distributed within the matrix as homogeinity of the structure is very important to ensure isotropy of the nanocomposites. Otherwise, the nanocomposite will be anisotropic, which is very dangerous in critical applications because failure could occur suddenly due to differences in material properties at different points or directions. Microscopic analysis is peculiar to structural investigation of the nanocomposites because of its ability to detect individual nanoparticles with their agglomerates and present their spatial configurations within the polymeric matrix. Two-dimensional geometric models in Figure 3.9 represent typical TEM images of agricultural waste nanoparticle-reinforced polymers at varied level of reinforcements. Even dispersion of agricultural waste nanoparticles within the polymeric matrix is proposed as observed in Figure 3.9. Such structure leads to property isotropy, that is the property of the nanocompoite is independent of a point or direction. Since each nanoparticle resists/delays deformation of the nanocomposite when subjected to loading, the essence of their presence in the matrix is to increase the load bearing capacity of the nanocompoites. That is, improvement in the strength, stiffness, and toughness of the nanocomposites and degree of improvements increases as the amount of nanoparticles (Figure 3.9a–c) in the polymeric matrix increases until the nanoparticle additions reaches the binding limit of the polymeric matrix. The binding limit is described as the saturation point. Above the saturation point, not all nanoparticles added to the matrix are connected, implying that there are freely existing nanoparticles creating regions of discontinuity within the nanocomposite which act as stress raisers and cause local deformation of the nanocomposite at a stress level below the expectations. Below the matrix saturation point, improvement in the strength, toughness, and rigidity with maintenance of hardness values has been reported and attributed to even dispersion of the nanoparticles with the polymeric matrix [32]. At every point within the matrix, there is a good interaction of the nanoparticle and the polymer chain leading to strong adhesion of the nanoparticles to the polymeric molecules. Therefore, a nanocomposite blend is achievable at every point within the nanocomposite structure. This is not possible in a random dispersion of nanoparticles within the matrix as proposed in Figure 3.9d, which causes variations in materials' properties at different locations within the nanocomposites. Poor processing resulting from inadequate stirring of the nanoparticle and polymers is a possible cause of the random dispersion. Therefore, it is very pertinent to break up nanoparticle agglomerates before adding to the polymer during processing. Agitation of nanoparticle in organic solvent has been employed to dissolve particle agglomerates prior to addition to the polymer [33]. This is followed by evaporation while the polymer-nanoparticle mixture is agitated. After evaporation, oil is added to the mixture for binding nanoparticles to the polymeric noodles before further processing. Although there is a possibility of the random reinforcement dispersion with nanoparticle-reinforced epoxy due to poor processing technique, the problem is common with fibre- or nanotube-reinforced polymers. For the fibre-reinforced polymer, the problem has been reduced through stacking of different layers of fibre laminate reinforced polymers in different directions [34] and sophisticated stirring techniques (ball mill, sonications) are employed to improve dispersion of nanotube in the polymers [35–38]. Uses of TEM as a tool to examine dispersion of nano reinforcement in polymers have been reported. A report on polystyrene butylacrylate in [25] reveals application of the TEM to examine dipersion of – nano-ZnO particles at different wt% in the polystyrene butylacrylate. Similarly, TEM was used to study distribution of the carbon nanotubes in epoxy as reported in [35,38] and polyamides as read in [39], silver nanoparticles in polylactic acid, iron nanoparticle in 4-polyvinyl pyridine, calcium carbonate nanoparticles in polyethylene terephthalate, and silica nanoparticles in polymaleimide [9]. Moreover, other microscopic equipment like scanning electron microcope (SEM), atomic force microscope (AFM), and optical microcope have been reported for studying microstructural properties of the polymeric nanocomposites. For instance, microstructural properties of carbon nanotube-reinforced epoxy [40], eggshell nanoparticle-reinforced recycled low density polyethylene [33], and tensile fractured surfaces of glass fibre-reinforced epoxy [41] were studied using SEM. Moreover, dispersion of carbon nanotube in epoxy was presented in a review article by Bauhofer et al. (2009) using an optical micrograph. Microstructural fractography, studying the pullout mechanism of glass fibre and multiwall carbon nanotube-reinforced epoxy, was reported in [41] using AFM. This implies that AFM, OM, SEM, and TEM are already established for microstructural property investigations, showing a degree of adhesion of nano-reinforcement and polymeric matrix. However, presentation of one equipment is different from another.

3.4.5.5 Dimensional Stability

Dimensional stability is very important for many engineering applications. Incorporating small amounts of nanoparticles into polymers led to a reduction in the coefficient of thermal expansion (CTE), which helped to make the thermal stability better.

When SiO_2 nanoparticles were introduced to the epoxy matrix as read in [25], the coefficient of thermal expansion was greatly reduced because silica nanoparticles have much lower CTE compared to epoxy. This lowering of the CTE as a result of SiO_2 addition is needed for applications where cryogenic temperatures are involved.

3.4.6 APPLICATION OF POLYMER NANOCOMPOSITES

The potential of nanocomposites is so significant that they are applied to an array of industries ranging from packaging to biomedical applications [16]. They are being used for packaging of materials, in self-cleaning nanocoatings, as nanosensors, regenerative medicine, tissue engineering, drug-delivery as carriers, and wound-dressing systems [17]. Moreover, Polycaprolactone-SiO_2 is used for manufacturing bone-bioerodible for skeletal tissue repair, polyimide-SiO_2 for microelectronic devices, PMMA-SiO_2 nanocomposite for dental and optical applications, particulate SiO_2 reinforced polyethylacrylate for catalyst support and stationary phase for chromatography, poly(ρ-phenylene vinylene)-SiO_2 as non-linear material for optical waveguides [9], Ajwa date seed-PMMA nanocomposite for bone cement application, *parquetina nigrescens* fibre-reinforced polylactic acid for food packaging and dental applications, poly(3,4-ethylene-dioxythiophene)-V_2O_5 as cathode material for lithium ion batteries, aluminium nanoparticle reinforced epoxy for automobile bumper application [32], eggshell nanoparticle-reinforced recycled low-density polyethylene for producing automobile bumpers [33] and TiO_2 nanoparticle-reinforced poly(amide-imide) for gas separation applications.

One of the major differences between nanocomposites and normal composites is that with minimal impact on the weight of the material, they offer much better properties. These improved properties of polymer nanocomposites have led to the development of devices like gas sensors, solar cells, light-emitting diodes, and synthetic methods like spin-coating, lamination, etc. The discovery of conductive polymer matrix materials like polyaniline and poly(thiophenes) using semiconducting nanoparticles such as CdS, CdTe, and ZnS are some of the highlights of the applications. The possibility for conduction to take place through these semiconductor chains of nanoparticles is fundamental for the manufacture of some very sensitive sensors where the properties of the polymer matrix and the nanoparticle are combined.

Some of the advantages of agricultural wastes in polymer composites include biodegradability, renewability, cost-effectiveness, economically feasible, and low density with a high specific strength. The use of polymer composites is mainly due to their ability to be easily prepared at a high production rate with a reduced cost compared to conventional materials. As polymeric nanocomposites from agricultural wastes have been employed for productions of artefacts like automobile bumpers and bone cements for biomedical applications, more of such nanocomposites are emerging at reduced costs for advanced applications.

3.5 ADVANTAGES AND DISADVANTAGES OF PARTICULATE NANOCOMPOSITES BASED ON AGRICULTURAL WASTES

Polymer nanocomposites from agricultural waste nanoparticles as the reinforcement have characteristics that make them stand out and others that make them unattractive for industrial applications.

3.5.1 ADVANTAGES

Some of the advantages include:

  i. **High surface area:** Nanoparticles have a very high surface-area-to-volume ratio, resulting in more interaction with the matrix. The interfacial region can sufficiently react with the matrix molecules leading to highly toughened polymer nanocomposites [42]. Nanoparticles make it possible for low loading amounts to be used while achieving optimal performance [43].
  ii. **Cost reduction:** when agricultural wastes are used to produce nanoparticles, they can reduce the cost of the nanocomposite.
  iii. **Improved processing:** the very small nature of nanoparticles can make it easier for the processing of polymer nanocomposites.
  iv. **Density control:** the density of nanoparticles can be effectively controlled with the use of nanoparticles since they have similar dimensions.
  v. **Optical effects like translucency:** due to the low nanoparticle loading used in producing nanocomposites, the transmittance is slightly reduced while the optical clarity is better.
  vi. **Thermal conductivity:** the thermal conductivity of polymer nanocomposites can be effectively controlled with the use of nanoparticles.
  vii. **Controlled thermal expansion coefficient:** the nature of the nanoparticle used in preparing nanocomposites will determine the thermal expansion characteristics. The ability to synthesize nanoparticles of different materials using the many different agricultural wastes available ensures that nanocomposites with different thermal expansion coefficients can be produced.
  viii. **Electrical properties:** The nature of the nanoparticle used to reinforce the nanocomposite affects the electrical properties. The ability to obtain different nanoparticles from these agricultural wastes ensures that the desired electrical characteristics can be gotten. Also, when nanoparticles are present, the electrical breakdown strength and endurance of nanocomposites increase.
  ix. **Magnetic properties:** Nanoparticles from agricultural wastes that show magnetic properties can be used to enhance the properties of the nanocomposites.
  x. **Flame retardancy:** Nanoparticles from agricultural wastes can be used to reduce the susceptibility of nanocomposites to attack by fire.
  xi. **Improved mechanical properties notably hardness, stiffness:** A lot of reports in the literature have already shown that nanoparticles do improve the mechanical properties of polymer nanocomposites. The level of improvement now depends on the particle loading and the chemical makeup of the nanoparticle.

3.5.2 DISADVANTAGES

There are certain disadvantages with using nanoparticles in the production of polymer nanocomposites, some of which include:

i. Polymer nanocomposites whose source of reinforcements include layered fillers such as carbon nanotubes, graphene, graphite, etc. are already very well established in the market because they have successfully been commercialised. This makes their cost more competitive in the market compared to those from agricultural waste particles whose processing methods and steps are still being developed [44].
ii. There is currently the absence of cost-effective methods to control the dispersion of the nanoparticles, leading to aggregation of the nanoparticles and thereby canceling out the advantages.
iii. The accurate prediction of the properties of nanocomposites based on the amount and type of nanoparticles and polymer used remains a major challenge.

3.6 COST COMPARISON BETWEEN CONVENTIONAL AND AGRICULTURAL PARTICLE POLYMERIC NANOCOMPOSITES

The success or otherwise of polymer nanocomposites based on particulates from agricultural wastes depends on their cost compared to that used in preparing similar nanocomposites using other types of fillers. Fillers (nanoparticles) increase the cost of processing because of the need for different equipment, higher energy cost as well as lower throughput. Due to the steps involved in converting agricultural wastes into nanoparticles, a lot of costs are associated with such nanocomposites. Since the process of producing the nanoparticles from these waste materials is not well established, there is a need for new equipment, and a lot of energy is involved while the output will be low since the process is still being developed. The colour of the nanoparticle, its purity level, and the size of its particles, shape, and surface treatment are factors that determine the cost of polymer nanocomposites.

A cost comparison has been made for different types of nanofillers as shown in Table 3.2. Due to the recent nature of research on agricultural wastes for nanoparticle production, not much information about the cost can be found. The table however indicates the possible costs associated with polymer nanocomposites from agricultural wastes especially for nano-silica. Carbon nanotubes (single wall or multiple wall) remain the most expensive type of nanofillers because of the stringent measures applied in their production. Carbon nanofibers are also expensive costing between $95 and $1500 for every kilogram produced. The cost of nanoparticles from agricultural waste will closely resemble that of nano-silica but due to the extra processing involved, the cost will be higher than $8.50 per kilogram that have been reported for nano-silica [42].

Clays and carbonates cost $57.5/tonne while specialist ones cost more than 100 times more. The cost of the nanocomposites is determined by properties like colour, purity, particle size, shape, and surface treatment. The particles are two to three times heavier when compared to polymers.

3.7 FUTURE DIRECTIONS OF PARTICULATE POLYMER NANOCOMPOSITES

Polymer nanocomposites remain the future of packaging as well as many other industries.

As the prices continue to decrease, more companies will begin to invest in the area of polymer nanocomposites. More processes will be developed leading to even lower costs in due course. The opportunities that abound with polymer nanocomposites are very huge thereby attracting tremendous worldwide interest. Due to the potentials of polymer nanocomposites (especially from nanoparticles), many governments around the world are partnering with research and development-based organisations and businesses intending to tap the benefits of the development [45]. The global market for advanced polymeric composite is expected to grow from $9.34 billion in 2017 to a forecast amount of $16.83 billion by 2025 [46]. Biodegradable polymer nanocomposites hold a very great promise for use in biodegradable materials especially in the human body due to easy decomposition into CO_2, H_2O, etc. through microbial activities. Plants are increasingly being used as polymer sources because of the abundance of lignin, cellulose, and hemicellulose.

The volume of sales by companies that deal with polymer nanocomposites will continue to increase. There is hardly any sector of the economy that will not benefit from the burgeoning polymer nanocomposites technology. It is expected to be dominant in all aspects of our lives. Currently, its effects are being felt in nanomaterials used to produce cosmetics, automotive batteries, etc. In the medium term, biosensors, lighting, and memory devices will be produced from polymer nanocomposites. In the long term, aerospace, biotechnology devices, etc. will increasingly be produced using polymer nanocomposites.

In terms of cost, based on the report by market and market [47], the forecast for polymer nanocomposites in terms of value based on the information obtained for the period between 2014 and 2019 is expected to have exceeded $5.1 billion by 2020, which is in line with the report of the FIOR market [46]. Globally, the polymer nanocomposites market has seen a lot of growth in the last few years. Emerging economies, especially China, Japan, and India, are the major drivers of this growth. Technological advancement, bio-raw materials development, and innovative techniques are expected to be the key influencing factors for global growth in this area. Applications to drive this growth include packaging, electronics, semiconductors, and automotives.

3.8 CONCLUSION

Polymer nanocomposites reinforced with nanoparticles derived from agricultural waste products have been reviewed in this report. Although some of these wastes are used for feeding animals, as manures and other low-level activities, a lot is left unutilised. Nanoparticles, especially those based on silica, have been produced from these wastes and are suitable for producing different types of nanocomposites. The properties of nanocomposites such as mechanical properties, barrier properties, thermal properties, and microstructural properties have been proven to improve greatly compared to those of normal composites because of the surface area to volume advantages in the former.

Both thermoplastic and thermosetting polymers are used as matrix materials for making polymer nanocomposites and the application areas for these nanocomposites are vast, ranging from packaging to even biomedical areas. The cost of polymer nanocomposites derived from agricultural wastes is, however, still high because the necessary technology is still being developed. The future of polymer nanocomposites derived from agricultural wastes is very bright because as the demand for the products keeps increasing, the technology becomes cheaper while the production volume increases.

REFERENCES

[1] UNEP United Nations Environment Programme; Division of Technology, I.a.E.I.E.T.C. (2009). Converting Waste Agricultural Biomass into a Resource – Compendium of Technologies. 1–441.
[2] Mohlala, L.M., Bodunrin, M.O., Awosusi, A.A., Daramola, M.O., Cele, N.P., and Olubambi, P.A. (2016). Beneficiation of Corbncob and Sugarcane Bagasse for Energy Generation and Materials Development in Nigeria and South Africa: A Short Overview. *Alexandria Engineering Journal*, **55**, 3025–3036.
[3] Maraveas, C. (2020). Production of Sustainable and Biodegradable Polymers from Agriculturalwaste. *Polymers*, **12**, 1127. 10.3390/polym12051127
[4] Ghosh, S., Dairkee, U.K., Chowdhury, R., and Bhattachary, P. (2017). Hydrogen from Food Processing Wastes Via Photofermentation Using Purple Non-Sulfur Bacteria (Pnsb) – a Review. *Energy Conversion and Management*, **141**(1), 299–314. 10.1016/j.enconman.2016.09.001
[5] Abdelbasir, S.M., McCourt, K.M., Lee, C.M., and Vanegas, D.C. (2020). Waste-Derived Nanoparticles: Synthesis Approaches, Environmental Applications, and Sustainability Considerations. *Frontiers in Chemistry*, **31**. 10.3389/fchem.2020.00782
[6] Abdel-Shafy, H.I. and Mansour, M.S.M. (2018). Solid Waste Issue: Sources, Composition, Disposal, Recycling, and Valorization. *Egyptian Journal of Petroleum*, **27**(4), 1275–1290. 10.1016/j.ejpe.2018.07.003
[7] Arifin, B., Aprilia, S., Alam, P.N., Mulana, F., Amin, A., Anaska, D.M., and Putri, D.E. (2018). Characterization Nanofillers from Agriculture Waste for Polymer Nanocomposites Reinforcement. *MATEC Web of Conferences*, **156**(05020), 1–4. 10.1051/matecconf/201815605020
[8] Dungani, R., Aditiawati, P., Aprilia, S., Yuniarti, K., Karliati, T., Suwandhi, I., and Sumardi, I. (2018). Biomaterial from Oil Palm Waste: Properties, Characterization and Applications, in *Oil Palm*. V. Waisundara, Editor, IntechOpen. 31–51.
[9] Hsieh, Y.-Y., Tsai, Y.-C., He, J.-R., Yang, P.-F., Lin, H.-P., Hsu, C.-H., and Loganathan, A. (2017). Rice Husk Agricultural Waste-Derived Low Ionic Content Carbon–Silica Nanocomposite for Green Reinforced Epoxy Resin Electronic Packaging Material. *Journal of the Taiwan Institute of Chemical Engineers*, **78**, 493–499. 10.1016/j.jtice.2017.06.010

[10] Obi, F.O., Ugwuishiwu, B.O., and Nwakaire, J.N. (2016). Agricultural Waste Concept, Generation, Utilization and Management. *Nigerian Journal of Technology (NIJOTECH)*, **35**(4), 957–964. 10.4314/njt.v35i4.34

[11] Pelissari, F.M., Andrade-Mahecha, M.M., Sobral, P.J.D.A., and Menegalli, F.C. (2017). Nanocomposites Based on Banana Starch Reinforced with Cellulose Nanofibers Isolated from Banana Peels. *Journal of Colloid and Interface Science*, **505**, 154–167. 10.1016/j.jcis.2017.05.106

[12] Wahi, R., Aziz, S.M.A., Hamdan, S., and Ngaini, Z. (2015). Biochar Production from Agricultural Wastes Via Low-Temperature Microwave Carbonization. 1–4. 10.1109/RFM.2015.7587754

[13] Valavala, P.K. and Odegard, G.M. (2005). Modelling Techniques for Determination of Mechanical Properties of Polymer Nanocomposites. *Reviews on Advanced Materials Science*, **9**, 34–44.

[14] Kaviya, S., Santhanalakshmi, J., Viswanathan, B., Muthumary, J., and Srinivasan, K. (2011). Biosynthesis of Silver Nanoparticles Using Citrus Sinensis Peel Extract and Its Antibacterial Activity. *Spectrochim Acta A: Mol Biomol Spectrosc*, **79**(3), 594–598. doi: 10.1016/j.saa.2011.03.040

[15] Armstrong, G. (2015). An Introduction to Polymer Nanocomposites. *European Journal of Physics*, **36**(6), 063001. 10.1088/0143-0807/36/6/063001

[16] Neto, W.P.F., Mariano, M., Silvaa, I.S.V.D., Silvérioa, H.A., Putauxe, J.-L., Otaguroa, H., Pasquinia, D., and Dufresnec, A. (2016). Mechanical Properties of Natural Rubber Nanocomposites Reinforcedwith High Aspect Ratio Cellulose Nanocrystals Isolated from Soy Hulls. *Carbohydrate Polymers*, **153**, 143–152. 10.1016/j.carbpol.2016.07.073

[17] Ciolacu, D.E. and Darie, R.N. (2016). Nanocomposites Based on Cellulose, Hemicelluloses, and Lignin, in *Nanomaterials and Nanocomposites: Zero- Tothree-Dimensional Materials and Their Composites*, P.M. Visakh and M.J.M. Morlanes, Editors, Wiley-VCH Verlag. pp. 391–423.

[18] Camargo, P.H.C., Satyanarayana, K.G., and Wypych, F. (2009). Nanocomposites: Synthesis, Structure, Properties and New Application Opportunities. *Materials Research*, **12**(1), 1–39.

[19] Müller, K., Bugnicourt, E., Latorre, M., Jorda, M., Sanz, Y.E., Lagaron, J.M., Miesbauer, O., Bianchin, A., Hankin, S., Bölz, U., Pérez, G., Jesdinszki, M., Lindner, M., Scheuerer, Z., Castelló, S., and Schmid, M. (2017). Review on the Processing and Properties of Polymer Nanocomposites and Nanocoatings and Their Applications in the Packaging, Automotive and Solar Energy Fields. *Nanomaterials*, **7**(4), 74.

[20] Thomas, S.P.R.S., Bandyopadhyay, S., and Thomas, S. (2007). Polymer Nanocomposites: Preparation Properties and Applications. *Nanocomposites*, **2**, 49–56.

[21] Oliveira, A.D.D. and Beatrice, C.A.G. (2018). Polymer Nanocomposites with Different Types of Nanofiller, in *Nanocomposites – Recent Evolutions*. S. Sivasankaran, Editor, IntechOpen. pp. 103–127. 10.5772/intechopen.81329

[22] Oliveira, M. and Machado, A.V. (2013). Preparation of Polymer-Based Nanocomposites by Different Routes. Nova Science Publishers. pp. 1–22.

[23] Machado, A.V., Araújo, A., and Oliveira, M. (2015). Assessment of Polymer-Based Nanocomposites Biodegradability. Nova Science Publishers. pp. 1–28.

[24] Kaur, M., Mubarak, N.M., Chin, B.L.F., Khalid, M., Karri, R.R., Walvekar, R., Abdullah, E.C., and Tanjung, F.A. (2020). Extraction of Reinforced Epoxy Nanocomposite Using Agricultural Waste Biomass. in *2nd International Conference on Materials Technology and Energy*. IOP Conference Series.

[25] Fu, S., Sun, Z., Huang, P., Li, Y., and Hu, N. (2019). Some Basic Aspects of Polymer Nanocomposites: A Critical Review. *Nano Materials Science*, **1**, 2–30. 10.1016/j.nanoms.2019.02.006

[26] Chan, C.-M., Wu, J., Li, J.-X., and Cheung, Y.-K. (2002). Polypropylenecalcium Carbonate Nanocomposites. *Polymer*, **43**(10), 2981–2992.

[27] Petrović, Z.S., Javni, I., Waddon, A., and Bánhegyi, G. (2000). Structure and Properties of Polyurethane–Silica Nanocomposites. *Journal of Applied Polymer Science*, **76**, 133–151.

[28] Rong, M.Z., Ji, Q.L., Zhang, M.Q., and Friedrich, K. (2002). Graft Polymerization of Vinyl Monomers onto Nanosized Alumina Particles. *European Polymer Journal*, **38**(8), 1573–1582.

[29] Liu, J., Zong, G., He, L., Zhang, Y., Liu, C., and Wang, L. (2015). Effects of Fumed and Mesoporous Silica Nanoparticles on the Properties of Sylgard 184 Polydimethylsiloxane. *Micromachines*, **6**, 855–864.

[30] Uddin, F. (2008). Clays, Nano Clays, and Montmorillonite Minerals. *Metallurgical and Materials Transactions A*, **39**(12), 2804–2814.

[31] Zhang, L., Xiong, Y., Ou, E., Chen, Z., Xiong, Y., and Xu, W. (2011). Preparation and Properties of Nylon 6/Carboxylic Silica Nanocomposites Via in Situ Polymerization. *Applied polymer science*, **122**(2), 1316–1324. 10.1002/app.33967

[32] Bello, S.A., Agunsoye, J.O., Adebisi, J.A., Adeyemo, R.G., and Hassan, S.B. (2020). Optimization of Tensile Properties of Epoxy Aluminum Particulate Composites Using Regression Models. *Journal of King Saud University – Science*, **32**(1), 402–411. 10.1016/j.jksus.2018.06.002

[33] Bello, S.A., Raji, N.K., Kolawole, M.Y., Adebayo, M.K., Adebisi, J.A., Okunola, K.A., and AbdulSalaam, M.O. (2021). Eggshell Nanoparticle Reinforced Recycled Low-Density Polyethylene: A New Material for Automobile Application. *Journal of King Saud University – Engineering Sciences*. 10.1016/j.jksues.2021.04.008

[34] Bello, S.A. (2020). Carbon-Fiber Composites: Development, Structure, Properties, and Applications, in *Handbook of Nanomaterials and Nanocomposites for Energy and Environmental Applications*, O.V. Kharissova, L.M.T. Martínez, and B.I. Kharisov, Editors, Springer International Publishing. pp. 1–22. 10.1007/978-3-030-11155-7_86-1

[35] Chen, H., Jacobs, O., Wu, W., Rüdiger, G., and Schädel, B. (2007). Effect of Dispersion Method on Tribological Properties of Carbon Nanotube Reinforced Epoxy Resin Composites. *Polymer Testing*, **26**(3), 351–360. 10.1016/j.polymertesting. 2006.11.004

[36] Cui, S., Canet, R., Derre, A., Couzi, M., and Delhaes, P. (2003). Characterisation of Multiwall Carbon Nanotubes and Influence of Surfactant in the Nanocomposite Processing. *Carbon*, **41**, 797–809.

[37] Gojny, F.H. and Schulte, K. (2004). Functionalisation Effect on the Thermo-Mechanical Behaviour of Multi-Wall Carbon Nanotube/Epoxy-Composites. *Composites Science and Technology*, **64**(15), 2303–2308. 10.1016/j.compscitech.2004.01.024

[38] Gojny, F.H., Wichmann, M.H.G., Fiedler, B., and Schulte, K. (2005). Influence of Different Carbon Nanotubes on the Mechanical Properties of Epoxy Matrix Composites – a Comparative Study. *Composites Science and Technology*, **65**(15–16), 2300–2313. 10.1016/j.compscitech.2005.04.021

[39] Krause, B., Pötschke, P., and Häußler, L. (2009). Influence of Small Scale Melt Mixing Conditions on Electrical Resistivity of Carbon Nanotube-Polyamide Composites. *Composites Science and Technology*, **69**(10), 1505–1515. 10.1016/ j.compscitech.2008.07.007

[40] Saadatyar, S., Beheshty, M.H., and Sahraeian, R. (2021). Mechanical Properties of Multiwall Carbon Nanotubes/Unidirectional Carbon Fiber-Reinforced Epoxy Hybrid Nanocomposites in Transverse and Longitudinal Fiber Directions. *Polymers and Polymer Composites*, **29**(9_suppl), S74–S84. 10.1177/0967391120986516

[41] Dhilipkumar, T. and Rajesh, M. (2021). Effect of Using Multiwall Carbon Nanotube Reinforced Epoxy Adhesive in Enhancing Glass Fiber Reinforced Polymer Composite through Cocure Manufacturing Technique. *Polymer Composites*, **42**(8), 3758–3772. 10.1002/pc.26091

[42] Zaman, I., Manshoor, B., Khalid, A., and Araby, S. (2014). From Clay to Graphene for Polymer Nanocomposites—A Survey. *Journal of Polymer Research*, **21**(429), 1–11. DOI 10.1007/s10965-014-0429-0. (DOI 10.1007/s10965-014-0429-0)

[43] Downing-Perrault, A. (2005). Polymer Nanocomposites Are the Future. 1–13.

[44] Li, S., Lin, M.M., Toprak, M.S., Kim, D.K., and Muhammed, M. (2010). Nanocomposites of Polymer and Inorganic Nanoparticles for Optical and Magnetic Applications. *Nano Reviews*, **1**(1), 5214, 1–19. DOI: 10.3402/nano.v1i0.5214

[45] Hussain, F., Hojjati, M., Okamoto, M., and Gorga, R.E. (2006). Review Article: Polymer-Matrix Nanocomposites, Processing, Manufacturing, and Application: An Overview. *Journal of Composite Materials*, **40**(17), 1511–1575.

[46] FIOR Market. (2020). Global Advanced Polymer Composites Market Is Expected to Reach Usd 16.83 Billion by 2025: Fior Markets. https://www.globenewswire.com/news-release/2020/02/20/1988157/0/en/Global-Advanced-Polymer-Composites-Market-is-Expected-to-Reach-USD-16-83-Billion-by-2025-Fior-Markets.html

[47] https://www.marketsandmarkets.com. 2021.

4 Organic and Inorganic Nanoparticles from Agricultural Waste

Jeleel Adekunle Adebisi
Department of Metallurgical and Materials Engineering,
University of Ilorin, Kwara State, Nigeria

CONTENTS

4.1 INTRODUCTION

Nanoparticles are particles with an average dimension or size distribution between 1 and 100 nm [1]. The advent of nanotechnology has improved manufacturing of several products with higher efficiency because of improved material properties [2]. Nanotechnology has been deployed in automobile, structural, medical, aeronautical, and energy applications. It is noteworthy to state that agricultural wastes have been utilised for providing feedstocks in most of these applications.

Agricultural wastes are wastes generated annually in many parts of the world from various agricultural operations with little or no value. This includes any form of wastes from farms (poultry, livestock shelters, abattoir, or fields). The wastes may include animal wastes, manures (from both plants and animals), crop/plant residues, run-off from fields, pesticides (that enter water, air or soil) and even salt or

DOI: 10.1201/9781003170549-5

silt drained from fields. Recycling of agricultural wastes attracted the interest of researchers because of the volume generated and thus used for energy production [3], adsorbents of heavy metals [4], activated carbon [5], refractory and ceramic products [6], livestock feed [7], construction materials [8], silica [9], silicon [10], and composite reinforcement [11]. Renewability, biodegradability, pollution inhibitory, and economic amelioration are some of the merits that make utilisation of wastes from this source appealing.

4.2 METHODS OF SYNTHESIS

Nanoparticles can be produced through various known methods categorised as either "bottom-up" or "top down". This depends on whether the materials are made from elemental basis via chemical processes (bottom-up) or reduced from its bulky state via mechanical milling or chemical attrition processes (top-down). Bottom-up nanoparticle synthesis are achieved via pyrolysis, solvothermal reaction, inert gas condensation, sol-gel fabrication, and structured media. The top-down methods are accomplished majorly by milling and attrition. Several literatures have touched on the general concepts of these methods [2,12]. They can also be classified as physical, chemical, or biological [13]. Utilisation of different nanoparticles synthesis methods from various sources have been reported [13–20]. Comprehensive, different mechanical methods used in organic polymer nanofibers have been reviewed by Abdul Khalil and Davoudpour [21]. However, some examples of the methods employed in synthesis/production of nanoparticles from agricultural wastes are described.

4.2.1 TOP-DOWN APPROACH

4.2.1.1 Mechanical Milling

This is one of the basic methods of impacting size reduction up to the nano range in materials. The material to be milled is usually a solid phase of bulky mass or at least sizeable agricultural material like leaf, bark, seed, flower, husk, or even the shell of animals. Milling usually involves disintegration of the material, which usually passes through the micro-state before attaining the desired nano size. Size reduction may require initial crushing before grinding operation. An instance is production of nanoparticles from coconut shells. Crushing is usually carried out in a dry state, unlike grinding, that may require either dry or wet depending on the nature of the material.

Milling is usually carried out in vessels (usually a metallic shell but sometimes polymer composites) that contains grinding medium and the charge. A milling machine could be stirred or tumbling type based on mode of motion. Milling is influenced by varied factors depending on the mechanism of breakage. Mode of breakage can be through compression, chipping, or abrasion, as presented in Figure 4.1. The grinding medium can be rod, vibratory, shaker, planetary ball, and attritor ball mills.

Size reduction is caused by impact and attrition of the milling media against the material. It has been employed for metals, ceramics, and several agricultural wastes at either dry or wet condition [2]. The materials milled are typical of brittle nature but ductile materials are possible under stipulated conditions [1]. Agricultural

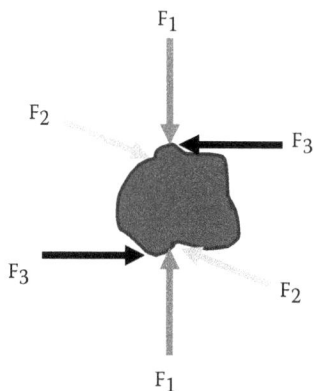

FIGURE 4.1   Modes of application of force on solid mass to impact size reduction during the crushing or milling process. F_1 is by compression, F_2 is chipping, and F_3 is abrasion.

wastes (both from plants and animals) usually recycled for nanoparticles production are natural composites. They contain a mixture of polymers (cellulose, hemi-cellulose, lignin) and ceramics (oxides, carbides, silicates, and complexes of elements usually metals). The polymer phase makes them ductile and tough depending on lignocellulose contents. Examples include residues from plant origins and animal residues like bones and shells (from crustaceans, molluscs, and eggs). Thermal treatments such as carbonisation and calcination have been employed to induce brittleness and sometimes make them stronger and thermally stable. This involves heating the agricultural wastes above 575°C in the absence, limited supply or presence of air (oxygen). Ball milling was employed by Bello, Agunsoye [22] to produce coconut shell nanoparticles. It has been used for eggshell and cassava leaf nanoparticle production [23,24].

The major advantage of milling is high production rate and has been employed in the production of nanocomposite and nano-grained bulk materials. Nonetheless, they are usually produced with defects due to a large amount of strain imparted to the particles during the process. Nanoparticles produced have varied particles sizes (geometrically) and yield broad particles sizes. Hence, particle size analysis may be required if close range particle sizes are required. Moreover, mechanical milling is usually applied to brille and mostly crystalline materials. Recent studies have shown that other classes of materials such as polymers and very ductile, FCC metals can be milled mechanically. Also, agglomeration of nanoparticles after long milling has been reported.

4.2.1.2   Combustion

A simple nonisothermal decomposition (at different heating rate of 5, 10, 15, and 20 K/min between 300 and 1000 K) of HCl pre-treated rice husk in a controlled oxidative environment produced uniformly sized ultrafine nanoparticle with an average size of 60 nm [25]. In a simple combustion of acid-treated rice husk, nano silica has been produced with a particle size of 7 nm at an instance [26]. In order to produce silicon nanoparticles from some plants using metallothermic reduction, silica nanoparticles were extracted using a similar method [27,28]. Ma, Liu [29] used microwave-assisted calcination of hydrolysed treated corn tissues for synthesis of silica incorporated in bio-templated hierarchical titanium oxide for composite structure. Pretreatment and/or post-calcination processes (chemically or biologically) are usually carried out to improve on purity of the products [30].

4.2.1.3 Ionotropic Gelation

This is a procedure used in nanoparticle production between ionic species (one of which must be polymeric) by electrostatic interactions between them under predetermined certain conditions. It has been used in the production of chitosan from crustaceans, molluscs, insects, and fungi that contain chitin [31]. Preparation of chitin from the shells of the organisms involves washing, drying, and crushing before chemical or biological reactions. Chemical route involves three steps of deproteination, demineralisation, and decolouration. Biological route involves using microorganisms and enzymes for chitin extraction. Production of chitosan from chitin is commonly achieved via deacetylation process, which requires acids or alkalis application [32,33]. Chitosan nanoparticles are usually produced from the reaction by ultrasonic disassembly.

4.2.1.4 Sonochemical

Blue agave wastes (bagasse and leaves) were used for the production of cellulose nanocrystal and nanofiber using a combination of mechanical and chemical procedures [34]. The procedures involve organosolv pulping (for removal of lignin and hemicellulose), milling (mechanical milling, sonication and defibrillation), and bleaching (total chlorine free sequence). The main physical process comes from the application of high ultrasound radiation to form acoustic cavitation in solutions, which usually initiate or enhance a chemical activity.

4.2.2 Bottom-Up Approach

4.2.2.1 Wet Chemical/Precipitation

This refers to any scenario in which a solid separate from a solution via a chemical change process. Precipitation is quite like gelation except in the form in which the particles separate from the solution. In the study carried out by Mohanraj, Kannan [35], silica nanoparticles with average size of 22 nm were produced from maize cob. Extraction of the silica follows calcination, solation and precipitation (unlike gelation in sol-gel method) using 1% polyvinyl alcohol as a dispersing medium. Extracts from plants have been used as bioreducing agent for synthesis of silver, copper oxide, titanium oxide, zinc oxide, iron, and gold nanoparticles [36–40]. Biogenic processes for synthesis of important nanoparticles from horticultural and aquacultural food wastes have been enunciated in the review by Ghosh and Fawcett [41]. These processes usually occur at ambient conditions with no toxic substance which makes them eco-friendly but reproducibility and low in-dept scientific concept challenges. Walnut shell extracts has also been employed as reducing agent and catalyst for $NaBH_4$ in Cu/Cu_2O nanocomposites [42]. In-situ copper nanoparticles synthesis into tamarind fruit shell powder (TFSP) has also been reported by Ashok and Hariram [43]. The synthesis involves a single step of bioreduction of 0.25 mol/L aqueous $CuSO_4 \cdot 5H_2O$ solution by dried TFSP at 80°C. In the study carried out by Oruç and Ergüt [44], activated carbon was prepared from lemon waste. The activated carbon was then mixed with $FeSO_4.7H_2O$ and $ZnCl_2.4H_2O$ solutions before the addition of lemon leaves the extract as a bioreductant. Activated carbon/Fe-Zn bimetallic nanoparticles were then separated using centrifugation before washing and drying.

4.2.2.2   Sol-Gel

Sol-gel is a method that involves a set of chemical reactions that irreversibly convert a homogeneous solution (a sol) of molecular reactant precursors into a three-dimensional polymer (a gel), forming an elastic solid, usually filling the same volume as the solution [2]. It has been utilised for synthesis of products from several agricultural wastes [45–48]. El-Sayed and El-Samni [49] reported various silica content obtained from different sources: sorghum, wheat, corn, bamboo, bagasse, lantana, sunflower, rice husk, rice straw, and bread fruit tree. Vaibhav and Vijayalakshmi [50] produced nano silica from bamboo leaves, rice husk, groundnut shell, and sugarcane bagasse using the pyro-chemical method.

The process involves extraction of silicon oxide present in the plant cell directly or after calcination of the biomass. An alkaline medium is used for silicon oxide extraction from the feedstock to form sodium silicate that will thereafter be reduced to precipitate the silica gel. The precipitation/gelation process may occur by purging the sodium silicate solution with carbon dioxide or use acid to effect neutralisation reaction [51,52]. The product formed is to age for about 18 hours. The gel is then filtered and dried to obtain silica nanoparticles. Figure 4.2 presents procedures used in preparing sodium silicate solutions from cassava periderm ash (UCPA), maize stalk ash (UMSA), and maize cob ash (UMCA).

Gelation was done using dilute hydrochloric acid at different pHs between 3 and 7. The solution was left to age for 18 hours, which leaves a clear salt solution and gel, as shown in Figure 4.3. Decantation of the clear solution and filtration of the silica sol produce silica gel that can be dried to obtain silica nanoparticles, as presented in Figure 4.4. Sometimes refluxing of the silica occurs in order to improve the purity of the silica nanoparticles produced, especially in biomedical applications [53].

Basically, the sol-gel process can be applied on any agricultural wastes with inherent silica content. Silica has been reported as one of the constituents of most agricultural wastes/residues, usually of a plant background. Silica content in plant residues is enriched when calcined at a temperature below 700°C for leachable amorphous silica. The ash obtained is highly impure and the silica content is usually low without any chemical treatments. Acid leaching may result in removal some compounds, thus improving silica content, but the silica that will be obtained will still contain acid-resistant compounds like aluminium oxide that is amphoteric in nature.

However, leaching of agricultural waste ashes with alkaline solution selectively dissolves silica to form sodium silica as indicated in Equation 4.1. This process is dependent majorly on temperature and concentration of the alkaline medium, with little influence of time. The mixture is filtered to obtain a clear filtrate solution and solid residue is retain on the filter. Neutralisation of sodium silicate solution with acid (in this case HCl) as presented in Equation 4.2 yields silica gel with salt solution. Neutralisation is a spontaneous reaction but aging of silica gel network is time and

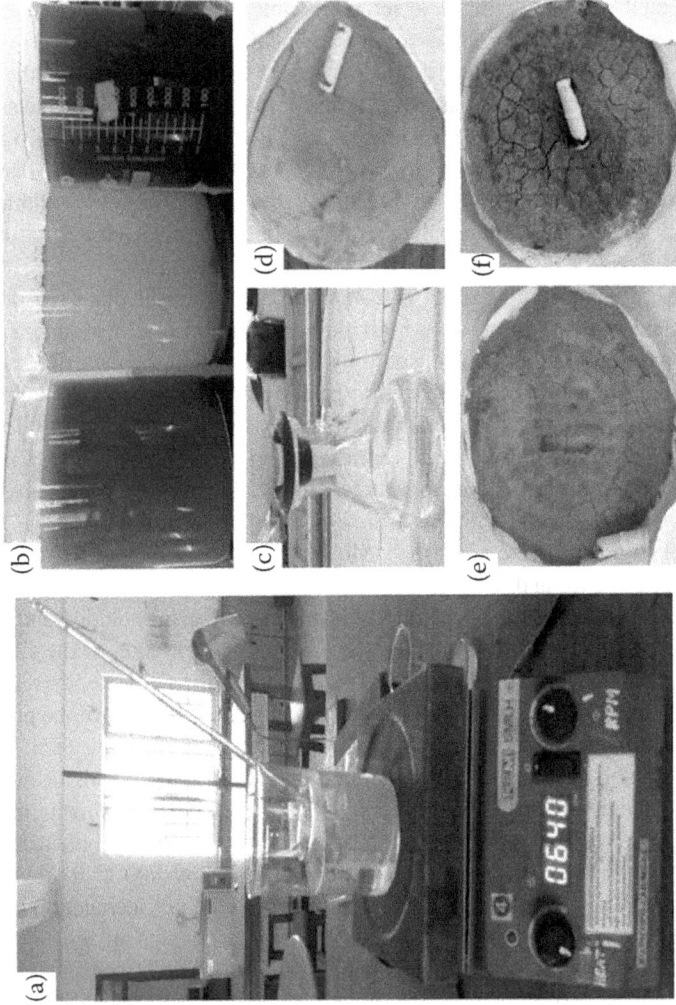

FIGURE 4.2 Solation and filtration of UCPA, UMSA, and UMCA. (a) Solation of UMSA magnetically stirred at 640 rpm; (b) cooled mixture of silicate and residues for UCPA, UMSA, and UMCA; (c) filtered silicate solution; (d) UCPA residue; (e) UMSA residue; and (f) UMCA residue.

FIGURE 4.3 Amount of silica precipitated after 18 hours of aging at different pH for (a1) UCPA, (a2) TCPA, (b1) UMSA, (b2) TMSA, (c1) UMCA, and (c2) TMCA.

(a)   (b)

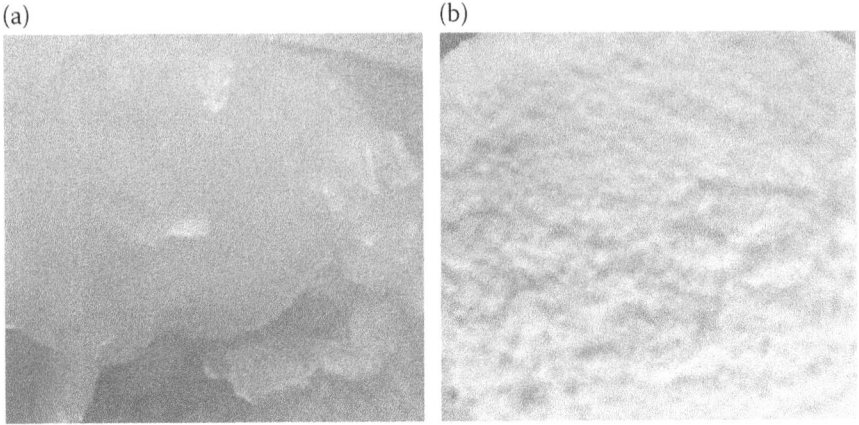

FIGURE 4.4 (a) Silica gel and (b) silica nanoparticles produced from cassava periderm.

concentration dependent. This is called gelation and the waiting period is termed *aging*. Process of gelation is hydrolysis-polycondensation reaction. The mixture is then washed and rinsed to obtain a salt solution filtrate and silica gel as residue.

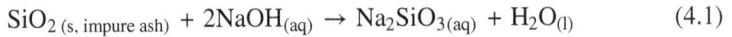

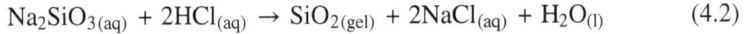

$$SiO_{2 \text{ (s, impure ash)}} + 2NaOH_{(aq)} \rightarrow Na_2SiO_{3(aq)} + H_2O_{(l)} \quad\quad (4.1)$$

$$Na_2SiO_{3(aq)} + 2HCl_{(aq)} \rightarrow SiO_{2(gel)} + 2NaCl_{(aq)} + H_2O_{(l)} \quad\quad (4.2)$$

In-situ bacterial-cellulose-silica composite was produced in two stages of silica synthesis from rice husk ash and then permeation of refluxed silica in bacteria cellulose via sol-gel method [54]. This approach enables formation of a silica framework in cellulosic matrix and the resultant composite was observed to be more thermally stable.

4.2.2.2.1 Metallothermic Reduction

The procedures for processing agricultural wastes to silicon nanoparticles are summarised in Figure 4.5. It involves silica nanoparticle synthesis as described in a sol-gel approach and then reducing the silica using magnesium powder. The produced silicon nanoparticles, though revealed to some degree of agglomeration, but the majority of the particles produced still fell within the nano size, as presented in Figure 4.6.

4.2.2.3   Hydrothermal

Hydrolysis is a reaction of a substance with water molecules leading to breaking down large molecules into smaller ones. It usually involves catalysis by proton or hydroxide (and sometimes inorganic ions such as phosphate ions) present in the aqueous solution, which plays the role of acid-base catalysis. Kumar and Negi [55] synthesised cellulose nanocrystals from sugarcane bagasse at two stages of chemically purified cellulose isolation from the wastes and then extraction of cellulose nanocrystals from

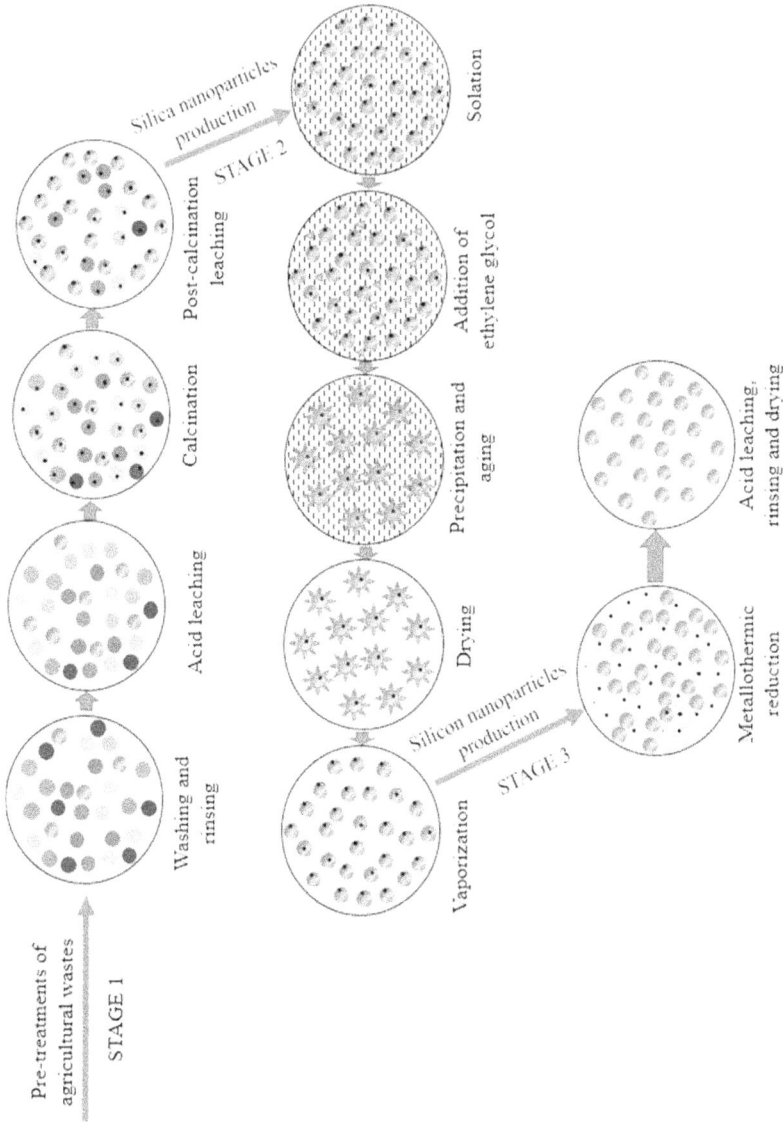

FIGURE 4.5 Schematic procedure of preparing silicon nanoparticles from agricultural wastes.

FIGURE 4.6 TEM image of silicon nanoparticles showing agglomerates.

the purified cellulose. Hydroxyapatite nanoparticle is an important biomaterial used in orthopaedics, dentistry, and trauma surgery. It has been synthesised from animal shells and bones [56–58]. The processes involve cleaning (washing, etching, and drying between 110°C–120°C), crushing and grinding, calcination to form CaO, hydrothermal reaction for form hydroxyapatite, and finally size control using the reverse microemulsion method.

4.3 CHARACTERISATION AND SIZE DETERMINATION

Nanoparticles obtained from different methods using agricultural wastes are usually characterised for chemical composition, functional groups, phase identification, morphological characteristics, microstructural properties, surface area, and size distribution. The characterisation techniques reported from previous works are energy dispersive X-ray spectroscopy (EDS), X-ray fluorescence (XRF), atomic emission spectroscopy (AES), Fourier transform infrared spectroscopy (FT-IR), dual polarisation interferometry (DPI), nuclear magnetic resonance (NMR), Raman spectroscopy, ultraviolet-visible spectroscopy, matrix-assisted laser desorption/ionisation time-of-flight mass spectrometry (MALDI-TOF), Rutherford backscattering spectrometry (RBS), X-ray diffractometry (XRD), particle size analyser (PSA), dynamic light scattering (DLS), nanoparticle tracking analysis (NTA), tunable resistive pulse sensing (TRPS), optical microscopy, scanning electron microscopy (SEM), transmission electron microscopy (TEM), atomic force microscopy (AFM), and X-ray photoelectron spectroscopy (XPS) [24,28,37,55].

Energy dispersive X-ray spectroscopy makes use of the X-ray spectrum emitted by a solid sample bombarded with a focused beam of electrons to obtain a localised chemical species. High-level accuracy of elemental composition for powdered nanoparticles can be determined using an X-ray fluorescence (XRF) analytical technique. Atomic emission spectroscopy (AES) can be effectively used to find the chemical composition up to ppb of solution samples.

The determination of chemical species with functional groups in nanoparticles is usually examined (both qualitative and quantitative) using Fourier transform infrared spectroscopy (FT-IR) analysis. Dual polarisation interferometry (DPI) is used to measure conformational change nanoparticles (protein or other biomolecules) that are adsorbed to surface of a material. Nuclear magnetic resonance (NMR) is a powerful non-selective analytical tool that enables ascertain molecular structure including relative configuration, relative and absolute concentrations, and even intermolecular interactions without the destruction of the analyte. Raman spectroscopy is used for the examination of chemical bonding, structural phases, and molecular interactions in matter at reduced dimensions (up to a few mm) with high sensitivity. The ultraviolet-visible spectroscopy technique offers the ability to study the charge transfer between individual atoms/ions and molecules which are independent of surrounding vibrational bands. It is usually used for determination of structure and composition of materials.

Matrix-assisted laser desorption/ionisation time-of-flight mass spectrometry (MALDI-TOF) is used to analyse biomolecules and various organic molecules that may ionise by conventional ionisation methods due to their fragile nature. Rutherford backscattering spectrometry (RBS) is an ion scattering technique used for analysing thin film with reference standards. The structure and sometimes crystal particle sizes of nanoparticles can be estimated using X-ray diffractometry (XRD).

Particle size distribution of nanoparticles can be determined by particle size analyser (PSA). Dynamic light scattering (DLS) for measurement of average particle size of nanoparticles. Nanoparticle tracking analysis (NTA) allows direct tracking of Brownian motion of individual nanoparticles thus providing means of measuring particle sizes of individual entities in solution. Tunable resistive pulse sensing (TRPS) enables the simultaneous measurement of size, concentration, and surface charge for a wide variety of nanoparticles.

Distribution, morphology, and particle size analysis are quantified using electron microscopy (SEM and TEM). Atomic force microscopy (AFM) is used for force measurement, topographic imaging, and manipulation. X-ray photoelectron spectroscopy (XPS) is a characterisation method employed to map particle topography up to 10 nm depth and provide chemical composition and chemical bonding states of a surface.

4.4 INORGANIC NANOPARTICLES: STRUCTURES, PROPERTIES, AND APPLICATIONS

Silica and silicon have been synthesised from rice husk, sugarcane bagasse, cassava periderm, maize stalk, maize cob, and several others as earlier reported [10]. Nanoparticles are employed as reinforcing phase in composite materials [59,60].

Ma and Liu [29] fabricated hierarchical porous TiO_2–SiO_2 with a size between 13.8 and 20.3 nm for photocatalysis. The efficiency of harvesting visible light is enhanced from 400 to 750 nm range and has a larger specific surface area of 2.5 times the commercial P25. A review by Zamani and Marjani [61] presented metal oxide nanostructures synthesis and applications from agricultural wastes in a concise manner including recent momentous improvements. Reduction of silver in silver nitrate was achieved using extract from *Annona squamosa* peel and banana peel [62,63]. The particles were observed to be nanoparticles with an average particle size of 35 ± 5 nm. In another instance, the extract from banana peels was used as reductant for palladium chloride to synthesise palladium nanoparticles [64].

Silicon nanoparticles have been reviewed to be used in agriculture as nanopesticides, nanofertilizers and alleviating biotic and abiotic toxic effects arising from environmental stresses in plants [65,66]. Various inorganic nanoparticles (Au, Ag, Fe, Cu, Ni, Pd, CoO, FeO, Fe_2O_3, Fe_3O_4, SiO_2 and TiO_2) as reported in the review by Shanmugam, Hari [67] have been investigated with positive outcomes in biohydrogen production. Functionalised activated carbon Fe-Zn nanocomposite has been reported to be useful in removal of toxic organic compounds from water sources [44].

4.5 ORGANIC NANOPARTICLES: STRUCTURES, PROPERTIES, AND APPLICATIONS

Nanofibrillated cellulose is used in surface modification, security papers, nanopapers, food packaging, and coating additives. Cellulose nanocrystal was extracted from barley and *Agave tequilana* by-products from food industry using alkalibleaching treatments [68]. Methylcellulose has been synthesised from sugarcane bagasse via two stages of acid-hydrolysis [55]. Crystallinity of 72.5% was obtained that is higher than chemically purified cellulose with crystallinity of 63.5%. The nanocrystals presented nanorod morphology with a diameter and length of 20–60 nm and 250–480 nm, respectively. The cellulose obtained was proposed for biocomposite production.

Carbon nanotubes and nano-activated carbon have been reported to increase biohydrogen production according to Shanmugam and Hari [67]. Carbon nanotube is an excellent material discovered in 1991 with reported excellent properties in terms of mechanical, optical, thermal, electrical, aspect ratio, and surface area characteristics. It has found applications in various fields such as energy, water treatment, medicine, and engineering. Carbon nanotubes with diameters between 0.6–100 nm and length up to 130 μm have been reportedly prepared from rice straw, olive oil, turpentine oil, coconut oil, palm oil, and sesame oil using catalysed spray pyrolysis or combustion vapour deposition (CVD) assisted pyrolysis [69]. A very high temperature that was bane for carbon nanotube cost implications has been drastically reduced from more than 3000°C to temperatures below 900°C and time below 60 minutes.

4.6 CONCLUSIONS

Synthesis of nanoparticles from agricultural wastes applied similar procedures used for synthetic feedstocks. Not all methods used for the synthesis of nanoparticles from synthetic materials have been exploited for agricultural wastes. Both bottom-up and top-down approaches reported in literature have been enumerated. They include mechanical milling, combustion, ionotropic gelation, and sonochemical are among those reported for top-down approaches. Bottom-up methods reviewed are precipitation, sol-gel, hydrothermal, and metallothermic reduction. Techniques employed in characterisation of nanoparticles have been highlighted, stating their purposes. Nanoparticles produced have been classified into organic and inorganic covering their applications. Nanoparticles produced from agricultural wastes have been observed to be suitable, mostly in biomedical applications due to their mild biocompatibility. Production of nanoparticles from these sources is appealing due to the merits of encouraging 3R principles on waste management. They are also found to be cheaper and environmentally friendly. It is desirable to explore more nanoparticle synthesis from agricultural wastes.

REFERENCES

[1] De Castro, C.L. and Mitchell, B.S. (2003). *Nanoparticles from Mechanical Attrition*, in *Synthesis, Functionalization and Surface Treatment of Nanoparticles*, M.I. Baraton, Editor. American Scientific Publishers. pp. 1–15.

[2] Pokropivny, V., Lohmus, R., Hussainova, I., Pokropivny, A., and Vlassov, S. (2007). *Introduction to Nanomaterials and Nanotechnology*. Tartu University, Tallinn University, Frantsevich Institute for Problems of Materials Science of NASU. p. 192.

[3] Zhao, P., Shen, Y., Ge, S., Chen, Z., and Yoshikawa, K. (2014). Clean Solid Biofuel Production from High Moisture Content Waste Biomass Employing Hydrothermal Treatment. *Applied Energy*, **131**, 345–367. http://www.sciencedirect.com/science/article/pii/S0306261914006205

[4] Simate, G.S. and Ndlovu, S. (2015). The Removal of Heavy Metals in a Packed Bed Column Using Immobilized Cassava Peel Waste Biomass. *Journal of Industrial and Engineering Chemistry*, **25**, 635–643. http://www.sciencedirect.com/science/article/pii/S1226086X14001853

[5] Kalderis, D., Koutoulakis, D., Paraskeva, P., Diamadopoulos, E., Otal, E., Valle, J.O.D., and Fernández-Pereira, C. (2008). Adsorption of Polluting Substances on Activated Carbons Prepared from Rice Husk and Sugarcane Bagasse. *Chemical Engineering Journal*, **144**(1), 42–50. http://www.sciencedirect.com/science/article/pii/S1385894708000223

[6] Aigbodion, V.S., Hassan, S.B., Ause, T., and Nyior, G.B. (2010). Potential Utilization of Solid Waste (Bagasse Ash). *Journal of Minerals & Materials Characterization & Engineering*, **9**(1), 67–77.

[7] Heuzé, V., Tran, G., Bastianelli, D., Archimède, H., Lebas, F., and Régnier, C. (2014). *Cassava Peels, Cassava Pomace and Other Cassava By-products*. April 11, 2016, 15:24 [cited 2014 07/09/2018]; Feedipedia.org. [A programme by INRA, CIRAD, AFZ and FAO.]. Available from: http://www.feedipedia.org/node/526

[8] Villamizar, M.C.N., Araque, V.S., Reyes, C.A.R., and Silva, R.S. (2012). Effect of the Addition of Coal-ash and Cassava Peels on the Engineering Properties of Compressed Earth Blocks. *Construction and Building Materials*, **36**, 276–286. http://www.sciencedirect.com/science/article/pii/S0950061812002589

[9] Gu, S., Zhou, J., Luo, Z., Wang, Q., and Shi, Z. (2015). Kinetic Study on the Preparation of Silica from Rice Husk Under Various Pretreatments. *J Therm Anal Calorim*, **119**, 2159–2169.

[10] Venkateswaran, S., Yuvakkumar, R., and Rajendran, V. (2013). Nano Silicon from Nano Silica Using Natural Resource (Rha) for Solar Cell Fabrication. *Phosphorus, Sulfur, and Silicon and the Related Elements*, **188**(9), 1178–1193. 10.1080/10426507. 2012.740106

[11] Siqueira, G., Bras, J., Follain, N., Belbekhouche, S., Marais, S., and Dufresne, A. (2013). Thermal and Mechanical Properties of Bio-nanocomposites Reinforced by Luffa Cylindrica Cellulose Nanocrystals. *Carbohydrate Polymers*, **91**(2), 711–717. http://www.sciencedirect.com/science/article/pii/S0144861712008314

[12] Jesus, M. and Grazu, V. (2012). *Nanobiotechnology: Inorganic Nanoparticles vs Organic Nanoparticles*. Elsevier.

[13] Iravani, S., Korbekandi, H., Mirmohammadi, S.V., and Zolfaghari, B. (2014). Synthesis of Silver Nanoparticles: Chemical, Physical and Biological Methods. *Research in Pharmaceutical Sciences*, **9**(6), 385–406. https://www.ncbi.nlm.nih. gov/pmc/articles/PMC4326978/

[14] Narayanan, K.B. and Sakthivel, N. (2010). Biological Synthesis of Metal Nanoparticles by Microbes. *Advances in Colloid and Interface Science*, **156**(1), 1–13. http://www.sciencedirect.com/science/article/pii/S0001868610000254

[15] Kavitha, K., Baker, S., Rakshith, D., Kavitha, H., Yashwantha Rao, H., Harini, B., and Satish, S. (2013). Plants as Green Source towards Synthesis of Nanoparticles. *International Research Journal* of *Biological Sciences*, **2**(6), 66–76.

[16] Mittal, A.K., Chisti, Y., and Banerjee, U.C. (2013). Synthesis of Metallic Nanoparticles Using Plant Extracts. *Biotechnology Advances*, **31**(2), 346–356. http://www.sciencedirect.com/science/article/pii/S0734975013000050

[17] Wu, S.-H., Mou, C.-Y., and Lin, H.-P. (2013). Synthesis of Mesoporous Silica Nanoparticles. *Chemical Society Reviews*, **42**(9), 3862–3875. 10.1039/C3CS35405A

[18] Dhand, C., Dwivedi, N., Loh, X.J., Jie Ying, A.N., Verma, N.K., Beuerman, R.W., Lakshminarayanan, R., and Ramakrishna, S. (2015). Methods and Strategies for the Synthesis of Diverse Nanoparticles and Their Applications: A Comprehensive Overview. *RSC Advances*, **5**(127), 105003–105037. 10.1039/C5RA19388E

[19] Singh, P., Kim, Y.-J., Zhang, D., and Yang, D.-C. (2016). Biological Synthesis of Nanoparticles from Plants and Microorganisms. *Trends in Biotechnology*, **34**(7), 588–599. http://www.sciencedirect.com/science/article/pii/S0167779916000408

[20] Anu Mary Ealia, S. and Saravanakumar, M.P. (2017). A Review on the Classification, Characterisation, Synthesis of Nanoparticles and Their Application. *IOP Conference Series: Materials Science and Engineering*, **263**, 032019. 10.1088/ 1757-899X/263/3/032019

[21] Abdul Khalil, H.P.S., Davoudpour, Y., Islam, M.N., Mustapha, A., Sudesh, K., Dungani, R., and Jawaid, M. (2014). Production and Modification of Nanofibrillated Cellulose Using Various Mechanical Processes: A Review. *Carbohydrate Polymers*, **99**, 649–665. http://www.sciencedirect.com/science/article/pii/S0144861713008539

[22] Bello, S.A., Agunsoye, J.O., Adebisi, J.A., Kolawole, F.O., and Hassan, S.B. (2016). Physical Properties of Coconut Shell Nanoparticles. *Journal of Science, Engineering and Technology*, **12**(1), 63–79. https://www.nepjol.info/index.php/ KUSET/article/view/21566

[23] Kolawole, F.O., Kolawole, S.K., Agunsoye, J.O., Bello, S.A., Adebisi, J.A., Soboyejo, W.O., and Hassan, S.B. (2018). Synthesis and Characterization of Cassava Bark Nanoparticles. *MRS Advances*, **3**(42-43), 2519–2526. https://www. cambridge.org/core/article/synthesis-and-characterization-of-cassava-bark-nanoparticles/7E249BE60AEDD0DE7B03989EEA815899

[24] Bello, S.A., Raji, N.K., Kolawole, M.Y., Adebayo, M.K., Adebisi, J.A., Okunola, K.A., and AbdulSalaam, M.O. (2021). Eggshell Nanoparticle Reinforced Recycled

Low-Density Polyethylene: A New Material for Automobile Application. *Journal of King Saud University-Engineering Sciences*.

[25] Liou, T.-H. (2004). Preparation and Characterization of Nano-Structured Silica from Rice Husk. *Materials Science and Engineering*, **A**(364), 313–323.

[26] Carmona, V.B., Oliveira, R.M., Silva, W.T.L., Mattoso, L.H.C., and Marconcini, J.M. (2013). Nanosilica from Rice Husk: Extraction and Characterization. *Industrial Crops and Products*, **43**, 291–296. http://www.sciencedirect.com/science/article/pii/S0926669012003962

[27] Zemnukhova, L.A., Panasenko, A.E., Tsoi, E.A., Fedorishcheva, G.A., Shapkin, N.P., Artem'yanov, A.P., and Maiorov, V.Y. (2014). Composition and Structure of Amorphous Silica Produced from Rice Husk and Straw. *Inorganic Materials*, **50**(1), 75–81. 10.1134/S0020168514010208

[28] Adebisi, J.A., Agunsoye, J.O., Ahmed, I.I., Bello, S.A., Haris, M., Ramakokovhu, M.M., and Hassan, S.B. (2021). Production of Silicon Nanoparticles from Selected Agricultural Wastes. *Materials Today: Proceedings*, **38**, 669–674. 10.1016/j.matpr.2020.03.658

[29] Ma, H., Liu, W.-W., Zhu, S.-W., Ma, Q., Fan, Y.-S., and Cheng, B.-J. (2013). Biotemplated Hierarchical TiO2–SiO2 Composites Derived from Zea Mays Linn. for Efficient Dye Photodegradation. *Journal of Porous Materials*, **20**, 1205–1215.

[30] Espíndola-Gonzalez, A., Martínez-Hernández, A.L., Angeles-Chávez, C., Castaño, V.M., and Velasco-Santos, C. (2010). Novel Crystalline SiO2 Nanoparticles via Annelids Bioprocessing of Agro-Industrial Wastes. *Nanoscale Research Letters*, **5**(9), 1408–1417. 10.1007/s11671-010-9654-6

[31] Yadav, M., Goswami, P., Paritosh, K., Kumar, M., Pareek, N., and Vivekanand, V. (2019). Seafood Waste: A Source for Preparation of Commercially Employable Chitin/Chitosan Materials. *Bioresources and Bioprocessing*, **6**(1), 8. 10.1186/s40643-019-0243-y

[32] Oh, J.-W., Chun, S.C., and Chandrasekaran, M. (2019). Preparation and In Vitro Characterization of Chitosan Nanoparticles and Their Broad-Spectrum Antifungal Action Compared to Antibacterial Activities against Phytopathogens of Tomato. *Agronomy*, **9**(1), 21. https://www.mdpi.com/2073-4395/9/1/21

[33] Divya, K. and Jisha, M.S. (2018). Chitosan Nanoparticles Preparation and Applications. *Environmental Chemistry Letters*, **16**(1), 101–112. 10.1007/s10311-017-0670-y

[34] Robles, E., Fernández-Rodríguez, J., Barbosa, A.M., Gordobil, O., Carreño, N.L.V., and Labidi, J. (2018). Production of Cellulose Nanoparticles from Blue Agave Waste Treated with Environmentally Friendly Processes. *Carbohydrate Polymers*, **183**, 294–302. http://www.sciencedirect.com/science/article/pii/S0144861718300158

[35] Mohanraj, K., Kannan, S., Barathan, S., and Sivakumar, G. (2012). Preparation and Characterization of Nano SiO_2 from Corn Cob Ash by Precipitation Method. *Optoelectronics and Advanced Materials – Rapid Communications*, **6**(3–4), 394–397.

[36] Sasidharan, D., Namitha, T., Johnson, S.P., Jose, V., and Mathew, P. (2020). Synthesis of Silver and Copper Oxide Nanoparticles Using Myristica Fragrans Fruit Extract: Antimicrobial and Catalytic Applications. *Sustainable Chemistry and Pharmacy*, **16**, 100255.

[37] Sk, I., Khan, M.A., Haque, A., Ghosh, S., Roy, D., Homechuadhuri, S., and Alam, M.A. (2020). Synthesis of Gold and Silver Nanoparticles Using Malva verticillata Leaves Extract: Study of Gold Nanoparticles Catalysed Reduction of Nitro-schiff Bases and Antibacterial Activities of Silver Nanoparticles. *Current Research in Green and Sustainable Chemistry*, **3**, 100006.

[38] El-Remaily, M.A.E.A.A.A., Abu-Dief, A.M., and Elhady, O. (2019). Green Synthesis of TiO_2 Nanoparticles as an Efficient Heterogeneous Catalyst with High Reusability for

Synthesis of 1,2-Dihydroquinoline Derivatives. *Applied Organometallic Chemistry*, **33**(8), e5005. https://onlinelibrary.wiley.com/doi/abs/10.1002/aoc.5005

[39] Król, A., Railean-Plugaru, V., Pomastowski, P., and Buszewski, B. (2019). Phytochemical Investigation of Medicago sativa L. Extract and Its Potential as a Safe Source for the Synthesis of ZnO Nanoparticles: The Proposed Mechanism of Formation and Antimicrobial Activity. *Phytochemistry Letters*, **31**, 170–180. http://www.sciencedirect.com/science/article/pii/S1874390018307183

[40] Pan, Z., Lin, Y., Sarkar, B., Owens, G., and Chen, Z. (2020). Green Synthesis of Iron Nanoparticles Using Red Peanut Skin Extract: Synthesis Mechanism, Characterization and Effect of Conditions on Chromium Removal. *Journal of Colloid and Interface Science*, **558**, 106–114. http://www.sciencedirect.com/science/article/pii/S0021979719311518

[41] Ghosh, P.R., Fawcett, D., Sharma, S.B., and Poinern, G.E. (2017). Production of High-value Nanoparticles via Biogenic Processes Using Aquacultural and Horticultural Food Waste. *Materials*, **10**(8), 852.

[42] Saadati F.L.V., Zamani A., J. and Serb. Chem. Soc., 82, 1-12. (2017). Environmentally Benign Copper Nanoparticles Supported on Walnut Shell as a Highly Durable Nanocatalyst for the Synthesis of Propargylamines. *Journal of Serbian Chemical Society*, **82**, 1–12.

[43] Ashok, B., Hariram, N., Siengchin, S., and Rajulu, A.V. (2020). Modification of Tamarind Fruit Shell Powder with In situ Generated Copper Nanoparticles by Single Step Hydrothermal Method. *Journal of Bioresources and Bioproducts*, **5**(3), 180–185. http://www.sciencedirect.com/science/article/pii/S2369969820300955

[44] Oruç, Z., Ergüt, M., Uzunoğlu, D., and Özer, A. (2019). Green Synthesis of Biomass-derived Activated Carbon/Fe-Zn Bimetallic Nanoparticles from Lemon (Citrus limon (L.) Burm. f.) Wastes for Heterogeneous Fenton-Like Decolorization of Reactive Red 2. *Journal of Environmental Chemical Engineering*, **7**(4), 103231. http://www.sciencedirect.com/science/article/pii/S2213343719303549

[45] Adepoju, A.D., Adebisi, J.A., Odusote, J.K., Ahmed, I.I., and Hassan, S.B. (2016). Preparation of Silica from Cassava Periderm. *The Journal of Solid Waste Technology and Management*, **42**(3), 216–221. 10.5276/JSWTM.2016.216

[46] Adebisi, J.A., Agunsoye, J.O., Bello, S.A., Haris, M., Ramakokovhu, M.M., Daramola, M.O., and Hassan, S.B. (Year). Production of Amorphous silica Nanoparticles from Cassava periderm Using Modified Sol-gel Approach. in *16th Annual International Conference of Nigeria International Materials Congress [NIMACON - 2017]*. University of Benin, Benin City, Edo State, Nigeria.

[47] Adebisi, J.A., Agunsoye, J.O., Bello, S.A., Haris, M., Ramakokovhu, M.M., Daramola, M.O., and Hassan, S.B. (2020). Green Production of Silica Nanoparticles from Maize Stalk. *Particulate Science and Technology*, **38**(6), 667–675.

[48] Adebisi, J.A., Agunsoye, J.O., Bello, S.A., Haris, M., Senthilkumar, M., Rherari, A., and Hassan, S.B. (2020). Silica Nanoparticles Synthesized from Maize Cob Using Modified Sol-Gel Method. *Nigerian Journal of Technological Research*, **15**(2), 22–29.

[49] El-Sayed, M.A. and El-Samni, T.M. (2006). Physical and Chemical Properties of Rice Straw Ash and Its Effect on the Cement Paste Produced from Different Cement Types. *Journal King Saud University*, **19**(1), 21–30.

[50] Vaibhav, V., Vijayalakshmi, U., and Roopan, S.M. (2015). Agricultural Waste as a Source for the Production of Silica Nanoparticles. *Spectrochimica Acta Part A: Molecular and Biomolecular Spectroscopy*, **139**, 515–520. http://www.sciencedirect.com/science/article/pii/S1386142514018630

[51] Tang, Q. and Wang, T. (2005). Preparation of Silica Aerogel from Rice Hull Ash by Supercritical Carbon Dioxide Drying. *The Journal of Supercritical Fluids*, **35**(1), 91–94. http://www.sciencedirect.com/science/article/pii/S0896844604003213

[52] Sudirmana, Anggaravidya, M., Budianto, E., and Gunawan, I. (2012). Synthesis and Characterization of Polyester-Based Nanocomposite. *Procedia Chemistry*, **4**, 107–113.

[53] Imoisili, P.E., Ukoba, K.O., and Jen, T.-C. (2020). Green Technology Extraction and Characterisation of Silica Nanoparticles from Palm Kernel Shell Ash Via Sol–gel. *Journal of Materials Research and Technology*, **9**(1), 307–313. http://www.sciencedirect.com/science/article/pii/S2238785419314139

[54] Soemphol, W., Charee, P., Audtarat, S., Sompech, S., Hongsachart, P., and Dasri, T. (2020). Characterization of a Bacterial Cellulose-silica Nanocomposite Prepared from Agricultural Waste Products. *Materials Research Express*, **7**(1), 015085. 10.1088/2053-1591/ab6c25

[55] Kumar, A., Negi, Y.S., Bhardwaj, N.K., and Choudhary, V. (2012). Synthesis and Characterization of Methylcellulose/PVA Based Porous Composite. *Carbohydrate Polymers*, **88**(4), 1364–1372. http://www.sciencedirect.com/science/article/pii/S0144861712001464

[56] Malla, K.P., Regmi, S., Nepal, A., Bhattarai, S., Yadav, R.J., Sakurai, S., and Adhikari, R. (2020). Extraction and Characterization of Novel Natural Hydroxyapatite Bioceramic by Thermal Decomposition of Waste Ostrich Bone. *International Journal of Biomaterials*, **2020**, 1–10.

[57] Odusote, J.K., Danyuo, Y., Baruwa, A.D., and Azeez, A.A. (2019). Synthesis and Characterization of Hydroxyapatite from Bovine Bone for Production of Dental Implants. *Journal of Applied Biomaterials & Functional Materials*, **17**(2), 1–7. https://journals.sagepub.com/doi/abs/10.1177/2280800019836829

[58] Yinka, K.M., Olayiwola, A.J., Sulaiman, A., Ali, A., and Iqbal, F. (2020). Preparation and Characterization of Hydroxyapatite Powder for Biomedical Applications from Giant African Land Snail Shell Using a Hydrothermal Technique. *Engineering and Applied Science Research*, **47**(3), 275–286.

[59] Aigbodion, V.S. (2019). Bean Pod Ash Nanoparticles a Promising Reinforcement for Aluminium Matrix Biocomposites. *Journal of Materials Research and Technology*, **8**(6), 6011–6020. http://www.sciencedirect.com/science/article/pii/S223878541930095X

[60] Agunsoye, J.O., Odumosu, A.K., and Dada, O. (2019). Novel Epoxy-carbonized Coconut Shell Nanoparticles Composites for Car Bumper Application. *The International Journal of Advanced Manufacturing Technology*, **102**(1), 893–899. 10.1007/s00170-018-3206-0

[61] Zamani, A., Marjani, A.P., and Mousavi, Z. (2019). Agricultural Waste Biomass-assisted Nanostructures: Synthesis and Application. *Green Processing and Synthesis*, **8**(1), 421–429. https://www.degruyter.com/view/journals/gps/8/1/article-p421.xml

[62] Kumar, R., Roopan, S.M., Prabhakarn, A., Khanna, V.G., and Chakroborty, S. (2012). Agricultural Waste Annona squamosa Peel Extract: Biosynthesis of Silver Nanoparticles. *Spectrochimica Acta Part A: Molecular and Biomolecular Spectroscopy*, **90**, 173–176. http://www.sciencedirect.com/science/article/pii/S138614251200042X

[63] Bankar, A., Joshi, B., Kumar, A.R., and Zinjarde, S. (2010). Banana Peel Extract Mediated Novel Route for the Synthesis of Silver Nanoparticles. *Colloids and Surfaces A: Physicochemical and Engineering Aspects*, **368**(1), 58–63. http://www.sciencedirect.com/science/article/pii/S0927775710004164

[64] Bankar, A., Joshi, B., Kumar, A.R., and Zinjarde, S. (2010). Banana Peel Extract Mediated Novel Route for the Synthesis of Palladium Nanoparticles. *Materials Letters*, **64**(18), 1951–1953. http://www.sciencedirect.com/science/article/pii/S0167577X10004738

[65] Roychoudhury, A. (2020). Silicon-Nanoparticles in Crop Improvement and Agriculture. *International Journal on Recent Advancement in Biotechnology & Nanotechnology [ISSN: 2582-1571 (online)]*, **3**(1).

[66] Saratale, R.G., Saratale, G.D., Shin, H.S., Jacob, J.M., Pugazhendhi, A., Bhaisare, M., and Kumar, G. (2018). New Insights on the Green Synthesis of Metallic Nanoparticles Using Plant and Waste Biomaterials: Current Knowledge, Their Agricultural and Environmental Applications. *Environmental Science and Pollution Research*, **25**(11), 10164–10183. 10.1007/s11356-017-9912-6

[67] Shanmugam, S., Hari, A., Pandey, A., Mathimani, T., Felix, L., and Pugazhendhi, A. (2020). Comprehensive Review on the Application of Inorganic and Organic Nanoparticles for Enhancing Biohydrogen Production. *Fuel*, **270**, 117453. http://www.sciencedirect.com/science/article/pii/S0016236120304488

[68] Espino, E., Cakir, M., Domenek, S., Román-Gutiérrez, A.D., Belgacem, N., and Bras, J. (2014). Isolation and Characterization of Cellulose Nanocrystals from Industrial By-products of Agave tequilana and Barley. *Industrial Crops and Products*, **62**(0), 552–559. http://www.sciencedirect.com/science/article/pii/S0926669014005627

[69] Makgabutlane, B., Nthunya, L.N., Maubane-Nkadimeng, M.S., and Mhlanga, S.D. (2021). Green Synthesis of Carbon Nanotubes to Address the Water-energy-food Nexus: A Critical Review. *Journal of Environmental Chemical Engineering*, **9**(1), 104736. https://www.sciencedirect.com/science/article/pii/S221334372031085X

5 Computational Approaches to Polymeric Nanocomposites

Saheed Olalekan Ojo
Bernal Institute, School of Engineering, University of Limerick, Castletroy, Ireland

Sikiru Oluwarotimi Ismail
Centre for Engineering Research, Department of Engineering, School of Physics, Engineering and Computer Science, University of Hertfordshire, England, UK

CONTENTS

DOI: 10.1201/9781003170549-6

5.1 INTRODUCTION

The emergence of nanocomposites as a new class of materials has triggered interest in their development and applications in the last few decades. The unique properties of nanoparticles when introduced into a continuous polymer matrix have resulted in new phenomena and exceptional properties with promising applications in engineering industries, such as automotive, packaging, and medical devices. For specific applications, prediction of properties is necessary for efficient design and synthesis of new materials. However, the poor understanding of fundamental issues bordering on prediction of nanoparticle structure (specifically, the effects of nanofiller size and architecture on the nanocomposite morphology), dynamics (that is, the effect of nanofillers on the rheological characteristics of the melt), solid-state properties, and processing methods and conditions constitute challenges hindering further developments in the field of nanocomposites.

To a large extent, the quality of dispersion of nanoscale fillers in the polymer matrix significantly affects the final properties of polymeric nanocomposites (PNCs), yet it is challenging to achieve optimal dispersion of nanoparticles owing to their tendency to form nanoparticle aggregates and platelet stacks or due to uncertainties in the properties of nanoparticles and dissimilarity between the chemical properties of matrix and nanofillers [1].

Considering these observations, the experimental approach creates some challenges, which limit the ability to characterise the structure and control the process of fabricating PNCs. Therefore, for efficient design of polymeric nanocomposite (PNC) materials, predictive tools that can sufficiently capture these nanoscale phenomena must be employed to guide material synthesis. Fundamentally, predictive models address issues, including thermodynamics and kinetics formation of PNCs; the effect of nanoparticles on the polymer rheological behaviour; hierarchical features of the structure and dynamics of PNCs from the molecular scale to the macroscale; and the reinforcement mechanisms of nanoparticles in PNCs [2]. Furthermore, modelling of nanocomposite properties is essential to eliminate the need to synthesise every composite to determine their properties. Practically, predictive models use realistic assumptions that are capable of accurately and efficiently simulating experimental observations in the nanoscale to generate information needed for design purposes in the macroscale. From the multiscale perspective, a computational approach to modelling PNCs involves three procedures depending on the structural level [2]:

• Molecular scale methods: These refer to modelling and simulation methods which employ atoms, molecules, or their clusters as the basic unit of the simulation cell. Popular methods in this category include Monte Carlo (MC) and molecular dynamics (MD) methods.

- Microscale methods: These represent modelling methods which combine the merits of molecular and continuum methods to investigate the microscopic structure and phase separation of PNCs. Examples in this context include dissipative particle dynamics (DPD), dynamic density functional theory (DDFT), lattice Boltzmann (LB), Brownian dynamics (BD), and time-dependent Ginzburg-Lanau method.
- Mesoscale and macroscale methods: These are methods which rely on the fundamental laws of equilibrium together with the laws of conservation of energy, mass, moment, and entropy to describe the mesoscale and macroscopic behaviours of PNC structure represented by an equivalent continuum material. Examples include micromechanics, finite element method, equivalent-continuum, and self-similar approaches.

These methods are generally classified into analytical and numerical methods and are applicable at different length and time scales and at different levels of complexity.

5.2 ANALYTICAL METHODS

For some simplified structures, analytical models give satisfactory outcomes in terms of accuracy, computational efficiency, and predictive capacity. Analytical methods rely on micromechanical approach by using a representative volume element (RVE) to statistically describe the local continuum properties [1]. In this context, the properties of the nanocomposite can be determined based on the properties of the constituents and their volume fractions or geometry of the reinforcements and their dispersion characteristics in the matrix. Some examples of micromechanical analytical models include the Einstein [3], rule of mixtures, cox (shear-lag), Halpin-Tsai, Nielson, as well as Mori-Tanaka models [1]. For didactic purpose, this chapter will focus on analytical methods for the prediction of mechanical and thermal properties of PNCs.

5.2.1 MECHANICAL PROPERTIES

5.2.1.1 Rule of Mixtures

Rule of mixtures (ROM) is the simplest approach to predict the properties of composites. The basic assumption in this approach is that the properties of composites are dependent on the volume fraction of the constituent phases in the composite. Essentially, the properties of the filler and matrix constituents are used to estimate the macroscopic properties of composites [1].

In line with the Voigt's isostrain assumption, specifically the strain parallel to continuous parallel fibre must be equal in both filler and matrix components of the fibre, the longitudinal modulus, E_{11} of a composite with aligned continuous fillers is expressed as Equation (5.1) [4].

$$E_{11} = E_f V_f + E_m V_m \qquad (5.1)$$

where E_f and E_m represent the moduli of fibre and matrix respectively, while V_f and V_m denote the volume fractions of fibre and matrix respectively. Conversely, the transverse modulus, E_{22}, is determined according to Reuss's assumption in which transverse stress in the filler and the matrix is equal, leading to Equation (5.2) [4].

$$\frac{1}{E_{22}} = \frac{V_f}{E_f} + \frac{V_m}{E_m} \qquad (5.2)$$

Similarly, the in-plane shear modulus, G_{12}, can be derived according to the equal stress assumption, that is expressed in Equation 5.3.

$$\frac{1}{G_{12}} = \frac{V_f}{G_f} + \frac{V_m}{G_m} \qquad (5.3)$$

where G_f and G_m are shear moduli of the fibre and matrix, respectively. In case the matrix is filled with randomly distributed fillers, the elastic and shear moduli of the nanocomposite are approximated as Equations (5.4) and (5.5), respectively [4].

$$E_c = \frac{3}{8}E_{11} + \frac{5}{8}E_{22} \qquad (5.4)$$

$$G_c = \frac{1}{8}E_{11} + \frac{1}{4}E_{22} \qquad (5.5)$$

It is noted that other properties of the composite can be predicted by the ROM theory as summarised and presented in Table 5.1 [1].

Due to the isostrain condition, the fibres fail before the matrix reaches its tensile strength. Thus, the ROM model is inadequate to predict the strength of unidirectional composites [5]. In addition, ROM theory neglects the effect of the filler orientation and size in its estimate, leading to inaccuracies in the predictions.

5.2.1.2 Halpin-Tsai Model

The Halpin-Tsai (HT) model is a semi-empirical method that predicts the stiffness of unidirectional composites based on the geometry and the orientation of the filler and the elastic properties of filler and matrix. The general form of the longitudinal, E_{11}, and transverse, E_{22}, moduli is expressed as Equation (5.6) [4,6].

$$\frac{E_i}{E_m} = \frac{1 + \eta_i \xi_i V_f}{1 - \eta_i V_f} \qquad (5.6)$$

Parameter η_i is expressed in Equation 5.7 [4].

TABLE 5.1

Rule of Mixtures Formula for Property Prediction of Composite Materials

Property	Equation
Density (ρ)	$\rho = \rho_f V_f + \rho_m V_m$
Average stress ($\bar{\sigma}$)	$\bar{\sigma} = \bar{\sigma}_f V_f + \bar{\sigma}_m V_m$
Average strain ($\bar{\varepsilon}$)	$\bar{\varepsilon} = \bar{\varepsilon}_f V_f + \bar{\varepsilon}_m V_m$
Poisson's ratio (υ)	$\upsilon = \upsilon_f V_f + \upsilon_m V_m$
Thermal conductivity (κ)	$\kappa = \kappa_f V_f + \kappa_m V_m$
Thermal expansion coefficient (α)	$\alpha = \alpha_f V_f + \alpha_m V_m$
Electrical conductivity (μ)	$\mu = \mu_f V_f + \mu_m V_m$
Gas diffusivity (D)	$D = D_f V_f + D_m V_m$

Note: The subscripts, f and m, refer to filler (fibre) and matrix, respectively. Overbar terms refer to average values.

$$\eta_i = \frac{E_f/E_m - 1}{E_f/E_m + \xi_i} \quad i = 11,\ 22 \tag{5.7}$$

The shear modulus, G_{12}, can be determined as Equation (5.8) [4].

$$\frac{G_{12}}{G_m} = \frac{1}{1 - \eta_{12} V_f} \tag{5.8}$$

Parameter η_{12} is expressed in Equation 5.9 [4].

$$\eta_{12} = \frac{G_f/G_m - 1}{G_f/G_m} \tag{5.9}$$

In Equations (5.6–5.9), E_f and E_m are the Young's moduli of the filler and the matrix, respectively, while G_f and G_m correspond to the shear moduli of the fibre and matrix, respectively. In addition, V_f is the filler volume fraction and ξ is a geometry parameter, which is a measure of reinforcement geometry, packing geometry, and loading conditions, and is expressed for a rectangular-shaped (for example, platelets or lamellar-shaped) filler in the longitudinal and transverse directions, as presented in Equations (5.10) and (5.11), respectively.

$$\xi_{11} = \frac{2l}{t} \tag{5.10}$$

$$\xi_{22} = \frac{2w}{t} \qquad (5.11)$$

where l, w, and t are the filler length, width, and diameter, respectively. In the case of cylindrical-shaped filler with diameter, $d = t = w$, the reinforcement parameters in the longitudinal and transverse directions become $\xi_{11} = \frac{2l}{d}$ and $\xi_{22} = 2$, respectively. The classical HT mathematical model is not suitable for composites reinforced with more than one scale of reinforcements, for example, composites reinforced with randomly oriented fibres. In this regard, the modified HT model for elastic modulus of composite with randomly oriented fibres, E_c, takes the form of Equation (5.12) [4].

$$\frac{E_c}{E_m} = \frac{3}{8}\left(\frac{1 + \eta_L \xi V_f}{1 - \eta_L V_f}\right) + \frac{5}{8}\left(\frac{1 + 2\eta_T V_f}{1 - \eta_T V_f}\right) \qquad (5.12)$$

Parameters η_L and η_T, respectively, are expressed in Equations (5.13) and (5.14) [4].

$$\eta_L = \frac{E_f/E_m - 1}{E_f/E_m + 2l/d} \qquad (5.13)$$

$$\eta_T = \frac{E_f/E_m - 1}{E_f/E_m + 2} \qquad (5.14)$$

5.2.1.3   Mori-Tanaka Model

The Mori-Tanaka (MT) model is commonly used to model the behaviour of particle reinforced composites in which the elastic modulus and the Poisson's ratio are expressed as Equations (5.15) and (5.16), respectively [4].

$$E_c = \frac{9K_c G_c}{3K_c + G_c} \qquad (5.15)$$

$$\nu_c = \frac{3K_c - 2G_c}{6K_c + 2G_c} \qquad (5.16)$$

where K_c and G_c are the bulk and shear moduli of the composites, respectively, which are further expressed as Equations (5.17) and (5.18), respectively [4].

$$K_c = K_m + \frac{V_f(K_f - K_m)}{1 + (1 - V_f)[3(K_f - K_m)/(3K_m + 4G_m)]} \qquad (5.17)$$

$$G_c = G_m + \frac{V_f(G_f - G_m)}{1 + (1 - V_f)[(G_f - G_m)/(G_f + f_m)]} \qquad (5.18)$$

in which

$$K_m = \frac{E_m}{3(1 - 2v_m)}, \quad K_f = \frac{E_f}{3(1 - 2v_f)}, \quad G_m = \frac{E_m}{2(1 + v_m)},$$

$$G_f = \frac{E_f}{2(1 + v_f)}, \quad f_m = \frac{G_m(9K_m + 8G_m)}{6(K_m + 2G_m)}$$

A common phenomenon in particle-reinforced composites is the agglomeration of fillers to form inclusion-like constituents. In this regard, to predict an elastic field in and around a spherical inclusion in an isotropic matrix, a modified MT model which combines MT theory with the principle of Eshelby's inclusion was proposed by Tandon and Weng [7], with analytical expression for the longitudinal and transverse moduli, given as Equations (5.19) and (5.20), respectively [4].

$$\frac{E_{11}}{E_m} = \frac{A_0}{A_0 + V_f(A_1 + 2v_m A_2)} \qquad (5.19)$$

$$\frac{E_{22}}{E_m} = \frac{2A_0}{2A_0 + V_f[-2v_m A_3 + (1 - v_m)A_4 + (1 + v_m)A_0 A_5]} \qquad (5.20)$$

where v_m represents the Poisson's ratio of the matrix and the parameters, A_0, ..., A_5, which depend on the Eshelby's tensor as well as the properties of the filler and the matrix, are given in [7].

5.2.2 THERMAL PROPERTIES

For applications that involve thermal insulation of materials, polymer processing, smart coatings, or thermoelectric devices, modelling of thermal conductivity of PNCs is important, since the polymer composites may contain pores, metals, carbon materials, ceramics, and semiconductors [8]. In general, four approaches can be used to model the effective thermal conductivity of PNCs: (i) micromechanical analogy, (ii) effective medium approximation (EMA), (iii) numerical methods, and (iv) statistical approach. This section focuses on EMA and micromechanical analogy since they are analytically based.

5.2.2.1 Micromechanical Approach

In the micromechanical approach, thermal conductivity is considered analogous with elastic stiffness or elastic shear modulus such that the flux field disturbance becomes analogous to stress field disturbance by the dispersed filler. For highly conductive particles, such as carbon nanotubes (CNT), the thermal conductivities of the resulting PNC exhibit significant disparity with the prediction obtainable by

ROM [9]. Reasons for such discrepancies could be misalignment of CNTs, waviness of CNT shapes, CNT/matrix interfacial thermal resistance, void formation in CNT-based nanocomposites, or lattice defects in CNT structure [10]. To explore the effect of these factors, thermal conductivity models presented in this section focus on CNT-based nanocomposites, but the principle also applies to other highly conductive particles.

Classically, to estimate the axial thermal conductivity of an aligned straight CNT-reinforced nanocomposite considering the CNT volume fraction, length, and diameter, micromechanical analogy of elastic stiffness is extended to thermal conductivity to produce semi-empirical HT formula in Equation (5.21) [11]:

$$\kappa = \kappa_m \left(\frac{1 + \alpha\beta V_{CNT}}{1 - \beta\vartheta V_{CNT}} \right) \tag{5.21}$$

where the parameters α, β, and ϑ are expressed in Equations (5.22), (5.23), and (5.24).

$$\alpha = \frac{2l}{d} \tag{5.22}$$

$$\beta = \frac{\kappa_{CNT}/K_m - 1}{\kappa_{CNT}/K_m + \alpha} \tag{5.23}$$

$$\vartheta = 1 + V_{CNT} \left(\frac{1 - \vartheta_m}{\vartheta_m^2} \right) \tag{5.24}$$

Parameters κ_m and κ_{CNT} are the thermal conductivities of matrix and CNT, respectively. The terms l, d, and V_{CNT} denote length, diameter, and volume fraction of the CNT, respectively. It is noted that, CNT/polymer interface plays a key role in the prediction of thermal conductivities of CNT-based nanocomposite due to large surface area-to-volume ratio of CNTs. For a more realistic prediction of thermal conductivity of CNT-based PNC, CNT/polymer interfacial thermal resistance, random orientation, CNT alignment, CNT waviness should be accounted for, in addition to CNT volume fraction, length and diameter. For example, the effect of interfacial thermal resistance can be incorporated into Equation (5.21) by considering the effective CNT thermal conductivity, which assumes interfacial thermal barrier as an equivalent nanofibre based on the simple rule of mixture in Equation (5.25) [11].

$$\kappa_{CNT}^{eff} = \frac{\kappa_{CNT}}{1 + 2\kappa_{CNT}r_k/(\kappa_m L)}, \quad r_k = R_k\kappa_m \tag{5.25}$$

where r_k and R_k are the Kapitza radius and resistance, respectively. In addition, factors for CNT alignment, waviness, and random orientation can be included in the model through the expression in Equation (5.26) [11].

$$\beta = \frac{\gamma\delta\varrho\kappa_{CNT}/\kappa_m - 1}{\gamma\delta\varrho\kappa_{CNT}/\kappa_m + \alpha}, \quad \delta = 1 - \frac{A}{W} \tag{5.26}$$

Parameter γ is the CNT alignment factor, δ is the waviness efficiency factor and ϱ is the orientation factor, whereas A and W are the amplitude and half wavelength of a wavy CNT, respectively. In some cases, an aggregated state for CNT into the polymer matrix may be observed, due to the PNC fabrication process. Therefore, an aggregation efficiency factor can further be included in Equation (5.27) [11].

$$\beta = \frac{\gamma\delta\varrho\varphi\kappa_{CNT}/\kappa_m - 1}{\gamma\delta\varrho\varphi\kappa_{CNT}/\kappa_m + \alpha}, \quad \varphi = \exp(-V_{CNT}^\varepsilon) \tag{5.27}$$

where ε relates to the CNT aggregation degree, which depends on the fabrication process of the CNT-based nanocomposites.

In general, the choice of micromechanical model for good prediction of thermal conductivity depends on the degree of filler loading, nature of interface (specifically, interfacial resistance), shape and size of inclusions and filler distribution [10]. Therefore, for accurate thermal conductivity predictions, accurate input parameters must be supplied to the analytical models. One way to improve the accuracy of the predictions is the combination of numerical and analytical approaches. For example, in a study [9], atomistic MD simulations can be used to determine the interfacial thermal resistance between the nanoparticles and the polymer hydrocarbon chains, while an analytical approach is adopted to estimate the thermal conductivity, using the calculated resistance from MD simulation. Other examples of analytical models that can be applied for specific conditions of filler qualities are presented in Table 5.2.

5.2.2.2 Effective Medium Approach

In the effective medium approach (EMA), thermal conductivity is methodologically like electrical conductivity in terms of physical transport property. Therefore, the effective thermal resistance of PNCs can be estimated analogously to electrical resistance in which series and parallel arrangement can be derived, as shown in Figure 5.1. Theoretical assumptions in the derivation of the series and parallel models involve: (i) perfect interface between two phases in contact and (ii) in-dependent contribution of the phases to the overall thermal resistance and conductance.

In the case of series model, thermal resistance, R, is considered additive since the temperature drop along the heat flux direction is additive as demonstrated in Equation (5.28) [8].

$$\frac{1}{\kappa_e} = \frac{1}{L}\sum_i R_i = \sum_i \frac{V_{f_i}}{\kappa_i} \tag{5.28}$$

TABLE 5.2

Micromechanical-Based Thermal Models and Their Applications

Micromechanical Model	Applications	Equations
Benveniste-Miloh [12]	Composites with imperfect interfaces between constituents	Composites with prolate inclusions $$\frac{\kappa_{e,33}}{\kappa_m} = 1 + V_f\left(1 + \frac{\kappa_f}{\kappa_m}B_1\right)h(\xi_0)$$ $$\frac{\kappa_{e,11}}{\kappa_m} = \frac{\kappa_{e,22}}{\kappa_m} = 1 + V_f\left(1 + \frac{\kappa_f}{\kappa_m}D_1\right)g(\xi_0)$$ $$\kappa_e = \frac{2}{3}\kappa_{e,11} + \frac{1}{3}\kappa_{e,33}$$ $B_1, D_1, \xi_0, h,$ and g are defined in [10]. Composites with spherical inclusions $$\frac{\kappa_e}{\kappa_m} = \frac{2\kappa_m(1 - V_f) + r\beta\left[1 + 2V_f + \frac{2\kappa_m}{\kappa_f}(1 - V_f)\right]}{\kappa_m(2 + V_f) + r\beta\left[1 - V_f + \frac{\kappa_m}{\kappa_f}(2 + V_f)\right]}$$ β is the interfacial conductance.
Hasselman-Johnson [13]	Composites with uniform distribution and low loading of spherical inclusions	$$\frac{\kappa_e}{\kappa_m} = \frac{2V_f\left(\frac{\kappa_f}{\kappa_m} - \frac{\kappa_f}{rh} - 1\right) + \frac{\kappa_f}{\kappa_m} + \frac{2\kappa_f}{rh} + 2}{V_f\left(1 - \frac{\kappa_f}{\kappa_m} + \frac{\kappa_f}{rh}\right) + \frac{\kappa_f}{\kappa_m} + \frac{2\kappa_f}{rh} + 2}$$ r is the radius of the sphere and h is the interfacial conductance.

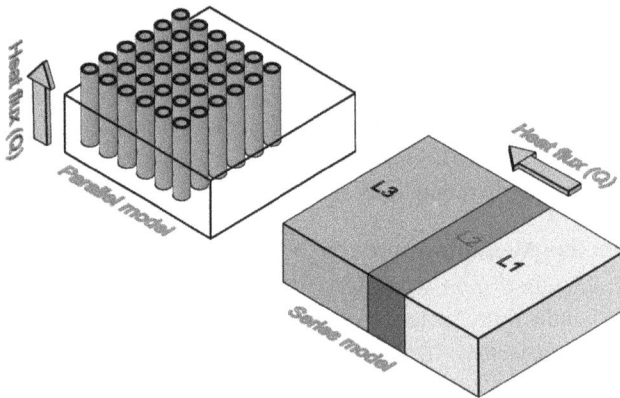

FIGURE 5.1 Parallel and series models.

The parameters are $R_i = \frac{L_i}{\kappa_i}$ and $V_{f_i} = \frac{L_i}{L}$, κ_e is the overall thermal conductivity, and L is the distance that the heat flux flows along. V_{f_i} is the volume fraction of nanofiller i. In the case of a parallel model, the effective thermal conductivity takes the form of Equation (5.29) [8].

$$\kappa_e = \sum_i V_{f_i}\kappa_i \tag{5.29}$$

Although the series model is suitable for laminated composites along the stacking direction, it underestimates the thermal conductivity of particulate composites as it ignores interaction between the fillers. On the other hand, the parallel model applies readily well to continuous-filled composites along the fibre direction but over-estimates the prediction of thermal conductivity of other types of composites. Consequently, the series and parallel models represent, respectively, the lower bound and upper bound for effective thermal conductivity of polymer-based composites. For PNCs, a more realistic prediction of thermal conductivity should account for randomness of particulate inclusions, interfacial thermal resistance of the matrix-particle blend, nanocomposites with multiphase components, and inclusion shape effect. For this purpose, a summary of improved thermal conductivity models for PNCs within EMA framework is provided in Table 5.3.

5.3 NUMERICAL METHODS

Due to significant scale difference between the components of nanocomposites, direct use of micromechanical analytical models may lead to loss of accuracy in the predictions. Numerical method provides accurate predictions of the properties of PNCs based on realistic physical assumptions of the system. Practically, the choice of numerical method to investigate thermodynamics and kinetic properties, mechanical properties and interfacial molecular structure of nanocomposites depends on the length and time scales of interest. For this reason, numerical modelling of PNCs will be discussed under three headings: molecular, microscale, and mesoscale/macroscale (continuum) approaches.

5.3.1 Molecular Scale Methods

Molecular scale methods (Figure 5.2) deal with interactions of atoms, molecules, or their clusters as basic units. The most popular techniques in this framework include molecular dynamics (MD), Monte Carlo (MC), and molecular mechanics (MM). In general, predicting the behaviour of PNC at this scale relies on thermodynamics and kinetics of the formation, molecular structure, and interactions.

5.3.1.1 Molecular Dynamics

Molecular dynamics (MD) is suitable to investigate the effect of fillers on polymer structures and the effect of polymer-filler interactions on the materials properties of polymer composites. In this context, MD is used to predict the physical properties and time evolution of system of interacting particles, such as atoms and molecules [21,22]. Such information as atomic positions, velocities, and forces are easily generated by MD, leading to derivation of macroscopic properties, such as heat capacities, pressure, and energy. Classical MD simulation is realised based on three components. Firstly, the initial positions and velocities of particles in a volume are defined. Secondly, the intermolecular or interatomic potential energy functions are defined. Finally, the evolution of the system in time is determined by solving Newton's equation of motion, given by Equation (5.30) [23].

TABLE 5.3

EMA Analytical Models and Their Applications

EMA Model	Applications	Expression
Russel [14]	Composites with discrete pores dispersed in the matrix.	$$\frac{\kappa_e}{\kappa_m} = \frac{V_f^{2/3} + \frac{\kappa_m}{\kappa_f}(1 - V_f^{2/3})}{V_f^{2/3} - V_f + \frac{\kappa_m}{\kappa_f}(1 + V_f - V_f^{2/3})}$$ κ_m is the thermal conductivity of continuous matrix phase.
Tsao [15]	Composites with particulate inclusions.	$$\frac{1}{\kappa_e} = \int_0^1 \frac{df_l}{\kappa_m + (\kappa_f - \kappa_m)\int_{f_l}^1 \frac{1}{\sigma\sqrt{2\pi}}e^{-0.5\left(\frac{f_l - \mu}{\sigma}\right)^2}df_l}$$ f_l is the one-dimensional porosity, μ is the mean of f_l, and σ is the standard deviation.
Cheng and Vachon [16]	Composites with particulate inclusions.	For $\kappa_f > \kappa_m$ $$\frac{1}{\kappa_e} = \frac{1}{\sqrt{C(\kappa_f - \kappa_m)[\kappa_m + B(\kappa_f - \kappa_m)]}} \ln \frac{\sqrt{\kappa_m + B(\kappa_f - \kappa_m)} + \frac{B}{2}C(\kappa_f - \kappa_m)}{\sqrt{\kappa_m + B(\kappa_f - \kappa_m)} - \frac{B}{2}C(\kappa_f - \kappa_m)} + \frac{1 - B}{\kappa_m}$$ For $\kappa_f < \kappa_m$ $$\frac{1}{\kappa_e} = \frac{2}{\sqrt{-C(\kappa_f - \kappa_m)[\kappa_m + B(\kappa_f - \kappa_m)]}} \tan^{-1}\sqrt{\frac{-C(\kappa_f - \kappa_m)}{\kappa_m + B(\kappa_f - \kappa_m)}} + \frac{1 - B}{\kappa_m}$$ $B = (3V_f/2)^{1/2}$, $C = -4/B$.
Maxwell [17]	Composites with: (i) spherical inclusions, (ii) very low filler loading (f), (iii), good dispersion, and (iv) no interfacial thermal resistance.	$$\frac{\kappa_e}{\kappa_m} = 1 + 3V_f \frac{\kappa_f - \kappa_m}{2\kappa_m + \kappa_f - f(\kappa_f - \kappa_m)}$$
Fricke [18]	Composites with: (i) arbitrarily shaped inclusions, (ii) very low filler loading (f), (iii) good dispersion, and (iv) no interfacial thermal resistance.	$$\kappa_e = \kappa_m + \frac{2V_f}{3(1 - V_f)}\sum_i^{a,b,c} \frac{\kappa_f - \kappa}{2 - abcL_i\left(1 - \frac{\kappa_f}{\kappa_m}\right)}$$ where $L_i = \int_0^\infty \frac{dt}{(i^2 + t)\sqrt{(a^2 + t)(b^2 + t)(c^2 + t)}}$ a, b, and c are the three major axes of an ellipsoidal inclusion.

Hamilton-Crosser [19]	Multiphase composites.

$$\frac{x_c}{x_m} = \frac{1 + \sum_{i=2}^{m} \frac{V_{f_i}(n_i - 1)(x_i - x_m)}{x_i + (n_i - 1)x_m}}{1 - \sum_{i=2}^{m} \frac{V_{f_i}(x_i - x_m)}{x_i + (n_i - 1)x_m}}$$

where $n = 3/\Psi$, Ψ being the sphericity defined as the ratio of the surface area of a sphere to the surface area of the particle. m represents the number of phases in the composite.

Hashin-Lin-Wong [8,20]	Particulate composites.

$$2\left[2 + c' + \frac{x_f}{x_m}(1 - c')\right]\left(\frac{x_c}{x_m}\right)^2 - \left[2(1 + 2c') + \frac{x_f}{x_m}(1 - 4c') + 9\left(\frac{x_f}{x_m} - 1\right)f\right]\frac{x_c}{x_m}$$
$$- \left[2(1 - c') + \frac{x_f}{x_m}(1 + 2c')\right] = 0$$

where $c' = \frac{a^3}{(a + l_k)^3}$, $l_k = R_k x_m$, R_k being the interfacial resistance.

Molecular Quantum
Monte Carlo (MC) dynamics mechanics

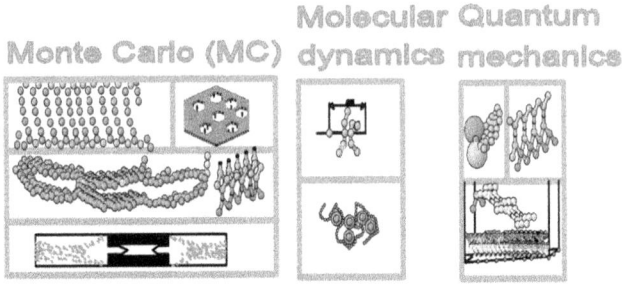

FIGURE 5.2   Molecular scale methods.

$$\sum_i^n \overrightarrow{F}_{ij} = m_j \frac{d\overrightarrow{r}_j}{dt^2} \tag{5.30}$$

where $\overrightarrow{F}_{ij}$ is the force exerted on the jth particle by the ith neighbouring particle at time t, while m_j and $\overrightarrow{r}_j$ are the mass and position of jth particle, respectively, and n is the number of neighbouring particles. To complete an MD simulation, proper consideration must be given to the choice of interatomic potentials, periodic boundary conditions, numerical integration scheme as well as pressure and temperature control to idealise meaningful thermodynamic setup [2]. The interactive forces between particles are calculated from interatomic potential energy function, which depends on the particle position, $\overrightarrow{r}$. Examples of methods that are used to determine the interactive force field include quantum method, empirical or quantum-empirical method. Typically, for an isolated system with n particles, the total energy, E, can be expressed as the Hamiltonian Equation (5.31) [23].

$$E = H = \sum_i^n \frac{\overrightarrow{p}_i^2}{2m_i} + U \tag{5.31}$$

The first term in Equation (5.31) is the kinetic energy of the particles, where $\overrightarrow{p}_i$ represents the momentum of particle, i, whereas U is the potential energy, due to interatomic interactions. The interatomic potential, U, depends on the nature of bonding among atoms. In this regard, different functions have been developed to characterise these bonds. Common examples of potential energy functions include:

- Lennard-Jones Potential [23]:

$$U(r) = \kappa\varepsilon \left[\left(\frac{\sigma}{r} \right)^n - \left(\frac{\sigma}{r} \right)^m \right], \tag{5.32}$$

where $\kappa = \frac{n}{n-m}\left(\frac{n}{m}\right)^{\frac{m}{n-m}}$, ε and σ are positive constants, while m and n are positive integers.

- Moorse Potential [23]:

$$U(r) = \Phi(r) = D\left[e^{-2\alpha(r-r_0)} - 2e^{-\alpha(r-r_0)}\right], \tag{5.33}$$

where D and α are constants and r_0 is the equilibrium distance between two atoms.

- Embedded Atom Potential [23]

$$U_{tot} = \sum_i E_e(\rho_i) + \frac{1}{2}\sum_{i,j(i\neq j)} \phi_{ij}(r), \tag{5.34}$$

in which ρ_i is the electron density of atom i, E_e is the embedding function, and ϕ_{ij} is the pair potential between particles i and j.

The force is related to the energy function through Equation (5.35) [23].

$$\overrightarrow{F_j} = m_j a_j = -\frac{dE}{dr_j} = -\frac{dU}{dr_j} \tag{5.35}$$

in which $a_j = \frac{d\overrightarrow{r}_j}{dt^2}$.

In a case where molecular interaction is involved, a pair potential is not sufficient to capture the particle interaction, since molecules are characterised by covalent bonds [2]. In this case, apart from energy change due to change of bond length (E_{bond}), force field contributions arising from energy changes associated with a change in the bond angle (E_{angle}), molecular rotations ($E_{torsion}$), out-of-plane interactions between molecules (E_{oop}), other types of interaction energies ($E_{nonbond}$), or cross terms between other interaction terms (E_{cross}) may be significant. Therefore, the total energy of a molecule is expressed as Equation (5.36) [2].

$$E = E_{bond} + E_{angle} + E_{torsion} + E_{oop} + E_{nonbond} + E_{cross} \tag{5.36}$$

In relation to Equation (5.35), the velocities are calculated from the accelerations while the positions are calculated from the velocities, using formulas in Equations (5.37) and (5.38), respectively.

$$a_j = \frac{dv_j}{dt} \tag{5.37}$$

$$v_j = \frac{dr_j}{dt} \qquad (5.38)$$

5.3.1.1.1 Time Integration

In MD simulations, the motion of a particle is evaluated over many time-steps. Therefore, a robust integration scheme which minimises accumulated error during the simulation is required to solve the system of differential equations. Examples of techniques commonly employed include Verlet algorithm, Leap-frog algorithm, Velocity verlet, Beeman's algorithm, Symplectic reversible integrators, and Gear predictor-corrector method [23,24]. The general criteria for choosing an algorithm include [24]:

- Computational efficiency.
- Accuracy for long time-step integration.
- Conservation of energy and momentum.
- Time-reversibility, that is, the system should go back to the original state, when $\delta t \rightarrow -\delta t$.

The Gear method [23] is explained here, due to its versatility. The Gear method uses prediction, evaluation, and correction steps to determine the particle position.

- Prediction Step

The particle position and its derivatives at time, $t + \Delta t$, are predicted using a truncated Taylor series in reference to the values of the position and its derivatives at time, t. The predicted position and its time-derivative are mathematically expressed in Equations (5.39)–(5.42) [23].

$$r_j^p(t + \Delta t) = r_j(t) + \dot{r}_j(t)\Delta t + \ddot{r}_j(t)\frac{(\Delta t)^2}{2!} + \dddot{r}_j(t)\frac{(\Delta t)^3}{3!} \qquad (5.39)$$

$$\dot{r}_j^p(t + \Delta t) = \dot{r}_j(t) + \ddot{r}_j(t)\Delta t + \dddot{r}_j(t)\frac{(\Delta t)^2}{2!} \qquad (5.40)$$

$$\ddot{r}_j^p(t + \Delta t) = \ddot{r}_j(t) + \dddot{r}_j(t)\Delta t \qquad (5.41)$$

$$\dddot{r}_j^p(t + \Delta t) = \dddot{r}_j(t) \qquad (5.42)$$

The superimposed dot in Equations (5.39)–(5.42) represents temporal derivatives of the position of the corresponding order, while the superscript p stands for the predicted value.

• Evaluation Step

The interatomic force at time, $t + \Delta t$, is calculated, using the predicted value of the position, r_j^p, as an input leading to Equation (5.43) [23].

$$\vec{F_j} = m_j a_j = -\frac{dU(r_j^p)}{dr_j^p} \tag{5.43}$$

• Correction Step

The discrepancy between the predicted value of the acceleration, $\ddot{r}_j^p$, and the evaluated value according to Equation (5.43), $\Delta\ddot{r}_j$, is used to correct the predicted values of the position and its derivatives, as shown in Equation (5.44) [23].

$$\Delta\ddot{r}_j = \left[\ddot{r}_j(t + \Delta t) - \ddot{r}_j^p(t + \Delta t) \right] \tag{5.44}$$

Parameter $\ddot{r}_j(t + \Delta t)$ is the acceleration obtained from Equation (5.43). The corrected position and its derivatives are then expressed in Equations (5.45)–(5.48) [23].

$$r_j^c(t + \Delta t) = r_j^p(t + \Delta t) + \frac{3}{32}\Delta\ddot{r}_j(\Delta t)^2 \tag{5.45}$$

$$\dot{r}_j^c(t + \Delta t) = \dot{r}_j^p(t + \Delta t) + \frac{251}{720}\Delta\ddot{r}_j(\Delta t)^2 \tag{5.46}$$

$$\ddot{r}_j^c(t + \Delta t) = \ddot{r}_j^p(t + \Delta t) + \Delta\ddot{r}_j \tag{5.47}$$

$$\dddot{r}_j^c(t + \Delta t) = \dddot{r}_j^p(t + \Delta t) + \frac{11}{6}\frac{\Delta\ddot{r}_j}{\Delta t} \tag{5.48}$$

Temperature and pressure control scheme is required in the MD simulation; therefore, an appropriate thermostat algorithm is typically included to re-scale the velocities of the particles such that unrealistic fluctuations of the kinetic energy are suppressed. The Berendsen thermostat [24] is a popular scheme that roughly yields correct canonical ensemble for large systems.

5.3.1.2 Monte Carlo

Monte Carlo (MC) is a stochastic method that relies on random sampling to compute the properties of a system. In the context of PNCs, MC techniques are suitable to investigate molecular response of nanoparticles under the influence of various factors. These factors include, but are not limited to, number of variables bounded to different constraints, accuracy of input parameters, and constraints [2].

Unlike deterministic models including MD, MC models consider risks associated with variation of input parameters. Typically, a statistical distribution is used as the source for each of the input parameters, followed by drawing random samples from each distribution to serve as the input variables. Then, output parameters corresponding to different input parameters are collected from several runs for statistical analysis and decision making. Summarily, to model a physical process, MC technique uses four steps [25]:

- Static model generation – a deterministic model characterising the real scenario is developed, using the most likely value of the input parameters.
- Input distribution identification – the deterministic model is translated into analogous statistical model based on stochastic nature of the input variables. Precisely, historical data of the input variables is used to identify the underlying distributions.
- Random variable generation – in this step, the statistical model generated from different input distributions is solved by numerical stochastic sampling experiment.
- Analysis and decision making – the sample of output variables collected from the simulation is statistically analysed to provide information for decision making.

Considering a canonical (NVT)[a] ensemble with N atoms at temperature T, a new configuration for an atom is formed by arbitrarily translating the atom position from point i to j, in which the change in Hamiltonian, H, is computed as Equation (5.49) [2].

$$\Delta H = H_j - H_i \tag{5.49}$$

where H_i and H_j are Hamiltonian corresponding to positions i and j, respectively. The new atomic configuration is accepted, depending on the direction of ΔH, which reduces the energy of the system. Typically, the new configuration is accepted if the condition in Equation (5.50) is satisfied.

$$\begin{cases} \Delta H < 0 \\ \Delta H \geq 0 \quad \text{subject to probability p} \end{cases} \tag{5.50}$$

The parameter p is given by Equation (5.51) [2].

$$p \propto \exp\left(-\frac{\Delta H}{k_B T}\right) \tag{5.51}$$

Parameter k_B is the Boltzmann constant and T is the absolute temperature of the particle.

According to [1], the new configuration may be accepted subject to the constraint in Equation (5.52).

$$\varsigma \le \exp\left(-\frac{\Delta H}{k_B T}\right) \qquad (5.52)$$

The parameter ς is a random number between 0 and 1; that is, $0 < \varsigma < 1$. In case Equation (5.52) is not satisfied, the new configuration is rejected, and the original position is retained. The process is then repeated for other arbitrarily selected atoms until equilibrium is attained.

Note[a]: NVT ensemble is a system whose internal states are controlled by thermodynamic variable (for example, absolute temperature, T) and mechanical variables (for example, number of particles, N, and volume of system, V).

5.3.2 Microscale Methods

Microscale techniques are aimed at combining the merits of molecular scale and macroscale (continuum) methods, while avoiding their disadvantages. Therefore, the microscale method is suitable to study the microscopic structure and phase separation of PNCs. In these methods, the polymer is characterised with a field representing microscopic particles in which molecular details are implicitly embedded. This approach is computationally beneficial given that the behaviour of nanoparticles can be simulated on length and time scales, which is inaccessible by molecular scale methods.

5.3.2.1 Lattice Boltzmann Method

The Lattice Boltzmann (LB) method is popular for efficient simulation of fluid flow, and it has been adopted for investigating phase separation of binary fluids in the presence of solid particles. One of the major advantages of the LB method is the convenience to incorporate microscopic physical interactions of the fluid particles into the numerical model [2]. Therefore, it is possible to couple the LB method with a molecular scale method, such as MD. The main disadvantage of the LB method is the numerical instability of the simulation, which may be associated with high interparticle interaction strength or high forcing rate.

The LB technique involves a collection of fictitious particles, residing on the nodes of regular shape lattices, interacting according to simple rules. To describe a particle occupation in the LB method, a distribution function of particle velocity along the i direction, $n_i(x, t)(i = 1, ..., N)$ is defined, where N denotes the number of directions of the particle velocities at each node, and the evolution rule for updating lattice site occupation by the particles in a given time step is defined based on the discrete velocity model for a multicomponent $(c = 1, ..., M)$ flow [24] as expressed in Equation 5.53.

$$f_i^c(x + e_i \delta_x, t + \delta_t) = f_i^c(x, t) + \Omega_i(f_i^c(x, t)) \quad i = 1, ..., N \qquad (5.53)$$

Parameter t is the time, e_i is the discrete velocity vector in the i direction, and Ω_i is the collision operator in the i direction representing changes, due to pairwise collisions. In the LB method, at each time step, particles at each lattice points move to the nearest nodal points in the lattice in accordance with the direction of velocity (a

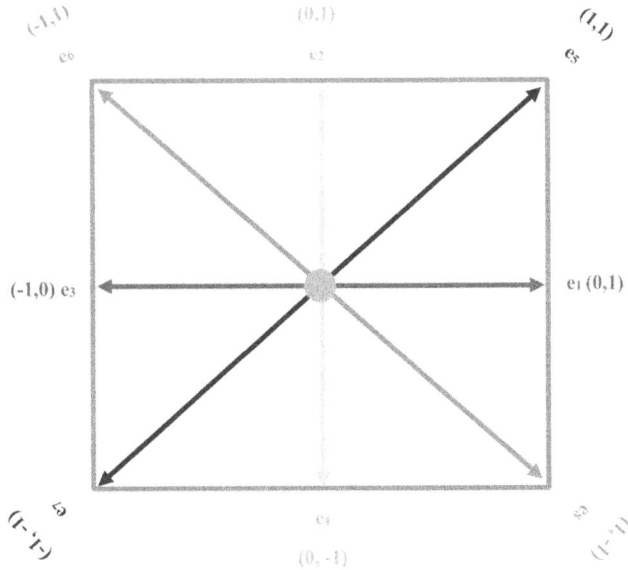

FIGURE 5.3 Nine velocity vectors of D2Q9 lattice.

process called streaming). Then, interaction of particles occurs, leading to change in the velocity directions of the particles (a process called collision or redistribution). A typical lattice structure is defined by the problem dimension, d, and the number of lattice vectors, n_v, and is compactly represented as DdQn_v. For example, Figure 5.3 depicts the discretised 2D with nine velocity vectors.

To characterise the pairwise collision, the collision operator can be expressed in different ways. For example, a simple operator in the form of a linear Bhatnagar-Gross-Krook, which uses single relaxation time, is expressed as Equation (5.54) [24].

$$\Omega_i = -\frac{1}{\tau^c}\left(f_i^c - f_i^{c,eq}\right) \tag{5.54}$$

where $f_i^{c,eq}$ corresponds to local equilibrium distribution function of f_i^c and τ^c is the viscosity-related dimensionless collision relaxation time of component c of the fluid (polymer). The relaxation parameter is related to the fluid kinematic viscosity of fluid component c, v^c, through Equation (5.55) [24].

$$\tau^c = \frac{v^c}{s^2} + \frac{1}{2} \tag{5.55}$$

Parameter $s = 1/\sqrt{3}$ is the lattice speed for unit spacing lattice structure.

In some applications, where stability issues are encountered due to a fixed Prandtl number associated with single-relaxation collision operator, the multiple-relaxation collision operator is preferred. The multiple-relaxation collision technique is suitable for varying kinematic and bulk viscosities and incorporates mechanisms for improved stability of the simulation.

5.3.2.2   Brownian Dynamics

The Brownian dynamics (BD) technique allows simulation of particles on the microsecond timescale, whereas the MD technique permits particle simulation up to a few nanoseconds. Unlike in MD where there is explicit description of solvent molecules, in BD an implicit description of solvent particles is applied in a continuum sense while ignoring internal motions of the molecules. The objective of the BD technique is achieved by accounting for the effect of solvent molecules on the polymer through dissipative and random force terms so that the governing equation of motion becomes Equation (5.56) [2].

$$F_i(t) = \sum_{i \ne j}^{n} F_{ij} - \alpha\mu_i + \beta\tau_i(t), \qquad (5.56)$$

Equation (5.56) is the Langevin equation, where F_{ij} is the force applied by particle, j on particle i, α and β are system-dependent constants, μ_i is the momentum of particle i,, and τ_i is a Gaussian random noise term. The approximation in Equation (5.56) leads to fluctuating forces such that energy and momentum are no longer conserved. Since Equation (5.56) does not obey the Navier-Stokes equations, the BD method is only suitable to simulate the diffusion properties but cannot reproduce the hydrodynamic flow properties of the system.

5.3.2.3   Dissipative Particle Dynamics

Dissipative particle dynamics (DPD) is a particle-based method, such as the MD and BD techniques, and is suitable for both Newtonian and non-Newtonian fluids on microscopic length and time scales. The basic unit of DPD is a molecular assembly (representing a particle) that is characterised by its mass, m_i, position, r_i, and momentum, μ_i. The force describing the interactions between two DPD particles may be expressed as a sum of the conservative, F_{ij}^C, dissipative, F_{ij}^D, and random, F_{ij}^R, forces in Equation (5.57) [2].

$$F_{ij} = F_{ij}^C + F_{ij}^D + F_{ij}^R, \qquad (5.57)$$

The total force, F_i, acting on a particle i at time t is given in Equation (5.58) [2].

$$F_i = \sum_{j \ne i} F_{ij}^C + F_{ij}^D + F_{ij}^R \qquad (5.58)$$

According to Equation (5.58), the macroscopic behaviour of DPD particles incorporates Navier-Stokes hydrodynamics, due to conservation of momentum. In comparison with MD, larger time steps are allowed in DPD simulation and hydrodynamic equilibrium is attained in a DPD system with far fewer particles.

5.3.3 MESOSCALE AND MACROSCOPIC METHODS

Notwithstanding the importance of molecular scale and microscale modelling, molecular properties of PNCs can be homogenised macroscopically according to different scales. As a result, in the macroscopic scale, the discrete atomic and molecular structure is ignored while continuity of material distributed throughout the control volume is adopted. The basis of macroscopic scale methods lies in the compliance with fundamental laws of continuity, equilibrium, momentum, conservation of energy, and conservation of entropy [2]. By employing appropriate constitutive relations and equation of state, the macroscale method measures the deformation of a continuum because of external forces by the resulting internal stresses and strains.

5.3.3.1 Micromechanics Approach

The micromechanics approach relies on the development of RVE to characterise local continuum properties, since the uniformity of continuum may not hold at the macroscopic scale (Figure 5.4). The RVE is periodically arranged in consistent with the smallest constituent that significantly affects the macroscopic behaviour. Therefore, micromechanics approach can assess discontinuities and interfacial profile of different constituents in a continuum. In addition, micromechanics

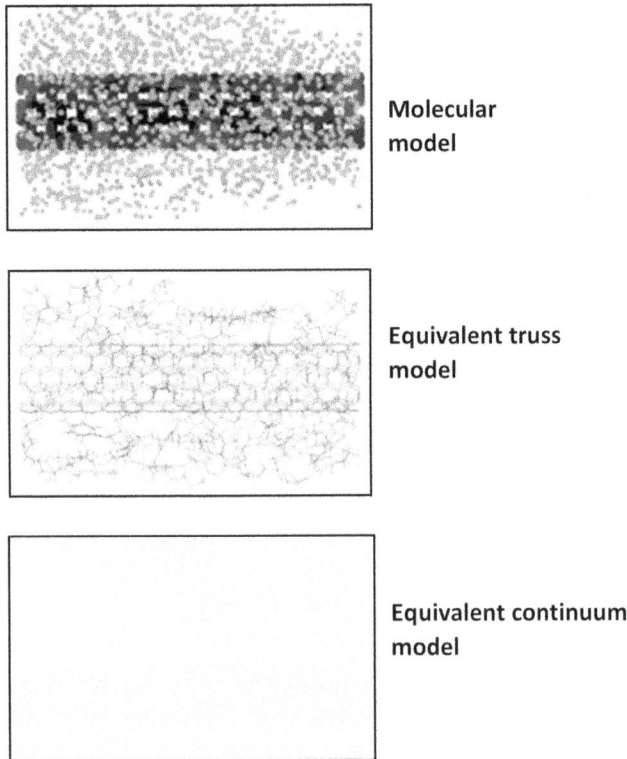

Molecular
model

Equivalent truss
model

Equivalent continuum
model

FIGURE 5.4   Representative volume elements of computational model at different length scale.

approach allows for coupling between mechanical and non-mechanical properties. In the context of PNCs, micromechanics methods tend to satisfy the following basic assumptions [2]:

- Linear elasticity of fillers and polymer matrix: This assumption suggests that a linear relationship exists between the total stress, σ, and the infinitesimal strain tensors, ε, for the matrix and nanophase constituents, as expressed in Equations (5.59) and (5.60).

$$\sigma^f = C^f \varepsilon^f \tag{5.59}$$

$$\sigma^m = C^m \varepsilon^m \tag{5.60}$$

C^m and C^f are the stiffness tensors for the matrix and fillers, respectively.

- Axisymmetric shape of fillers that are defined by aspect ratio: Since the local stress and strains are non-uniform throughout the PNC continuum, volume average stress and strain resultants are defined over the representative averaging volume, V, as presented by formulas in Equations (5.61) and (5.62), respectively.

$$\bar{\sigma} = \frac{1}{V} \int_V \sigma(x) dV \tag{5.61}$$

$$\bar{\varepsilon} = \frac{1}{V} \int_V \varepsilon(x) dV \tag{5.62}$$

The equivalent average stress and strain tensors of the fillers and matrix are expressed by formulas in Equations (5.63) and (5.64).

$$\bar{\sigma}^c = \frac{1}{V^c} \int_{V^c} \sigma(x) dV \tag{5.63}$$

$$\bar{\varepsilon}^c = \frac{1}{V^c} \int_{V^c} \varepsilon(x) dV \tag{5.64}$$

Parameter c denotes the filler, f or polymer matrix, and, m component. On this basis, the total average stress and strain are related to the filler and matrix averages according to Equations (5.65) and (5.66), respectively.

$$\bar{\sigma} = v^f \bar{\sigma}^f + v^m \bar{\sigma}^m \tag{5.65}$$

$$\bar{\varepsilon} = v^f \bar{\varepsilon}^f + v^m \bar{\varepsilon}^m \tag{5.66}$$

- Perfect bond between fillers and polymer matrix: The average properties of composite stiffness, $\mathbf{C}$, relate the average stress to the average strain tensors according to formula in Equation (5.67).

$$\bar{\sigma} = \mathbf{C}\bar{\varepsilon} \qquad (5.67a)$$

$\mathbf{C}$ is related to the filler and matrix stiffness tensors according to Equation (5.67b).

$$\mathbf{C} = \mathbf{C}^m + v^f(\mathbf{C}^f - \mathbf{C}^m)\mathbf{A} \qquad (5.67b)$$

Parameter $\mathbf{A}$ is the strain concentration tensor, defined as the ratio between the average filler strain, $\bar{\varepsilon}^f$, and the average strain, $\bar{\varepsilon}$, and can be determined by the formula in Equation (5.68).

$$\bar{\varepsilon}^f = \mathbf{A}\bar{\varepsilon} \qquad (5.68)$$

5.3.3.2   Equivalent-Continuum Approach

Classical micromechanics approach lacks the capacity to simulate the behaviours of nanotube-reinforced composites, due to a significant scale difference. The equivalent-continuum (EC) approach, proposed in [26], is a multiscale approach, which incorporates molecular, microscopic, and continuum scales to model nanotube-reinforced composites, such as single-walled carbon nanotube (SWCNT) composites.

In the first stage of EC simulation, the equilibrium structure of the polymer composite is generated with the MD technique in which the potential energy of the SWCNT-polymer composite system is defined according to MD principles. Then, an equivalent-truss model is used to construct the RVE, where each truss element represents an atomic interaction and each pin-joint of the truss corresponds to an atom in the molecular model (Figure 5.4).

Therefore, the equivalent-truss model represents the intermediate link between MD and EC systems. By equating the potential energy of the effective fibre, U^f, with the potential energies of MD and equivalent-truss models, U^{MD} and U^t, respectively, the relationship between the elastic stiffness tensor of the effective fibre and the force constants of the MD system can be established as Equation (5.69) [2].

$$U^f = U^{MD} = U^t \qquad (5.69)$$

By applying a set of loading conditions to Equation (5.69), the components of the elastic stiffness tensor of the effective fibre can be determined. Then, based on the mechanical properties of the effective fibre and the bulk polymer, the general constitutive properties of the SWCNT-based PNC can be determined, using an appropriate micromechanics model, such as MT and HT models [2].

5.3.3.3   Finite Element Method

The finite element method (FEM) is a numerical method developed to approximate solution of differential equations. FEM relies on the generation of mesh to idealise a structure and capture the response under complex loading, material, and geometric conditions. The procedure for the implementation of the FEM starts with the replacement of the continuum with finite elements constituting subdomains of the continuum. Finite elements are defined by shape functions, which are used to transform the element topology and approximate state variables at the element nodes. Appropriate constitutive laws are then selected to define the relationship between strain and stress fields, followed by application of the variational principle to describe the strain energy of the problem over each finite element. The total elastic energy in the continuum model, neglecting traction and body forces, is given by the sum of potential and kinetic energies in Equation 5.70 [2].

$$U = U^p + U^k \tag{5.70}$$

U^p denotes the potential energy expressed in the sample volume of the continuum and is given by Equation (5.71).

$$U^p = \frac{1}{2} \int \varepsilon^T C \varepsilon dV \tag{5.71}$$

in which ε represents the strain tensor, while $\mathbf{C}$ is the elastic stiffness tensor. The kinetic energy in the sample volume is denoted as U^k and is expressed by the formula in Equation (5.72).

$$U^k = \int \rho \frac{\partial^2 \mathbf{u}}{\partial t^2} dV \tag{5.72}$$

By replacing the continuum state variables in Equations (5.71) and (5.72) with discrete variables defined by a set of shape functions, a system of elemental equations is formed and subsequently assembled to form the global system of equations. The global system is numerically solved to determine the unknown state variables, which are consequently used to compute the stresses and strains of the PNC structure.

5.3.4 Multiscale Modelling

One of the main goals of simulation of PNCs is the accurate prediction of their hierarchical structures and properties. Towards enhancing the properties of PNCs for high-technology applications, structural characterisation, and precise manipulation of the fabrication of the nanostructured materials must be achieved. This implies that the nanocomposite structure as well as physical and chemical processes at the nanoscale level significantly influence the final properties of PNCs. Therefore, an adaptable modelling strategy is required to explore the design of PNC materials for optimised performance.

Prediction of the PNC behaviour based on application of analytical and numerical models at different independent scales is challenging considering length scales up to six orders of magnitudes or time scales spanning a dozen orders of magnitude (Figure 5.5), thus impeding the simulation efficiency [8]. Given this reality, it is impracticable for a single model to explore these length and time scales. Multiscale modelling is a practical way to bridge different length and time scales via combined computational methods that can simulate fundamental molecular processes and seamlessly transfer numerical parameters efficiently across wide scales to satisfactorily predict macroscale properties. In this context, two multiscale approaches that span from molecular to macroscopic levels are known as sequential and concurrent methods [2], as subsequently elucidated:

- Sequential Multiscale Method: This involves linking of hierarchical computational methods in a manner that allows the computation of operative parameters of a model on large scale from the calculated quantities obtained at a lower scale of the model [2]. An example of simulation tools designed to perform sequential multiscale modelling is OCTA, developed by Doi [27]. OCTA relies on multiple simulation engines to facilitate modelling of polymers from molecular scale up to the mesoscale.
- Concurrent Approach: This uses a combination of several computational methods linked together to bridge different scales of material behaviour concurrently [2]. Although this approach is computationally promising, considering the potential for seamless interaction between different scales of material behaviour, the method is in its early stage of development and, thus, it has a limited application.

Over the years, a variety of computational methods have been developed to efficiently capture phenomena at each length and time scale. For example, quantum mechanic methods are available to simulate systems containing atoms, while MD or MC methods are suitable for capturing material properties at atomic level. At the mesoscale level, BD or DPD approaches are examples of methods relevant to investigate mesoscale properties of polymer-based materials, while FEM possesses excellent potential to capture macroscopic properties of the system. The goal of multiscale modelling is to exploit the individual capacities of these methods, beginning from the quantum scale all the way to the process scale to predict the macroscopic properties of an engineering system of interest. Fundamentally, the key ingredients to construct a successful multiscale modelling are: (i) information about the basic processes that control the system at the lower scales and (ii) reliable strategies to link the degrees of freedom from the lower scale to the coarser scale [8].

In the context of polymer-based nanocomposites, the coarsening process at the lower scale, for example from QM to MD, is based on fundamental principles and can be generalised. However, the larger range of length and time scales that define the macromolecules constitutes specific challenges that complicate coarsening process at higher scales. As a result, scale integration in the specific context of PNCs can be achieved in different ways. At the lower scale, scale integration is based on force field application that retrieves information from the quantum

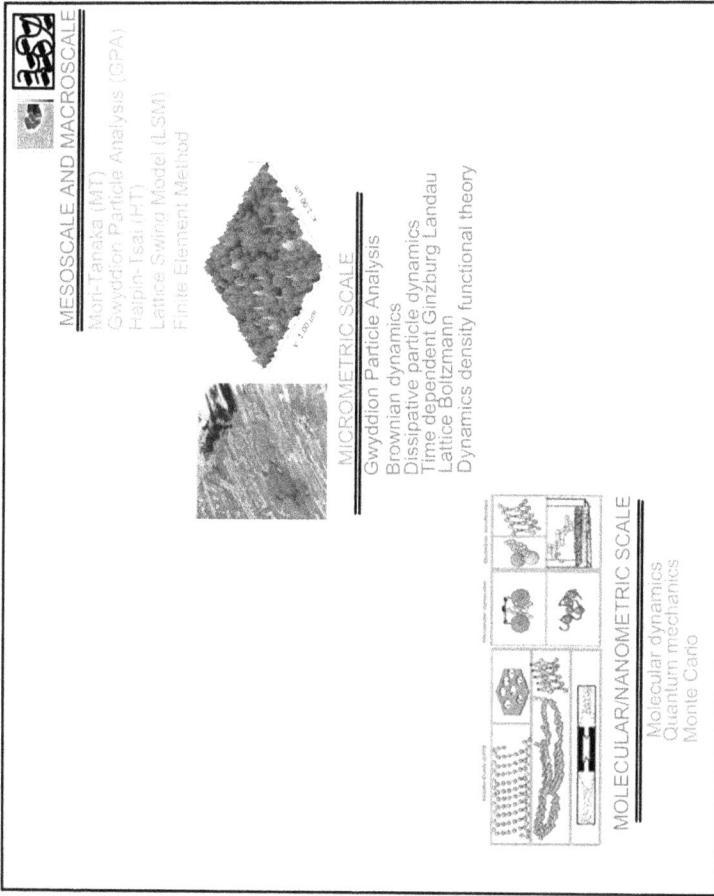

FIGURE 5.5 Hierarchical settings of PNC structure in time and length dimensions.

chemistry to the atomistic model. At the mesoscale level, essential features of the atomistic systems must be preserved structurally or thermodynamically, or both [28]. Finally, the most challenging aspect toward practical design application of material is linking to the mesoscale in which the microstructural features of PNCs are described to predict the property at the macroscopic level.

As an example of hierarchical multiscale computational process, montmorillonite (MMT)/poly (ethylene oxide) (PEO)–based PNC was investigated [29] to determine the effect of PEO molecular weight and the presence of water molecules on the interactions between polymer and clay platelets. Firstly, atomistic MD simulation of PEO-based PNC was carried out in a solvated environment to retrieve energetical (binding energy) and structural (interaction energy and chain conformation) information at the molecular level [30]. Then, the mesoscale property prediction was achieved by adopting the DPD approach in which the data gathered at the atomistic scale were projected to the corresponding energetical and structural information required to set up a coarse-grained mesoscale simulation. The mesoscale simulation produced information on the morphologies and density distribution of the system, which served as basic inputs to finite element simulation that was used to predict the property of the system at the macroscale level.

Stochastic multiscale approach is an example of the multiscale modelling. It is explained as follows. The random dispersion of reinforcing phases in PNCs significantly affects the microstructural properties which in turn influence the macroscopic behaviours. Other issues affecting the microstructural behaviours of PNCs include the size, orientation, and shape of the nanophases. Given these challenges, there is limited control in the manufacturing of tailored PNC materials, leading to uncertainties in the actual mechanical properties of the PNC constituents. In the context of modelling, critical experimental information may be omitted in the assumptions and approximations used to develop the PNC model. Besides, in the case of mixing phases that span multiple structural length scales, the load transfer mechanisms between the polymer matrix and nanophases are multiscale in nature. Consequently, the mechanical properties of PNCs are uncertain and substantial disparities may occur between experimentally measured mechanical properties and predictions by analytical or numerical calculations [31,32]. To incorporate the spatial randomness induced by the nonuniform distribution of constituent phases in polymers, the multiscale stochastic finite element method (MSFEM) has been proposed [33]. In MSFEM, the PNC is assumed to constitute a random heterogeneous media, due to random behaviours and uncertainties in the overall material behaviours. On this basis, a multiscale micromechanical approach is used to homogenise the system to obtain the estimates of local mechanical properties. This step is followed by Monte Carlo finite element scheme, which uses the local properties as inputs to compute the bulk properties of the PNC.

The procedure required to implement MSFEM can be summarised into four steps [33]:

- Definition of material region: A RVE that structurally captures the whole mixture on average and encapsulates sufficient number of inclusions is firstly defined. Other types of representative material regions include statistical volume element (SVE) [34] or statistically equivalent RVE (SERVE) [35], which incorporate separation of scale, have been proposed. According to [35], SERVE is suitable to compute local mechanical properties, as dictated by the actual randomness induced by the non-uniform dispersion of nanophase constituents in the PNC microstructure.

- Identification of spatial randomness: To capture heterogeneities, which induce randomness in the PNC structure, MSFEM statistically quantifies the variations of the volume fractions of nanophase constituents. The relative concentration of the nanophase constituents in the material region are assumed to be equivalent to the total amount of nanofillers in the PNC. According to Figure 5.6, variations observed in the volume fraction values of SWCNT are captured at the grid of material points in the material region through creation of a random field based on probability distribution functions [33].

- Homogenisation of multiscale properties: MSFEM determines the local properties of the PNC at the sub-element level of the material region (Figure 5.7). In MSFEM, the homogenisation process is performed in two stages based on the MT method to determine the stiffness tensor of homogenised spherical inclusions (IN) and the modified matrix (MM) in each FE in the material region. A two-phase media, which comprises a modified matrix and spherical inclusions, is adopted [36] to determine the effect of SWCNT dispersion and agglomeration on the mechanical properties of CNT-based PNC.

- Monte Carlo finite element model: MSFEM determines the overall mechanical properties of the PNC system by numerically solving elasticity problem in line with conventional FE principles, except that the constitutive parameters are random values obtained according to the Monte Carlo method.

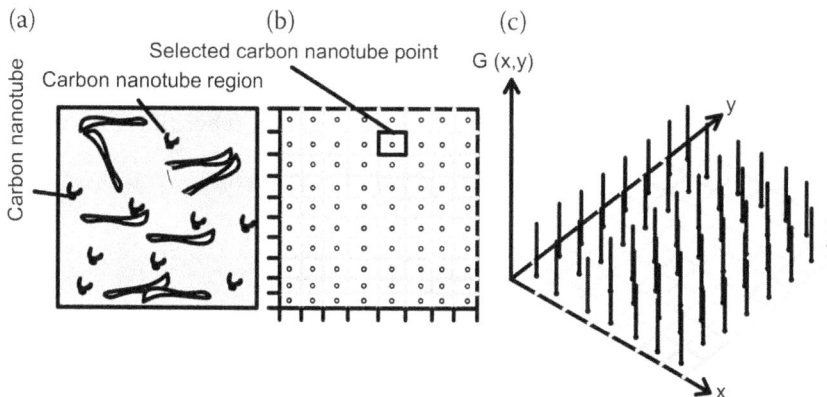

FIGURE 5.6 (a) Material region (MR), (b) material points with random CNT volume fraction, (c) discretised random field for variation of SWCNT in polymers.

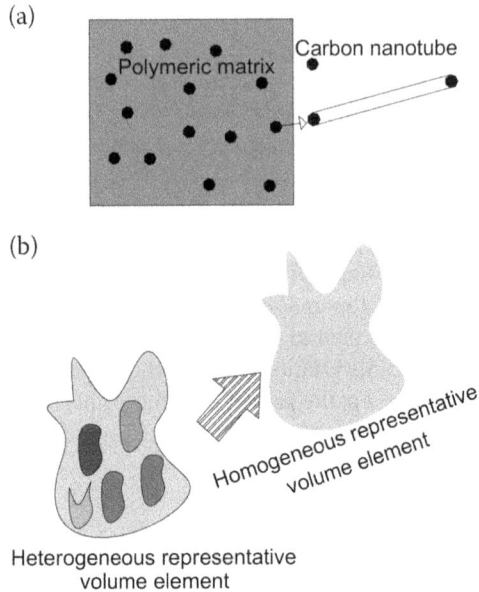

FIGURE 5.7 (a) Sub-element material structure in a finite element and (b) representative volume element homogenisation procedure.

5.4 COMPUTATIONAL APPROACH TO PNC PROPERTY PREDICTIONS

5.4.1 STIFFNESS AND STRENGTH

The analytical and numerical models describe above can be used to predict mechanical properties, such as strength and stiffness (elastic modulus), which are significantly influenced by the PNC filler size and polymer configuration [1]. In this context, extended micromechanical models have been adopted to predict the properties of PNCs [37,38]. For example, the micromechanical-based MT model has been combined with the probability distribution function (PDF) of the Doi-Hess hydrodynamic theory to predict the effective elastic properties of stationary, and flow-induced nanorod-based PNC [37]. Considering a perfect bonding arrangement between inclusions and matrix, an effective stiffness tensor which is valid for any volume fraction was derived according to the MT theory. The symmetry regarding the orientational distributions of the effective stiffness tensor of nano-rod composites are inherited from probability distribution functions (PDFs). The volume fraction dependent effective moduli can be consequently determined by implementing the MT formula using numerical databases for the PDFs [37].

Micromechanical models have also been used to predict the mechanical properties of nanotube-based PNC containing complex inclusions, where the Eshelby's equivalent tensor is combined with MT-type model with two aspect ratios to evaluate the longitudinal and shear moduli [38]. Given the constitutive equations for inclusion and matrix materials (see Equations 5.59–5.66), the volume-average stresses and strains were determined through integration of non-uniform local quantities over a large volume of the material filled with inclusions. The effective moduli of the PNC are eventually evaluated in terms of the average stresses and strains [38]. Moreover, to predict the interfacial strength of SWCNT-based PNC, a modified Kelly-Tyson approach (continuum-based method) has been proposed under the assumption of uniform interfacial shear and axial normal stresses [39]. The continuum mechanics approach, often combined with a suitable micromechanical method, is another strategy to predict the mechanical properties of PNCs.

In the aspect of numerical methods, Mokashi et al. [40] proposed a procedure based on molecular mechanics approach to obtain the tensile properties (specifically both elastic modulus and strength) of PNC, in which the atomic structure of the polymer matrix is first generated in the simulation cell using computer algorithm and relaxed to an equilibrium configuration whose parameters are recorded [40]. Then, the atomic structure of the nanofillers (for example, SWCNT) is generated and the coordinates merged with the polymer structure. Tensile boundary and loading conditions, like mechanical testing, are applied to the atoms at the boundaries of the simulation cell of the nanocomposite system. In each step of the incremental longitudinal displacement, the system is relaxed to an energetically stable state and this process is repeated until the polymer chains fracture. The tensile strength is determined from the total force, along the direction of the incremental displacement, acting on the boundary atoms per unit area of the cross-section.

A multiscale strategy comprising the FEM approach and a micromechanical method (for example, the MT method) has been extensively explored to predict the elastic properties of wavy or randomly oriented nanotube-based PNCs and silica nanoparticle-based PNC [41–43]. To investigate the effect of characteristic waviness of nanotubes embedded in polymers on the effective stiffness of nanotube-reinforced polymers, a 3D FEM model was used to numerically compute dilute strain concentration tensor. Thereafter, the concentration tensor was used to predict the effective modulus of the PNC with aligned or randomly oriented inclusions based on MT theory [42]. It is noted that for accurate prediction with this approach, especially in the case of high particle loading, it is important to account for the properties of the interface between matrix and the particles. In terms of accuracy of the FEM-micromechanical approach, fibre volume fraction is a significant factor to be considered when predicting stiffness properties of PNCs. For PNC with low-fibre volume fraction, MT method can accurately predict the stiffness of PNC [44]. Above a critical volume fraction, FEM-based models possess superior accuracy over MT micromechanical models, especially when handling complex fibre-polymer interactions.

5.4.2 Stress Transfer

The mechanical properties of PNC with high disparity in the modulus of its constituents (polymer and nanoparticles) are strongly influenced by stress-transfer mechanism. This is the case for clay/polymer nanocomposites, where clay nanoparticles exhibit much higher modulus than the polymer matrices, and there is presence of large interfacial area between the thin nanoclay platelets and the polymer. Therefore, due to the dominant role of the interfacial properties, stress transfer mechanism between rigid nanoparticles, and the polymer play a significant role in the modulus and strength properties [45].

Characterising the interfacial properties of PNCs is an effective way to determine the level of stress or load transfer (interfacial adhesion). This can be achieved using the MD [46], molecular mechanics (MM) [47], continuum mechanics (CM) [48], or an analytical approach [45]. Essentially, the interfacial shear stress is calculated based on the conditions of perfect or imperfect bonding, which determine the reinforcing efficiency of the fillers. Perfect bonding is a simple assumption that can lead to over-prediction of the mechanical properties since the aspect ratio and volume fraction may be overestimated. Nonetheless, for certain PNCs, such as nanoclay-based types, complete adhesion can be achieved at low concentration of clay platelets and a perfect bonding assumption can be justified. Under imperfect bonding condition, which applies for many PNCs, there is incomplete stress transfer at the polymer-filler interface, necessitating the estimation of effective parameters for the calculation of mechanical parameters. For example, to determine the modulus of PNC under imperfect bonding condition, the effective aspect ratio and effective volume fraction must be calculated to enhance the accuracy of the prediction [45].

The MD simulation has proved effective for estimating the shear strength and critical length required to enhance the load transfer capacity of nanotube-based composite [46]. The model system comprises a SWCNT embedded into a crystalline or amorphous matrix and subjected to a periodic boundary condition. In addition, the intramolecular interactions in the nanotube, polymer chains, and cross-links are characterised by a many-body potential function that allows the formation of chemical bonds between the nanotube and polymer matrix. The non-bonded interactions are captured by Lennard-Jones potentials. The polymer-nanotube system was first equilibrated using the MD simulation to create zero initial stress state. In the next step, further MD simulation was carried out with a uniform one-body force added to the atoms that comprise the nanotubes to determine the minimum shear strength required to pull the nanotube through the polymer matrices (i.e., crystalline, and amorphous). The shear strength, τ_c, is thus estimated as the total force at which the centre mass of the nanotube began to move freely independent of the matrix. Finally, the critical length of a nanotube that guarantees strong load transfer is estimated from τ_c according to Equation (5.73) [46].

$$l_c = \frac{\sigma_f d_f}{\tau_c} \qquad (5.73)$$

The parameter σ_f represents the fibre tensile strength and d_f denotes the fibre diameter.

5.4.3 Fatigue and Fracture

Studies have demonstrated that, without compromising the stiffness, nanoparticles can improve the fatigue resistance of polymer nanocomposites through crack-bridging and a frictional pull-out mechanism [1]. To determine the fatigue response of PNC, a good model should capture different mechanisms responsible for fatigue resistance, which include crack pinning, fibre bridging, crack tip deflection, and particle debonding (fibre pull-out). The dominating mechanism depends on the nature of polymer-nanofiller mix. Fibre pull-out is the main mechanism of fatigue failure in CNT-based nano-composites. A fracture mechanics model was proposed in [49] to determine the effective stress intensity factor amplitude to propagate crack in CNT-based PNC, under the assumption that the crack-opening displacement at a distance behind the crack tip equals the pull-out length, leading to Equation (5.74), where ΔK_I is the stress intensity factor amplitude in the absence of bridging, w_c is the work required to pull out a single nanotube, ρ is the number density of CNT penetrating the plane of the crack, and β ($\beta > 1$) is a parameter accounting for the number of fibres contributing to toughening, since not all fibres penetrating the crack plane participate in the bridging process.

$$\Delta K_I^{eff} = \sqrt{\Delta K_I^2 - \frac{2G\rho w_c}{(1-v)\beta}} \tag{5.74}$$

The modelling of a fracture in PNC is motivated by the fact that cracking is accompanied by bridging action of nanofillers, such as CNT [50]. Basically, in the idealisation of crack bridging, it is assumed that embedded nanotubes in a polymer matrix align perpendicular to the crack front in the matrix to provide resistance to crack propagation (Figure 5.8). In this context, the objective of fracture modelling in PNC is to relate the fracture energy of PNC to the nanoscale mechanical properties of the nanotube and the nanotube-polymer interface. Analytically, a cohesive zone model is an effective approach to characterise the fracture properties of nanotube-polymer interface, as applied in [50], where a shear opening model was derived based on cohesive zone potential, leading to Equation (5.75).

$$F_{max} = \frac{4\pi}{\sqrt{2e}} RL \frac{\Gamma_I}{\delta_{cr}} \tag{5.75}$$

where F_{max} represents the maximum force at which shear opening occurs at the interface; R and L are the nanotube outer radius and embedded length, respectively; Γ_I is the intrinsic fracture energy of the interface; and δ_{cr} is the critical crack opening displacement of the interface. According to [50], a good model for fracture energy should be able to correctly quantify the contribution of interfacial adhesion between

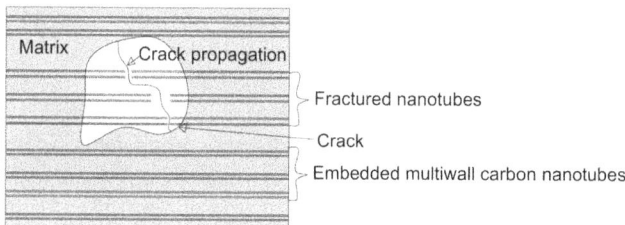

FIGURE 5.8 Schematic of crack bridging mechanism of carbon nanotube.

the nanotube and polymer, and at high velocities of nanotube pull-out, viscous dissipation of the polymer up to the time of pull-out must be accounted for, as the pull-out force is dependent on the velocity.

In general, to describe yielding or fracture behaviour of nanocomposites, detail atomistic observations involving chemical bond breaking must be accounted for [51]. Therefore, application of macroscopic or continuum modelling is limited in this context. For nanocomposites with hollow or filled nanotubes or nanowires, whose shapes resemble macroscopic thin-walled hollow columns or tubes, failure criteria can be described analogously to macroscopic conditions. As such, a multiscale approach may be an effective means to characterise fracture in these nanocomposites, considering the computational efficiency of macroscale methods and accuracy of atomistic modelling. An example of the multiscale method, which combined continuum mechanics with molecular dynamics to describe atomistic prediction of failure in CNT-based nanocomposites, has been presented in [51].

A CNT system under biaxial tensile-torsional loads was modelled using the MD technique in which short, covalent interactions are characterised with the aid of many-body and reactive empirical bond-order (REBO) potentials. Moreover, long-range van der Waals interactions are captured in the form of the Lennard-Jones potential. The CNT system, which is fixed at one end and subjected to applied biaxial tension-torsion at the other end, is subjected to incremental load to obtain the load paths. The stresses experienced by the CNTs were calculated by the continuum mechanics approach, which considers the CNTs as geometrically equivalent to macroscopic thin-walled columns or tubes. The calculated stresses are finally transformed to the principal stresses to determine the failure criteria.

In particle-filled nanocomposites, depending on the nature of load, a fracture may occur inter-grain in the embedding matrix. In such a case, a method that provides quantum mechanical description of the interactions is suggested to study the onset of fracture accurately, as demonstrated in [52], where a tight-binding MD method was used to investigate the fracture properties of tetrahedral amorphous carbon and nanodiamond-filled composites.

5.4.4 Creep

Modelling creep behaviour of materials is important because creep failure occurs at a stress lower than the yielding stress of the material. Since creep is a time-dependent plastic deformation over a long period of loading, viscoelasticity of the polymer matrix plays an important role and hence must be considered. Classically, the viscoelastic creep model known as the Burgers model (or four-element model) and an empirical model named the Findley model have been widely applied to PNCs [53]. However, these models are limited in application, considering the effect of elastic properties of nanocomposite constituents. Based on the principle of elastic-viscoelastic correspondence, an empirical formula for the creep strain of nanocomposites was proposed in [54], as expressed in Equation 5.76.

$$\varepsilon_{nc}(t, \bar{\sigma}, T) = \cfrac{\bar{\sigma}}{f\left(E_f, \left(S_{m_{exp}}(t, \bar{\sigma}, T)\left(\cfrac{S_{m_{exp}}(t_0, \bar{\sigma}, T)}{S_{m_{exp}}(t, \bar{\sigma}, T)}\right)^{(V_f k_c)}\right), V_f, r_e\right)} \quad (5.76)$$

where $S_{m_{exp}}$ represents the time-varying creep stiffness of the pure matrix, $\bar{\sigma}$ is applied stress, and T is the temperature. In addition, r_e is the exfoliation ratio, E_f is the elastic modulus of the nanofiller, and k_c represents the constraint factor which accounts for constraints imposed by the nanofillers on the movement of long chains and the molecule of the polymer.

In terms of a micromechanical approach, a realistic prediction of the creep performance of nanotube-based composites must consider the elastic properties of the nanotube, the viscoelastic attributes of the interface and the state of dispersion of the nanotubes in the polymer matrix [55]. In some cases, due to fabrication processes of the nanocomposites, it is essential to incorporate CNT agglomeration phenomena, such that the representative nanocomposite system is constituted by two fields: (i) pure matrix without CNT and (ii) inclusions containing CNT and rest of the matrix material (Figure 5.7a). In this way, the elastic stiffness of the randomly dispersed CNT-reinforced polymer nanocomposite can be described by the Eshelby principle [55]. Subsequently, the nanocomposite creep function is related to the CNT-polymer nanocomposite elastic stiffness tensor in the transform domain based on a simplified unit cell model.

5.5 CHALLENGES AND PROSPECTS

Simulating the behaviours of polymeric nanocomposites is a complex task which requires good understanding of nanoparticle structure (specifically, the effects of nanofiller size and architecture on the nanocomposite morphology), dynamics (that is, the effect of nanofiller on the rheological characteristics of the melt), solid-state properties, and processing methods and conditions. In addition, achieving optimal dispersion of nanoparticles in PNC is challenging owing to the tendency of nano-phase constituents to form nanoparticle aggregates and platelet stacks or due to uncertainties in the properties of nanoparticles and dissimilarity between the chemical properties of a matrix and nanofillers. In this context, theoretical models for PNC need to capture different phenomena based on a realistic assumption to generate accurate predictions. Considering the structure of a PNC system with length scales up to six orders of magnitude or time scale spanning a dozen orders of magnitude, application of analytical and numerical models at different independent scales is limited. This limitation is due to the impracticability of a single model to explore the wide length scales and time scales structurally and efficiently. As a result, modelling of PNCs demands bridging of different length and time scales through combined computational methods that can simulate fundamental molecular processes and seamlessly transfer numerical parameters efficiently across wide scales to satisfactorily predict bulk properties of PNC structures.

It is important to elucidate that availability, reliability, and accuracy of experimental data are challenges that significantly influence the performance of theoretical models for prediction of PNC behaviours. These challenges increase with the advent of more innovative PNCs through introduction of new bio-nanofillers and/or nano-particles, especially from agricultural wastes. Therefore, robust computational schemes, with capabilities for design, optimisation, and characterisation, are required to adequately explore the merits of PNC materials for enhanced engineering appli-cations. As the challenges in the field of PNCs evolve rapidly within the thriving field of composite science and technology, especially in the next decade, computational modelling of PNCs will remain one of the most active fields in the nearest future.

5.6 CONCLUDING REMARKS

Various theoretical models and methods/approaches have been extensively de-scribed with respect to the computational approaches to PNCs within this com-prehensive chapter. It is evident that nanoparticles improve both properties and applications of several PNCs. From the multiscale perspective, computational modelling of PNCs involves three approaches, depending on the structural levels: molecular, micro, and meso/macroscale. These approaches are generally classified into analytical and numerical methods. In addition, these approaches are applicable at different length, time scales and levels of complexity. Towards enhancing the properties of PNCs for high-technology applications, structural characterisation and precise manipulation of the fabrication of the nanostructured materials must be achieved. This implies that the nanocomposite structure as well as physical and chemical processes at the nanoscale level must be correctly characterised to control the final properties of PNCs. In this context, adaptable modelling strategies (using analytical and numerical computational tools) combined with advanced experi-mental techniques are required to effectively explore the design of PNC structures for optimised performances.

NOMENCLATURE

BD	Brownian dynamics
CM	continuum mechanics
CNT	carbon nanotube
DFT	density functional theory
DPD	dissipative particle dynamics
EC	Equivalent continuum
EMA	effective medium approximation/approach
FEM	finite element method
HT	Halpin-Tsai
LB	lattice Boltzmann
MC	Monte Carlo
MD	molecular dynamics
MM	molecular mechanics
MMT	montmorillonite
MR	material region
MSFEM	multiscale finite element method

MT	Mori-Tanaka
PDF	Probability distribution function
PEO	poly (ethylene oxide)
PNCs	polymeric nanocomposites
ROM	rule of mixtures
RVE	representative volume element
SERVE	statistically equivalent representative volume element
SVE	statistical volume element
SWCNT	single-walled carbon nanotube

REFERENCES

[1] Zeng, Q., and Yu, A. (2010). Prediction of the Mechanical Properties of Nanocomposites. in *Optimisation of Polymer Nanocomposite Properties*, V. Mittal, Editor, (pp. 301–331). Available at 10.1002/9783527629275.ch14

[2] Zeng, Q.H., Yu, A.B., and Lu, G.Q. (2008). Multiscale Modelling and Simulation of Polymer Nanocomposites. *Progress in Polymer Science*, **33**(2), 191–269.

[3] Loos, M. (2015). Fundamentals of Polymer Matrix Composites Containing CNTs. in *Carbon Nanotube Reinforced Composites*, M. Loos, Editor, (pp. 125–170). William Andrew Publishing. Available at 10.1016/B978-1-4557-3195-4.00005-9

[4] Zhao, S., Zhao, Z., Yang, Z., Ke, L., Kitipornchai, S., and Yang, J. (2020). Functionally Graded Graphene Reinforced Composite Structures: A Review. *Engineering Structures*, **210**, 110339.

[5] Hosford, W.R. (2013). Elementary Materials Science. Chapter 10. Available at https://www.asminternational.org/documents/10192/1849770/5373G_TOC.pdf

[6] Luo, Z., Li, X., Shang, J., Zhu, H., and Fang, D. (2018). Modified Rule of Mixtures and Halpin-Tsai Model for Prediction of Tensile Strength of Micron-Sized Reinforced Composites and Young's Modulus of Multiscale Reinforced Composites for Direct Extrusion Fabrication. *Advances in Mechanical Engineering*, **10**(7), 1–10.

[7] Tandon, G.P., and Weng, G.J. (1984). The Effect of Aspect Ratio of Inclusions on the Elastic Properties of Unidirectionally Aligned Composites. *Polymer Composites*, **5**(4), 327–333.

[8] Lin, W. (2013). Modelling of Thermal Conductivity of Polymer Nanocomposites. in *Modelling and Prediction of Polymer Nanocomposite Properties*, V. Mittal, Editor, (pp. 169–200).

[9] Clancy, T.C., Frankland, S.J.V., Hinkley, J.A., and Gates, T.S. (2010). Multiscale Modelling of Thermal Conductivity of Polymer/Carbon Nanocomposites. *International Journal of Thermal Sciences*, **49**(9), 1555–1560.

[10] Safi, M., Hassanzadeh-Aghdam, M.K., and Mahmoodi, M.J. (2020). A Semi-Empirical Model for Thermal Conductivity of Polymer Nanocomposites Containing Carbon Nanotubes. *Polymer Bulletin*, **77**(12), 6577–6590.

[11] Guthy, C., Du, F., Brand, S., Winey, K.I., and Fischer, J.E. (2007). Thermal Conductivity of Single-Walled Carbon Nanotube/PMMA Nanocomposites. *Journal of Heat Transfer*, **129**(8), 1096–1099.

[12] Benveniste, Y., and Miloh, T. (1986). The Effective Conductivity of Composites with Imperfect Thermal Contact at Constituent Interfaces. *International Journal of Engineering Science*, **24**(9), 1537–1552.

[13] Hasselman, D.P.H., and Johnson, L.F. (1987). Effective Thermal Conductivity of Composites with Interfacial Thermal Barrier Resistance. *Journal of Composite Materials*, **21**(6), 508–515.

[14] Russell, H.W. (1935). Principles of Heat Flow in Porous Insulators. *Journal of the American Ceramic Society*, **18**(1–12), 1–5.

[15] Tsao, G.T.N. (1961). Thermal Conductivity of Two-Phase Materials. *Industrial & Engineering Chemistry*, **53**(5), 395–397.

[16] Cheng, S.C., and Vachon, R.I. (1969). The Prediction of the Thermal Conductivity of Two and Three Phase Solid Heterogeneous Mixtures. *International Journal of Heat and Mass Transfer*, **12**(3), 249–264.

[17] Maxwell, J.C. (1873). *A Treatise on Electricity and Magnetism* (Vol. 1). Clarendon Press.

[18] Fricke, H. (1924). A Mathematical Treatment of the Electric Conductivity and Capacity of Disperse Systems I. The Electric Conductivity of a Suspension of Homogeneous Spheroids. *Physical Review*, **24**(5), 575–587.

[19] Hamilton, R.L., and Crosser, O.K. (1962). Thermal Conductivity of Heterogeneous Two-Component Systems. *Industrial & Engineering Chemistry Fundamentals*, **1**(3), 187–191.

[20] Hashin, Z. (1968). Assessment of The Self Consistent Scheme Approximation: Conductivity of Particulate Composites. *Journal of Composite Materials*, **2**(3), 284–300.

[21] Allen, M.P., and Tildesley, D.J. (2017). *Computer Simulation of Liquids*. Oxford University Press.

[22] Frenkel, D., and Smit, B. (2001). *Understanding Molecular Simulation: From Algorithms to Applications*, Vol. 1, Elsevier.

[23] Kwon, Y.W. (2015). Molecular Dynamics. *Multiphysics and Multiscale Modelling*, 145–186. doi:10.1201/b19098-5.

[24] Schneider R., Sharma A.R., Rai A. (2008). Introduction to Molecular Dynamics, in *Computational Many-Particle Physics. Lecture Notes in Physics*, Vol. 739, H. Fehske, R. Schneider, and A. Weiße, Editors, Springer. Available at 10.1007/978-3-540-74686-7_1

[25] Raychaudhuri, S. (2008, December). Introduction to Monte Carlo Simulation. In *2008 Winter Simulation Conference* (pp. 91–100). IEEE.

[26] Odegard, G.M., Gates, T.S., Wise, K.E., Park, C., and Siochi, E.J. (2003). Constitutive Modelling of Nanotube–Reinforced Polymer Composites. *Composites Science and Technology*, **63**(11), 1671–1687.

[27] Doi, M. (2002). Octa – A Free and Open Platform and Software of Multiscale Simulation for Soft Materials.

[28] Müller-Plathe, F. (2002). Coarse-Graining in Polymer Simulation: From the Atomistic to the Mesoscopic Scale and Back. *ChemPhysChem*, **3**(9), 754–769.

[29] Toth, R., Voorn, D.J., Handgraaf, J.W., Fraaije, J.G., Fermeglia, M., Pricl, S., and Posocco, P. (2009). Multiscale Computer Simulation Studies of Water-Based Montmorillonite/Poly (Ethylene Oxide) Nanocomposites. *Macromolecules*, **42**(21), 8260–8270.

[30] Kasemägi, H., Klintenberg, M., Aabloo, A., and Thomas, J.O. (2002). Molecular Dynamics Simulation of The $LiBF_4$–PEO System Containing Al_2O_3 Nanoparticles. *Solid State Ionics*, **147**(3–4), 367–375.

[31] Spanos, P.D., and Kontsos, A. (2008). A Multiscale Monte Carlo Finite Element Method for Determining Mechanical Properties of Polymer Nanocomposites. *Probabilistic Engineering Mechanics*, **23**(4), 456–470.

[32] Kontsos, A., and Spanos, P.D. (2009). Modelling of Nanoindentation Data and Characterisation of Polymer Nanocomposites by a Multiscale Stochastic Finite Element Method. *Journal of Computational and Theoretical Nanoscience*, **6**(10), 2273–2282.

[33] Kontsos, A., and Cuadra, J.A. (2013). Multiscale Stochastic Finite Elements Modelling of Polymer Nanocomposites. in *Modelling and Prediction of Polymer Nanocomposite Properties*, V. Mittal, Editor, (pp. 143–168). Wiley.

[34] Ostoja-Starzewski, M. (2006). Material Spatial Randomness: From Statistical to Representative Volume Element. *Probabilistic Engineering Mechanics*, **21**(2), 112–132.

[35] Swaminathan, S., Ghosh, S., and Pagano, N.J. (2006). Statistically Equivalent Representative Volume Elements for Unidirectional Composite Microstructures: Part I-Without Damage. *Journal of Composite Materials*, **40**(7), 583–604.

[36] Shi, D.L., Feng, X.Q., Huang, Y.Y., Hwang, K.C., and Gao, H. (2004). The Effect of Nanotube Waviness and Agglomeration on the Elastic Property of Carbon Nanotube-Reinforced Composites. *Journal of Engineering Materials and Technology*, **126**(3), 250–257.

[37] Zheng, X., Forest, M.G., Lipton, R., and Zhou, R. (2007). Nematic Polymer Mechanics: Flow-Induced Anisotropy. *Continuum Mechanics and Thermodynamics*, **18**(7), 377–394.

[38] Lee, K.Y., and Paul, D.R. (2005). A Model for Composites Containing Three-Dimensional Ellipsoidal Inclusions. *Polymer*, **46**(21), 9064–9080.

[39] Wagner, H.D. (2002). Nanotube–Polymer Adhesion: A Mechanics Approach. *Chemical Physics Letters*, **361**(1-2), 57–61.

[40] Mokashi, V.V., Qian, D., and Liu, Y. (2007). A Study on the Tensile Response and Fracture in Carbon Nanotube-Based Composites Using Molecular Mechanics. *Composites Science and Technology*, **67**(3-4), 530–540.

[41] Fisher, F.T., Bradshaw, R.D., and Brinson, L.C. (2003). Fibre Waviness in Nanotube-Reinforced Polymer Composites—I: Modulus Predictions Using Effective Nanotube Properties. *Composites Science and Technology*, **63**(11), 1689–1703.

[42] Bradshaw, R.D., Fisher, F.T., and Brinson, L.C. (2003). Fibre Waviness in Nanotube-Reinforced Polymer Composites—II: Modelling Via Numerical Approximation of the Dilute Strain Concentration Tensor. *Composites Science and Technology*, **63**(11), 1705–1722.

[43] Ashrafi, B., and Hubert, P. (2006). Modelling The Elastic Properties of Carbon Nanotube Array/Polymer Composites. *Composites Science and Technology*, **66**(3–4), 387–396.

[44] Hbaieb, K., Wang, Q.X., Chia, Y.H.J., and Cotterell, B. (2007). Modelling Stiffness of Polymer/Clay Nanocomposites. *Polymer*, **48**(3), 901–909.

[45] Zare, Y., Fasihi, M., and Rhee, K.Y. (2017). Efficiency of Stress Transfer between Polymer Matrix and Nanoplatelets in Clay/Polymer Nanocomposites. *Applied Clay Science*, **143**, 265–272.

[46] Frankland, S.J.V., Caglar, A., Brenner, D.W., and Griebel, M. (2002). Molecular Simulation of the Influence of Chemical Cross-Links on the Shear Strength of Carbon Nanotube – Polymer Interfaces. *The Journal of Physical Chemistry B*, **106**(12), 3046–3048.

[47] Wong, M., Paramsothy, M., Xu, X.J., Ren, Y., Li, S., and Liao, K. (2003). Physical Interactions at Carbon Nanotube-Polymer Interface. *Polymer*, **44**(25), 7757–7764.

[48] Li, C., and Chou, T.W. (2003). Multiscale Modelling of Carbon Nanotube Reinforced Polymer Composites. *Journal of Nanoscience and Nanotechnology*, **3**(5), 423–430.

[49] Zhang, W., Picu, R.C., and Koratkar, N. (2007). Suppression of Fatigue Crack Growth in Carbon Nanotube Composites. *Applied Physics Letters*, **91**(19), 1–4.

[50] Seshadri, M., and Saigal, S. (2007). Crack Bridging in Polymer Nanocomposites. *Journal of Engineering Mechanics*, **133**(8), 911–918.

[51] Jeong, B.W., Lim, J.K., and Sinnott, S.B. (2007). Multiscale-Failure Criteria of Carbon Nanotube Systems Under Biaxial Tension–Torsion. *Nanotechnology*, **18**(48), 1–7.

[52] Fyta, M.G., Remediakis, I.N., Kelires, P.C., and Papaconstantopoulos, D.A. (2006). Insights into the Fracture Mechanisms and Strength of Amorphous and Nanocomposite Carbon. *Physical Review Letters*, **96**(18), 1–4.

[53] Yang, J.L., Zhang, Z., Schlarb, A.K., and Friedrich, K. (2006). On The Characterisation of Tensile Creep Resistance of Polyamide 66 Nanocomposites. Part II: Modelling and Prediction of Long-Term Performance. *Polymer*, **47**(19), 6745–6758.

[54] Shokuhfar, A., Zare-Shahabadi, A., Atai, A.A., Ebrahimi-Nejad, S., and Termeh, M. (2012). Predictive Modelling of Creep in Polymer/Layered Silicate Nanocomposites. *Polymer Testing*, **31**(2), 345–354.

[55] Hassanzadeh-Aghdam, M.K., Mahmoodi, M.J., and Ansari, R. (2019). Creep Performance of Cnt Polymer Nanocomposites-An Emphasis on Viscoelastic Interphase and CNT Agglomeration. *Composites Part B: Engineering*, **168**, 274–281.

Section II

Agricultural Waste–Based Polyethylene-Based Nanocomposites

6 Agricultural Waste Reinforced Polyethylene-Based Hybrid Nanocomposites: Design Formulations and Mechanical Properties

Stephen Durowaye, Ademola Agbeleye,
Babatunde Bolasodun, and Eniola Apena
Department of Metallurgical and Materials Engineering,
University of Lagos, Lagos State, Nigeria

CONTENTS

DOI: 10.1201/9781003170549-8

6.1 INTRODUCTION

Thermoplastics are among the three types of polymers that are extensively used in various applications because of their unique and desirable characteristics. They melt at low temperatures and become solid when cooled. Examples of thermoplastics are polyethylene, nylon, polypropylene, etc. Thermoplastics are produced in large quantities because of their desirable qualities and increasing demand for them. For example, the production of polyethylene and nylons is above 80 million tonnes per year worldwide [1]. They exhibit linear or branched chain-like structure or may combine the two structures. Thermoplastics exhibit increased density, stiffness, tensile and yield strengths, and chemical and abrasion resistance with good dimensional properties. Despite possessing these desirable qualities, they have some limitations or disadvantages when compared with metals and ceramics. Some of the disadvantages are low tensile and compressive strengths and reduced/low modulus of elasticity [1]. Because they are widely used, many societies are confronted with the problem of indiscriminate disposal of thermoplastic waste. They are found in dumping sites, along the roads, gutters, canals, rivers, seashores, oceans, etc. The wastes are threats to both terrestrial and aquatic animals because of pollution. In many places, they are burnt, causing environmental and health hazards.

6.2 POLYETHYLENE AS A THERMOPLASTIC MATERIAL

Polyethylene is a thermoplastic polymer with long chain hydrocarbons. It is classified into several varieties based on density and branching. Some of the common varieties of polyethylene are low-density polyethylene (LDPE), high-density polyethylene (HDPE), and ultra-high molecular weight polyethylene (UHMWPE). LDPE is a branched thermoplastic with many relatively long branches of the main molecular chain. It is flexible and has low tensile and compressive strength compared to HDPE because of irregular packing of polymer chains. Generally, LDPE is used in food packaging materials, rigid containers, and plastic bags [2]. HDPE consists of chains without major branching. HDPE is more rigid than LDPE because of its high crystallinity content and it is used in packaging such as detergent bottles, milk jugs, garbage containers, and water pipes [2]. UHMWPE contains repeating units of polyethylene $[-(CH2-CH2)n-]$ and its molecular chain can be seen as a twisted string of wires. It is used in aerospace, automotive, production of safety materials, etc.

6.3 RECYCLING OF THERMOPLASTICS WASTE

Recycling of thermoplastic waste is regarded as a means of achieving useful products from plastic waste after their reprocessing/re-melting. It is an important means of solving the problem of environmental and ecological threats caused by thermoplastic waste which are not biodegradable and release harmful gases to the atmosphere when burnt [3]. Recycling has been reported to influence the environment positively as it mitigates pollution and works against climate change as there are direct and indirect impacts of global climate change on the environment, mankind, and economy [4]. In addition, it offers benefits like provision of raw materials for

manufacturing industry, decrease in energy consumption, and decrease in in-
cineration and landfill issues. It also provides employment for the people that are
into recycling [5]. There are various ways of recycling thermoplastic waste but
mechanical recycling is the most effective method. In addition, virgin or treated
thermoplastic waste can be recycled as a matrix with fillers and compatibilizers to
develop polymer matrix composites of enhanced characteristics [6]. Hence, miti-
gating the adverse effects of global climate change through recycling is a welcome
development and it should be seriously embraced and promoted. This is because of
its immense benefits among which are reducing the amount of waste sent to landfills
or burnt thereby reducing pollution, conserving the natural resources, and reducing
energy consumption, thereby making production economical.

6.4 FORMULATION, PROCESSING, AND CHARACTERISATION OF POLYETHYLENE MATRIX COMPOSITES

Generally, after determining the appropriate formulation of the input materials in
terms of weight percentage, the next step in producing polyethylene matrix composite is processing. During processing, the input materials can be modified/treated
for property enhancement. Formulation of materials can vary depending on the type
of materials selected, production method, and discretion of the researcher(s). In this
context, processing is converting the input materials to the desired composites by
mould casting, extrusion, injection moulding, etc. The composites produced are
usually subjected to various tests for properties assessment. Specifically, polyethylene matrix composites can be subjected to both tensile strength and modulus of
elasticity tests in accordance with ASTM D3039 [7] standard. The hardness test can
be conducted using the Brinell hardness tester in accordance with ASTM D2240 [8]
standard while impact energy test can be conducted by employing an Izod impact
tester in accordance with ASTM D256 [9] standard. Other tests that can be conducted are density, water absorption, thermal stability using thermogravimetric
analysis (TGA) and differential scanning calorimetry (DSC), microstructural examination using the scanning electron microscope (SEM) and transmission electron
microscope (TEM), X-ray diffraction (XRD), Fourier-transform infrared spectroscopy (FTIR), etc.

6.5 VIABILITY OF USING AGRICULTURAL WASTE PARTICLES TO DEVELOP POLYETHYLENE MATRIX COMPOSITES

Generally, the disposal of solid waste has been a problem in many societies because
of environmental and health issues. Agricultural and thermoplastic waste are
available in large quantities in many places. Agricultural wastes are the leftovers
from growing and processing of raw agricultural products such as fruits, plants,
dairy products, meat, and crops. Among the agricultural wastes are periwinkle
(*Turritela communis*) shells and cow bones. About 40.3 tonnes of periwinkles are
harvested yearly in Nigeria [10]. Likewise, large quantities of cow bones are
generated due to the high numbers of cows slaughtered daily for meat production.
In underdeveloped and some developing countries, cow bones are burnt openly

thereby polluting the environment. Effective management of these waste has become very imperative and most importantly, converting the ones that are not hazardous to useful engineering materials for industrial applications. Hence, efforts have been intensified via series of studies in developing polymer matrix composites (PMCs). Applying agricultural waste as fillers in the development of PMCs has gained much attention. Many of the agricultural wastes possess some valuable attributes like low density and appreciable tensile strength. They are not hazardous, exhibit low energy consumption, recyclable, and biodegradable [11]. In addition, they are easy and economical to generate in large quantities.

6.6   STATE-OF-THE-ART OF RESEARCH ON POLYETHYLENE MATRIX COMPOSITES: PROCESSING METHODS, PROPERTIES ENHANCEMENT, AND APPLICATIONS

Past and recent studies have been conducted by researchers in developing PMCs using the varieties of polyethylene (LDPE, HDPE, UHMWPE, etc.) as matrix materials with some agricultural wastes as fillers and the composites were subjected to various tests. Different processing methods were employed in the production of the composites with some modifications/treatments of the input materials and addition of a compatibilizer. These are primarily to ensure high output and property improvement to make the composites suitable for various applications. In the studies, the composites exhibited desirable physical, mechanical, thermal, and biodegradability characteristics. For example, [12] investigated the effect of agricultural waste (peanut husk) on low-density polyethylene (LDPE). The composites were produced by extrusion method and their physical, mechanical and biodegradability characteristics were examined. The results indicated that the composites exhibited reduced or lower mechanical and biodegradability characteristics when compared with the virgin LDPE. This is due to the poor compatibility between LDPE and peanut husk. However, the addition of maleic anhydride grafted polyethylene (MAPE), which acted as a compatibilizer, caused much better dispersion of agricultural filler in the matrix and homogeneity of the two phases. This caused a great improvement in the mechanical and biodegradability characteristics of the composites. Water absorption and thickness swelling indices increased with increasing filler content but reduced on addition of MAPE. The weight loss of the composites reduced with increasing MAPE addition. Furthermore, weight loss of composites via enzymatic degradation showed that the composites were biodegradable even at high levels of filler concentration.

Similarly, [13] studied the effect of varied particle sizes and weight percent of periwinkle shell on LDPE. The composites were produced by an injection moulding method with and without MAPE and their mechanical properties were examined. The results indicated that the composites exhibited increased mechanical characteristics such as tensile strength, modulus of elasticity, flexural strength, impact energy, and hardness with increasing periwinkle shell and compatibiliser concentrations at lower particle size. [14] investigated the influence of natural plantain fibres (NPFs) on high-density polyethylene (HDPE) using a compatibiliser. The composites were produced by an injection moulding method and their mechanical characteristics were examined.

The results indicated that NPF/HDPE composite exhibited a density of 993 kg/m^3 and improved tensile strength, yield strength, flexural strength, hardness, and impact energy compared to the unblended polyethylene. In addition, the composite was employed to produce automobile body fender with a lower/reduced density compared to that of steel indicating a comparative advantage, thereby serving as a very good substitute. [15] studied the effect of varied weight percent of bamboo fibres on polyethylene. The fibres were modified/treated with sodium hydroxide after which the input materials were used to produce the composites by a compression moulding method and they were subsequently characterised. The results revealed that water absorption by the composites increased with increasing fibres concentration indicating that the fibres had affinity for water. The microstructure revealed good distribution of fibres in the matrix at reduced/low fibre concentration, which improved the tensile strength, hardness, and impact energy of the composites. However, at high fibre concentrations, there was clustering of the fibres in the matrix, which reduced the mechanical properties of the composites.

[16] studied the influence of woven and randomly arranged sisal fibres on HDPE. The composites were produced by a hot compression moulding method and were subjected to mechanical tests. The results indicated that the composites exhibited improved or higher tensile strength and modulus of elasticity than the unblended/virgin HDPE. Likewise, [17] studied the effect of cellulose (cotton fabric waste) particles of varied weight percent on recycled polyethylene. The composites were produced by a mould casting method, using MAPE as a compatibiliser and were characterised. The microstructure of the composites revealed a very good dispersion of the cellulose in the polyethylene matrix with strong interfacial bonding of the two materials. This caused a great improvement in tensile strength and modulus of elasticity of the composites compared to unblended polyethylene. In addition, the composites were found to be thermally stable up to 300°C. [18] investigated the influence of varied weight percent of groundnut shell particles on HDPE. The composites were developed by a compression-moulding method and were characterised. The microstructure revealed a very good dispersion of the groundnut shell particles in the matrix with strong particles/matrix bonding. The storage modulus of all the composites decreased with an increase in temperature. The composite containing 25 wt% of groundnut shell exhibited the maximum storage modulus of 1158.47 MPa compared to 1033.58 MPa of the unblended HDPE, which is an indication that the composites are more thermally stable than unblended HDPE with an enhanced ability to bear load.

[19] studied the effect of glass fibres on ultra-high molecular weight polyethylene (UHMWPE) laminate mixed with epoxy resin and hardener. The composites were produced by a compression moulding method and the mechanical characteristics of the composites were examined. To ensure the compatibility of UHMWPE laminate with epoxy resin, surface modification of the laminate was done by immersing the laminate in chromic acid solution at room temperature for 30 minutes. This altered the surface chemistry and roughness of the laminate. The results showed that the flexural strength and flexural modulus of the glass fibres/UHMWPE composites greatly increased compared to the unblended UHMWPE laminate. Because of the enhanced and desirable properties exhibited by polyethylene matrix composites, they are widely used in various fields such as packaging, automotive, electrical, thermal energy

storage, biomedical, aerospace, etc. Specifically, UHMWPE fibre composites have witnessed numerous applications as safeguard materials like bulletproof vests, helmets, and armour. They are used as civil aeroplane cabins, cable products, fishing nets, and medical materials. Hence, this study is aimed at developing hybrid polymer composites using polyethylene as a matrix and agricultural waste (periwinkle shell and cow bone) as fillers.

6.7 MATERIALS PREPARATION AND PRODUCTION OF POLYETHYLENE COMPOSITES

Polyethylene waste generated on campus of the University of Lagos was collected, washed in liquid soap, rinsed with water, and sun-dried. Periwinkle shells and cow bones were sourced from the Oyingbo market in Lagos. The bones were washed in liquid soap, rinsed with water, and sun-dried for 5 hours. Alcohol was used to clean the bones after which they were kept for 3 hours in an electric oven set to 120°C. The periwinkle shells and cow bones were ground and sieved to 50 µm, as illustrated in Figure 6.1. A measured quantity of the polyethylene waste was placed in a crucible pot, charged into a L02-MAN crucible furnace (Figure 6.2) and heated until it melted at 130°C. Appropriate quantities (Table 6.1) of periwinkle shell particles were added to the melt. The mixture was stirred for 5 minutes with a stainless steel tong to ensure proper blending. The blend was poured into a wooden mould laced with paper tape to avoid sticking. The specimens were allowed to reach a semi-solid form by cooling after which they were removed from the moulds, pressed for 5 minutes at a pressure of 0.3 MPa, and recorded as the first batch. The same procedure was used for specimens blended with cow bone particles and

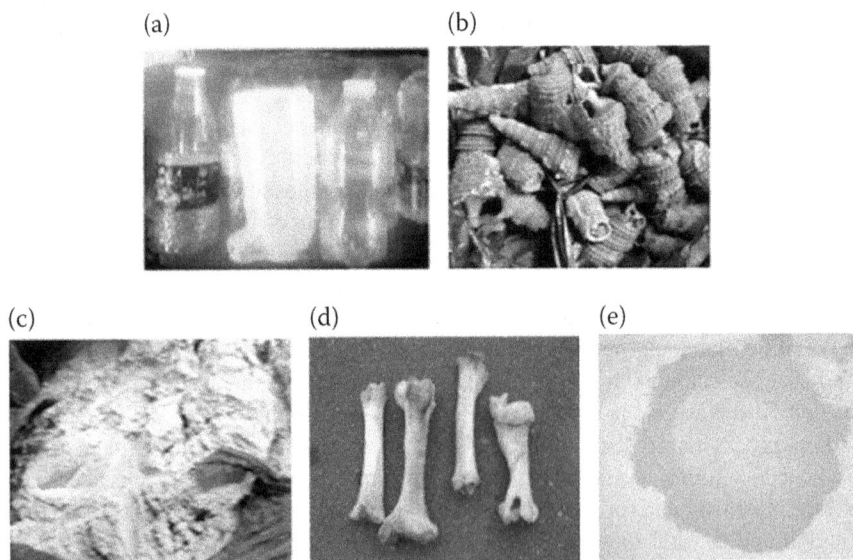

FIGURE 6.1 Photographs of input materials (a) polyethylene waste, (b) dried periwinkle shells, (c) 50-µm periwinkle shell, (d) cow bones, (e) 50-µm cow bone.

FIGURE 6.2 Melting of the polyethylene waste in a crucible furnace.

TABLE 6.1
Composite Design Formulation

Matrix (wt%)	Reinforcement (wt%)			
PET	50 μm PSP	50 μm CBP	50 μm Hybrid (PSP + CBP)	Total (wt%)
100 (control)	–	–	–	100
First Batch				
95	5	–	–	100
90	10	–	–	100
85	15	–	–	100
80	20	–	–	100
75	25	–	–	100
Second Batch				
95	–	5	–	100
90	–	10	–	100
85	–	15	–	100
80	–	20	–	100
75	–	25	–	100
Third Batch				
95	–	–	2.5 PSP + 2.5 CBP	100
90	–	–	5 PSP + 5 CBP	100
85	–	–	7.5 PSP + 7.5 CBP	100
80	–	–	10 PSP + 10 CBP	100
75	–	–	12.5 PSP + 12.5 CBP	100

PSP = Periwinkle shell particles; CBP = Cow bone particles.

TABLE 6.2
Composition of the Cow Bone

Elements	Ca	Fe	Au	Sr	Sn
Weight (%)	23.298	2.586	0	0	0

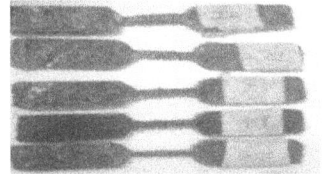

FIGURE 6.3 Enlarged photograph of some of the specimens for tensile strength test.

recorded as a second batch. The third batch was the hybrid specimens, which were polyethylene blended with equal amounts of periwinkle shell and cow bone. The unblended molten polyethylene specimens that were cooled to semi-solid form, removed from the mould, and pressed were taken as the control. Chemical composition of the cow bones is shown in Table 6.2, while a photograph of some of the composite specimens is shown in Figure 6.3.

6.8 CHARACTERISATION OF POLYETHYLENE MATRIX COMPOSITES

Microstructural examination of the composites was conducted using a scanning electron microscope (SEM) that has an energy dispersive X-ray spectroscope (EDS). The water absorption test was done by measuring the initial weight (W_1) of specimens after which they were separately immersed in water in beakers. After 24 hours of immersion, they were removed and weighed (W_2). This was repeated for 6 days (144 hours). The water absorption was determined using Equation (6.1) [20].

$$W_A(\%)\frac{W_2 - W_1}{W_1} \times 100 \qquad (6.1)$$

The tensile strength and modulus of elasticity tests were done in accordance with ASTM D3039 [7] standard. A hardness test was done in accordance with ASTM D2240 [8] standard while the impact energy test was done in accordance with ASTM D256 [9] standard using an Izod impact tester.

6.9 MICROSTRUCTURAL PROPERTIES OF POLYETHYLENE COMPOSITE

As shown in Figures 6.4–6.7, there are morphological differences (inhomogeneities) in the SEM micrographs of the specimens with dendritic (tree like) and oval shapes.

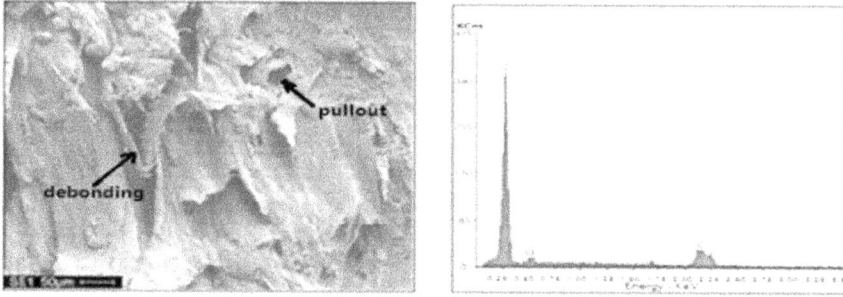

FIGURE 6.4   The SEM and EDS microstructure of the unreinforced polyethylene specimen after fracture.

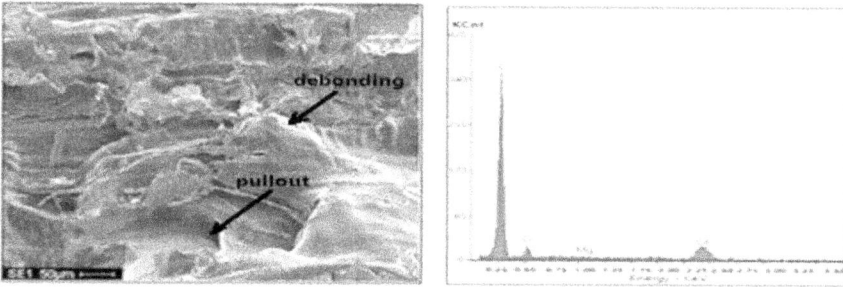

FIGURE 6.5   The SEM and EDS microstructure of 15 wt% periwinkle shell particles re-inforced polyethylene specimen after fracture.

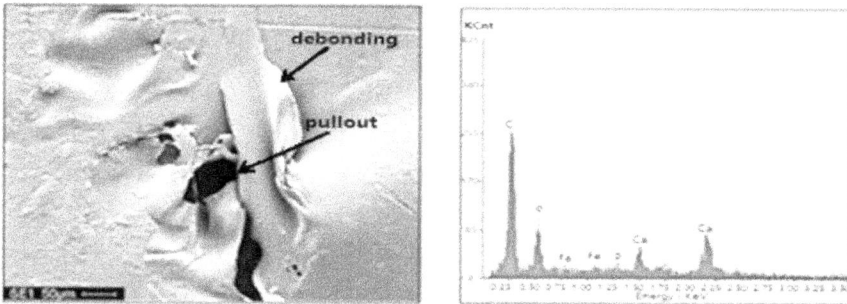

FIGURE 6.6   The SEM and EDS microstructure of the 15 wt% cow bone particles re-inforced polyethylene specimen after fracture.

The presence of carbon (C) in the spectrographs confirms that the polyethylene is a carbon-based organic material. The EDS shows that C, O, Fe, Na, Si, Ca, P, and Mg are contained in the hybrid specimen (Figure 6.7). There is evidence of fibres/particles pullout and debonding in the SEM images (Figures 6.4–6.7) due to impact fracture.

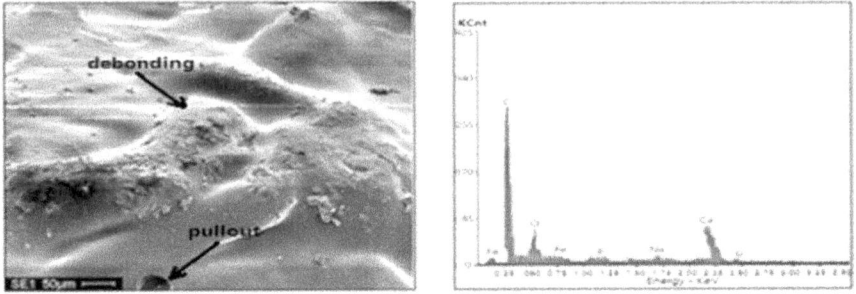

FIGURE 6.7 The SEM and EDS microstructure of the 15 wt% hybrid specimen after fracture.

6.10 WATER ABSORPTION OF POLYETHYLENE MATRIX COMPOSITES

The water absorbed by the specimens increased with time, as shown in Figure 6.8, indicating that pores existed in the microstructure, which agrees with the report by [21]. It was reported that water absorption had an adverse effect on the tensile strength of polymers as nylon, swelled and softened in water, leading to weak bonding, which reduced resistance to applied stress [20]. The hybrid specimen demonstrated the lowest water absorption of 0.3% when compared to other specimens. This is due to the interfacial bonding of periwinkle and cow bone particles with the polyethylene matrix which indicates reduced pores in the microstructure.

6.11 TENSILE STRENGTH OF POLYETHYLENE MATRIX COMPOSITES

The composites demonstrated increased tensile strength up to 15 wt% fillers additions but witnessed reduction in tensile strength when filler concentration was

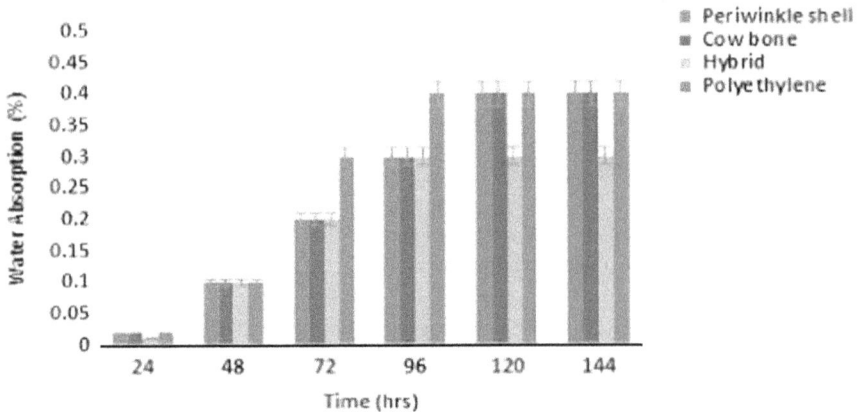

FIGURE 6.8 Graph of water absorption against time of the composites.

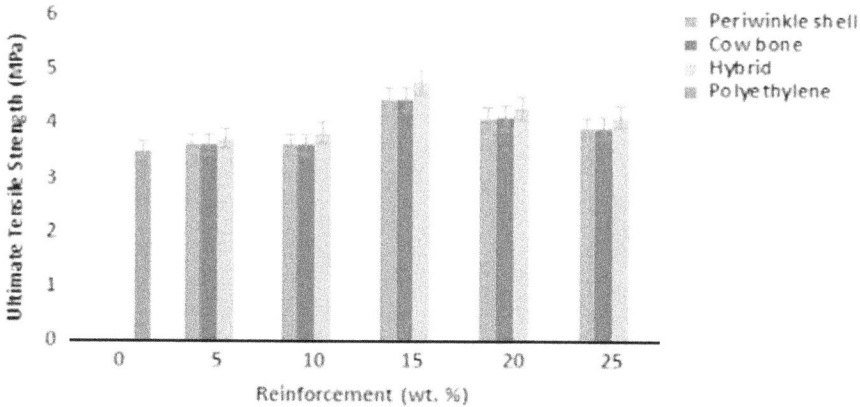

FIGURE 6.9 Effect of particles addition on tensile strength of the specimens.

higher than 15 wt%, as shown in Figure 6.9. However, all the composites demonstrated higher tensile strength than the unblended polyethylene. The hybrid composite exhibited the highest tensile strength of 4.76 MPa at 15 wt% filler additions. This is 36.4% higher than tensile strength of the unblended polyethylene of 3.49 MPa. The uniform dispersion of the fillers in the polymer matrix and strong interfacial bonding of the fillers and polyethylene caused increase in tensile strength. The characteristics of composites are greatly enhanced by the bonding between the filler and matrix as stress is transferred to the filler via this medium. The decrease in tensile strength is because of a reduction in inter-particle separation due to increased concentration of the particles in the polymer matrix, which increased the inter-particle stress concentration overlap that caused debonding under tension [22]. In addition, poor dispersion of fillers in polymer matrix may cause weak bonding of the particles with the polyethylene matrix. This ultimately had an adverse effect on load distribution.

6.12 MODULUS OF ELASTICITY OF POLYETHYLENE MATRIX COMPOSITES

As shown in Figure 6.10, the modulus of elasticity of the composites reduced with an increasing concentration of filler in the composites. During the application of tensile stress, there could be equality in the strain energy in the matrix with the bonding of particulates to a polymer matrix, which caused the particle-matrix interface to fracture and decreased the modulus of elasticity. This agrees with the report by [22].

6.13 HARDNESS OF POLYETHYLENE MATRIX COMPOSITES

As shown in Figure 6.11, the hardness of the composites increased as the concentration of fillers increased up to 15 wt%. The hard nature of the fillers must have enhanced the hardness. For example, an X-ray fluorescence test conducted on ash of

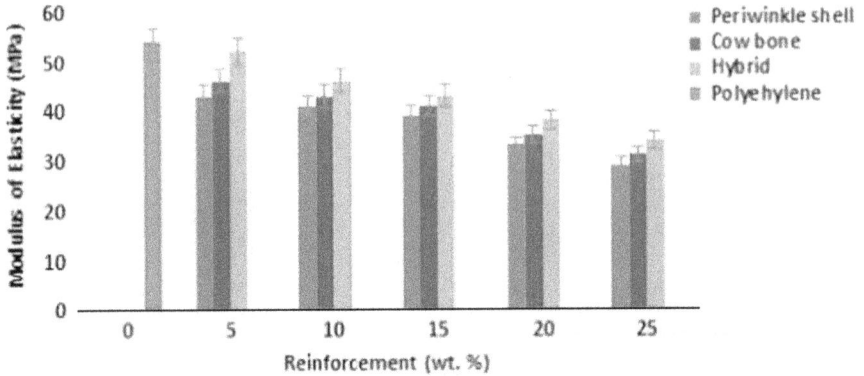

FIGURE 6.10   Effect of particles addition on modulus of elasticity of the specimens.

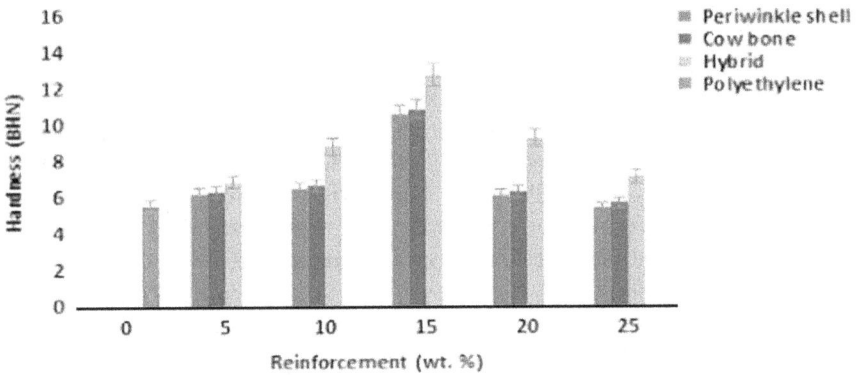

FIGURE 6.11   Effect of particles addition on hardness of the specimens.

periwinkle shell revealed that it contained hard materials like SiO_2, CaO, MgO, Cr_2O_3, and Fe_2O_3 [23]. The hybrid composite demonstrated the highest hardness of 12.78 BHN at 15 wt% filler concentration. Proper distribution of fillers in the matrix and strong bond between them caused the increase in hardness. This agrees with the report of [24]. However, increasing filler concentration beyond 15 wt% caused clustering of particles in the polyethylene. This led to weak bonding, which reduced the hardness.

6.14   IMPACT ENERGY OF POLYETHYLENE MATRIX COMPOSITES

The impact energy of the composites reduced as particle concentration increased, as shown in Figure 6.12. High hardness as concentration increased imparted brittleness to the specimens, which reduced the impact energy. There is evidence of fibre/particle pull out and debonding in the SEM images, as shown in Figures 6.4–6.7 due to impact fracture. Cracks formed during impact loading, fibre/particle pull out and debonding had an adverse effect on impact energy of the specimens. This agrees with the report of [25].

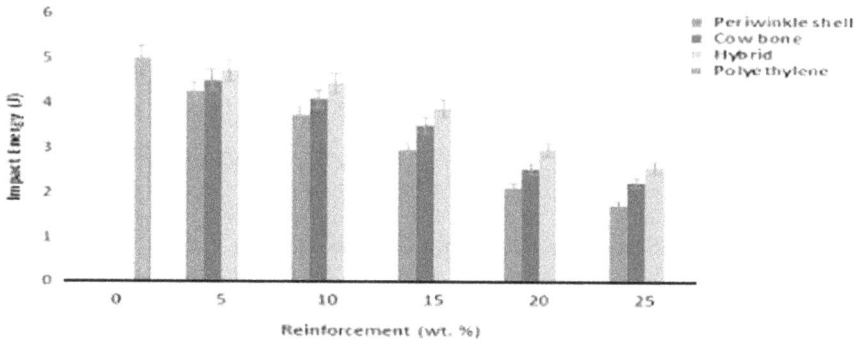

FIGURE 6.12   Effect of particles addition on impact energy of the specimens.

6.15   CONCLUSION/SUMMARY

Polyethylene waste was blended with agricultural waste (periwinkle shell and cow bone) particles to produce polyethylene matrix composites by mould casting and the composites were characterised to determine properties of the developed composites. The microstructure revealed an uninterrupted interlocking of the polymeric chains. Among the composites, the hybrid composite demonstrated appreciable characteristics. It exhibited the least water absorption of 0.3%. It also exhibited the greatest tensile strength of 4.76 MPa and hardness value of 12.75 BHN at 15 wt% filler concentrations. It also exhibited appreciable modulus of elasticity and impact energy of 43 MPa and 3.86 J, respectively, at 15 wt% reinforcement. The hybrid polyethylene matrix composite demonstrated the compatibility of the two-agricultural materials and their effectiveness as fillers. This composite could be a potential engineering material for applications in areas where low strength is required. Utilising agricultural and thermoplastic waste to develop composites for industrial application is laudable because of its economic, environmental, and health benefits. This will go a long way in putting an end to the indiscriminate disposal and open burning of the waste as it is being done in many underdeveloped and developing countries.

REFERENCES

[1] Groover, M.P. (2002). *Fundamentals of Modern Manufacturing*, 2nd Ed., John Wiley & Sons, New Jersey, USA.
[2] Khanam, P.N. and Al-Maadeed, M.A. (2015). Processing and Characterization of Polyethylene Based Composites. *Advanced Manufacturing, Polymer & Composites Science*, **1**, 63–79.
[3] Kehinde, O., Ramonu, O.J., Babaremu, K.O., and Justin, L.D. (2020). Plastic Wastes: Environmental Hazard and Instrument for Wealth Creation in Nigeria. *Elsevier – Heliyon*, **6**(10), 1–7.
[4] Grover, A., Gupta, A., Chandra, S., Kumari, A., and Khurana, S.M. (2015). Polyethylene and Environment. *International Journal of Environmental Sciences*, **5**(6), 1091–1105.
[5] Hopewell, J., Dvorak, R., and Kosior, E. (2009). Plastics Recycling Challenges and Opportunities. *Philosophical Transactions of the Royal Society B*, **364**, 2115–2126.

[6] Azeez, T.O., Onukwuli, O.D., Walter, P.E., and Menkiti, M.C. (2018). Influence of Chemical Surface Modifications on Mechanical Properties of Combretum Dolichopetalum Fibre High Density Polyethylene (HDPE) Composites. *Pakistan Journal of Scientific and Industrial Research Series A: Physical Sciences*, **61**(1), 28–34.

[7] ASTM D3039. (2017). *Standard Test Method for Tensile Properties of Polymer Matrix Composite Materials*. ASTM International, West Conshohocken, PA.

[8] ASTM D2240. (2015). *Standard Test Method for Rubber Property Durometer Hardness*. ASTM International, West Conshohocken, PA.

[9] ASTM D256. (2018). *Standard Test Methods for Determining the Izod Pendulum Impact Resistance of Plastics*. ASTM International, West Conshohocken, PA.

[10] Mmom, P.C. and Arokoya, S.B. (2010). Mangrove Forest Depletion, Biodiversity Loss and Traditional Resources Management Practices in the Niger Delta Nigeria. *Research Journal of Applied Sciences, Engineering and Technology*, **2**(1), 28–34.

[11] Malkapuram, R., Kumar, V., and Negi, Y.S. (2009). Recent Development in Natural Fiber Reinforced Polypropylene Composites. *Journal of Reinforced Plastics and Composites*, **28**(10), 1169–1189.

[12] Obasi, H.C. (2015). Peanut Husk Filled Polyethylene Composites: Effects of Filler Content and Compatibilizer on Properties. *Hindawi Publishing Corporation, Journal of Polymers*, Article ID 189289, 1–9.

[13] Nwanonenyi, S.C., Obidiegwu, M.U., and Onuegbu, G.C. (2013). Effects of Particle Sizes, Filler Contents and Compatibilization on the Properties of Linear Low Density Polyethylene Filled Periwinkle Shell Powder. *The International Journal of Engineering and Science (IJES)*, **2**(2), 1–8.

[14] Chukwutoo, I.C., Obiafudo, O.J., and Okafor, C.E. (2016). Characterization of Plantain Fiber Reinforced High Density Polyethylene Composite for Application in Design of Auto Body Fenders. *Journal of Innovative Research in Engineering and Sciences*, **4**(5), 574–587.

[15] Daramola, O.O., Akinwekomi, A.D., Adediran, A.A., Akindote-White, O., and Sadiku, E.R. (2019). Mechanical Performance and Water Uptake Behaviour of Treated Bamboo Fibre-Reinforced High-Density Polyethylene Composites. *Elsevier, Heliyon*, **5**, 1–6.

[16] Castro, B.D., Faria, P.E., Vieira, L.M., Rubio, C.V., Maziero, R., Rodrigues, P.C., and Rubio, J.C. (2020). Recycled Green PE Composites Reinforced with Woven and Randomly Arranged Sisal Fibres Processed by Hot Compression Moulding. *Acta Technologica Agriculturae*, **2**, 81–86.

[17] Mhumak, C. and Pechyen, C. (2017). Recycled Polyethylene and Waste Cellulose Composite: A Strategic Approach on Sustainable Plastic Packaging Application. *Journal of Waste Recycling*, **2**(2), 1–17.

[18] Jacob, J., Mamza, P.A., Ahmed, A.S., and Yaro, S.A. (2018). Effect of Groundnut Shell Powder on the Viscoelastic Properties of Recycled High Density Polyethylene Composites. *Bayero Journal of Pure and Applied Sciences*, **11**(1), 139–144.

[19] Shishupal, S., Yastuti, R.G., Ajeet, P.S., and Mukesh, K.V. (2018). Application of UHMWPE Fiber Based Composite Material. *International Journal of Research in Advent Technology*, **6**(7), 1768–1771.

[20] Mat-Shayuti, M.S., Abdullah, M.Z., and Megat-Yusoff, P.S.M. (2013).Water Absorption Properties and Morphology of Polypropylene/Polycarbonate/Polypropylene Graft Maleic Anhydride Blends. *Asian Journal of Scientific Research*, **6**(2), 167–176.

[21] Tewari, M., Singh, V.K., Gope, P.C., and Chaudhary, A.K. (2012). Evaluation of Mechanical Properties of Bagasse-Glass Fiber Reinforced Composite. *Journal of Materials and Environmental Science*, **3**(1), 171–184.

[22] Rutz, B.H. (2014). *Improvement of Mechanical Properties of Polymeric Composites*. PhD Thesis, Department of Chemical Engineering, University of Washington, USA.

[23] Aku, S.Y., Yawas, D.S., Madakson, P.B., and Amaren, S.G. (2012). Characterisation of Periwinkle Shell as Asbestos-Free Brake Pad Materials. *Pacific Journal of Science and Technology*, **13**(2), 57–63.

[24] Renner, K., Yang, M.S., Moczo, J., Choi, H.J., and Pukanszky, B. (2005). Analysis of the Debonding Process in Polypropylene Model Composites. *European Polymer Journal*, **41**, 2520–2529.

[25] Manikandan, A. and Rajkumar, A. (2016). Evaluation of Mechanical Properties of Synthetic Fiber Reinforced Polymer Composites by Mixture Design Analysis. *Polymers & Polymer Composites*, **24**, 455–462.

7 Delonix regia Pod Particles Reinforced Nanocomposites: Properties Comparison between Recycled and Virgin Low-Density Polyethylene

Sefiu Adekunle Bello
Department of Materials Science and Engineering, Kwara State University, Malete, Kwara State, Nigeria

Raphael Gboyega Adeyemo
Gateway (ICT) Polytechnic, Saapade, Ogun State, Nigeria

Sunday Wilson Balogun, Mohammed Kayode Adebayo, and Kehinde Adekunle Okunola
Department of Materials Science and Engineering, Kwara State University, Malete, Kwara State, Nigeria

CONTENTS

DOI: 10.1201/9781003170549-9

7.1 INTRODUCTION

Polyethylene is ductile, tough, and less expensive. It can serve as an alternate matrix
for developing advanced composite for specific applications, but their low strength is
a disadvantage. Their low strength can be improved through structural modification
including incorporation of reinforcement particles or fibres. Polyethylene has widely
been used for water packaging (sachet and bottle) and carrier bags for food packaging
because of its low cost. With those nonstructural applications, potentials of polyethylene as engineering plastics have not been fully utilised. Further applications of
polyethylene to structural applications can be realised by modification of its structure
through the addition of nanoparticles. This involves incorporation of particulate filler
into polyethylene, which their interaction engineers a polymeric composite having
superior properties to either the polyethylene or reinforcement itself. Particles

obtained from agricultural waste like *Delonix Regia* pods can be added to the polyethylene as a matrix to upgrade its properties to that level which could encompass specifications for automobile component applications.

Wide applications of polyethylene have consequences on environmental impacts. For instance, used sachet water bags and other expired polyethylene products are visible nearly everywhere, including drainages and waterways, dump sites, and ocean shorelines. Unfortunately, many people have not realised the income that can be generated from industrial waste through recycling and proper disposal.

Delonix regia trees produce seeds and pods annually and it is very common in Nigeria and some of the other African countries. The trees are planted for ornament and to prevent soil erosion. They produce large numbers of pods harbouring seeds annually. Both pods and seeds are rotting annually after falling from their trees because little or no value has been added to them. Production of *Delonix regia* pods/seeds is conservative and renewable and their harvest does not affect trees that protect our ecology. Their transformation to reinforcing particles/fibres for developing eco-friendly composite will add value to the waste and create new earnings from wastes.

Usages of both agricultural and industrial wastes will reduce problems associated with poor waste management. As the name implies, recycling is the new emerging technology that involves economic conversion of wastes into raw materials for the novel production of new materials for various engineering applications and for wealth creation. In 1992, the Commonwealth Government released the National Waste Minimisation and Recycling Strategy with a target of reducing the amount of solid waste going to landfill per capita by 50% from 1990 to 2000 [1]. The recycling of wastes reduces waste, saves energy, conserves natural resources, lessens the use of municipal landfills, and provides recyclers and municipalities with considerable revenue. The only solution to this problem is to harness wastes for novel production of engineering materials such as structural composites for various applications. Efforts have been made to find a substitute for polyethylene through synthesis of biodegradable polymers like polylactic acids (PLA) and polymethylmethacrylate (PMMA). The high cost of PLA or PMMA and differences in properties of the polyethylene-based composites with those of their substitutes are challenges for their perfect replacement in the services. Since polyethylene is a thermoplastic, it can be recycled for production of new materials to address its abundant wastes since perfect substitutes for polyethylene has not been fully successful. Therefore, recycling of polyethylene wastes for engineering new materials is important. This chapter reports a recent study on low-density polyethylene (LDPE) using the *Delonix regia* pod particles as reinforcement and benchmark properties of recycled LDPE-based composites with those of the virgin LDPE.

7.2 GREEN SYNTHESIS AND CHARACTERISATION

7.2.1 PREPARATION AND CHARACTERISATION OF *DELONIX REGIA* POD NANOPARTICLES

Delonix regia pods were rinsed separately in water, dried intermittently at 120°C for 6 hours a day for 5 days using the Shel lab vacuum oven model SVA S2E with serial no: 060 13 at Department of Materials Science and Engineering Laboratory, Kwara State University, Malete. The dried *Delonix regia* pods were broken manually into smaller pieces using mortar and pestle and pulverised after about 30 cycles of grinding with the aid of a locally fabricated disc grinder. Then, the pulverised pod powders were ball milled for 70 hours. *Delonix regia* pods particles were analysed to obtain quantitative measurement of particle sizes and their morphologies using a multipurpose field emission transmission electron microscope (TEM), Model: JEM 2100F. The samples for examination were prepared by drying separate particles of *Delonix regia* pods in a copper grid coated with a thin film of carbon, after which the transmission electron microscope operating at an accelerating voltage of 100 kV with an AMTXR41B 4-megapixel (2048 × 2048) bottom mount CCD camera was used to see a beam of electron through the sample and image of particles produced by transmitted electrons were captured.

7.2.2   FABRICATIONS OF *DELONIX REGIA* POD PARTICLE-REINFORCED LOW-DENSITY POLYETHYLENE

In a laboratory setup for developing low-density polyethylene composites using *Delonix regia* pod particles as additive fillers, a weighed amount equal to 2 weight percent (2 wt%) each of *Delonix regia* pod particles were first dissolved in absolute ethanol, agitated for 20 minutes in a 2 kg-capacity tumbler ball mill to break up the agglomerates, and added to the recycled low-density polyethylene (RLDPE). The mixture was stirred in the tumbler ball mill, after which it was dried until ethanol molecules were completely vapourised. Then, 4–5 drops of palm kernel oil were added, acting as binding agents/lubricants, and stirred again. The RLDPE compound was processed using a two-roll polymer mill setting to 140°C to soften and fabricate RLDPE sheet which was hot pressed at 90°C and 0.5–1.8 Nmm^{-2} using hydraulic compression moulding in line with [2] to produce various samples for analyses. The process was repeated with increasing wt% of *Delonix regia* pod particles up to 12 wt%. Other grades of nanocomposites were produced with virgin low-density polyethylene as the matrix using the same wt% of *Delonix regia* pod particles additions. The composite fabrication was carried out at a Materials Fabrication Laboratory.

7.2.3   STRUCTURAL CHARACTERISATIONS OF *DELONIX REGIA* POD PARTICLE-REINFORCED LOW-DENSITY POLYETHYLENE

7.2.3.1   X-Ray Diffraction Analysis

Representative samples of recycled and virgin polyethylene, virgin polyethylene reinforced with *Delonix regia* pod particles were examined by EMPYREAN PANalytical diffractometer with a Pixcel detector having Bragg-Bretano geometry in continuous operating mode and automatic divergence slit type using Cu Kα radiation emitted from Cu anode material at an accelerating voltage of 45 kV and current of 40 mA. Samples were scanned using a Gonio axis with 2Theta (°) position scale between 0 and 80° at a step size of 0.0260°. A chemical formula of phases/compounds present in the examined samples were identified using X'Pert HighScore Plus software in line with [3].

7.2.3.2 Scanning Electron Microscopic Characterisation of *Delonix regia* Pod Particle-Reinforced Low-Density Polyethylene

Spatial arrangement of phases that are present in the developed *Delonix regia* pod particle reinforced LDPE was studied using a scanning electron microscope ASPEX 3020 equipped with an energy-dispersive X-ray spectrometer (EDX). Scanning of the samples firmly fixed to the bottom of the holder using double-sided adhesive tape was done with an electron beam moving at 15 kV to produce signals and images due to secondary and back-scattered electrons, which gave information about the structure of the examined samples. The energy-dispersive spectroscopy gives the elemental composition of the polymeric composite samples.

7.2.4 MECHANICAL INVESTIGATION OF *DELONIX REGIA* POD PARTICLE-REINFORCED LOW-DENSITY POLYETHYLENE

7.2.4.1 Tensile Test

Standard samples of pristine LDPE and its composites having dog-bone shape of 80 (gauge length) $\times$ 6 $\times$ 6 mm^3 were subjected to uniaxial loading leading to gradual stretching at a strain rate of 0.001 per second until fracture occurred in each of the analysed samples. A testomeric testing machine (Machine no: 0500–10080) of Rect 5 kN was used for the analysis per ASTM D638-14 [4] with modifications at the NCAM Materials Testing Laboratory, Ganmo, Ilorin, Kwara State, Nigeria.

7.2.4.2 Flexural Test

A flexural property test was carried out on the rectangular LDPE composite samples of 120 mm long (total length), 10 mm thick, and 6 mm wide using a three-point bending method. The samples were placed horizontally on two pivots at 60 mm apart (span length) and loaded perpendicularly at the centre using the testomeric testing machine until the sample fractured. Three different samples were tested at each level of reinforcement, and average values of results obtained from the three examined samples at each level of reinforcement were plotted as functions of wt% of reinforcements additions and their standard deviations were presented as error bars. The analysis was carried out at room temperature of 18°C and 50% humidity per ASTM D7264/D7264M-15 [5] with slight modifications. The flexural strength, deflection, and modulus were estimated automatically by the tester.

7.2.4.3 Hardness Test

Hardness values of different grades of *Delonix regia* pod particle-reinforced LDPE composites were determined via microhardness approach using Vickers' hardness tester (Model MMT-X7A). A load of 50 gf was mounted on the examined polyethylene composites sample through an indenter for a dwelling time of 10 seconds. Diagonals of indentation produced in each of the *Delonix regia* pod particle-reinforced LDPE samples were measured and converted to the respective hardness value each of analysed sample automatically by the equipment. Prior to the hardness test, each sample surface was prepared using emery paper of 400 grit and polishing cloth. This analysis was performed at the Materials Laboratory, Department of Metallurgical and Materials Engineering, University of Lagos, Nigeria.

7.2.4.4 Impact Energy

The ability of *Delonix regia* pod particle-reinforced LDPE nanocomposites to absorb energy without fracture due to sudden impact of load was measured using the Hounsfield balanced Izod impact machine (Serial no: 3915) at the Materials Science and Engineering Department, Obafemi Awolowo University, Ile Ife, Osun State, Nigeria. $75 \times 6 \times 6$ mm^3 sample each of *Delonix regia* particle reinforced LDPE nanocomposites notched to a depth of 2 mm at an angle of 45° was stricken with a pendulum hammer released at a speed of 3.8 ms^{-1}. The impact energy absorbed by each sample was noted and recorded and the analysis was performed per ASTM D256-10 [6] with slight modifications. Before the test, the pendulum was released to set the scale to zero.

7.3 CHEMISTRY AND STRUCTURES

7.3.1 X-Ray Diffraction Profiles of *Delonix regia* Pod Particle-Reinforced Low-Density Polyethylene

Chemical properties of LDPE reported in this chapter as obtained from X-ray diffraction are presented in Figure 7.1. X-ray diffractograms of the pristine RLDPE and VLDPE have peaks having the same phases at corresponding angles (Figure 7.1), implying that their structural behaviour when no reinforcement is added to them is the same. For instance, ethylene urea ($C_3H_6N_2O$) observed at position of 20.5274 and 22.7464°, 35.4325, and 52.2759°; tri-methylene di-phosphonic acid ($C_3H_{10}O_6P_2$) at 20.5274 and 22.7464 and ethylenediamine hydrogen selenite ($C_2H_{10}N_2O_4Se$) at 20.5274, 22.7464, 35.4325, 42.1678, 46.1747, and 52.2759° are the phases observed with various XRD peaks of both RLDPE and VLDPE. In detail, phases observed with a peak at 20.5274° possess intensity of 10894.35 cts, full width half maximum (FWHM) of 0.4517°, and interplanar spacing (d-spacing) of 4.32677 Å. In addition, phases identified with the peak at 22.7464° have intensity of 2530.53 cts, FWHM of 0.6376° and d-spacing of 3.90945 Å; those observed at 35.4325° are characterised with an intensity of 625.66 cts, FWHM of 0.4251, and d-spacing of 2.53345 Å. Intensity of 358.86 cts, FWHM of 0.8502, and d-spacing of 2.14307 Å belong to the phases confirmed with the peak at 42.1678 and intensity of 182.26 cts, FWHM of 0.8502, and d-spacing of 1.96601 Å characterise the phases appeared on the peak at 46.1747°. Moreover, phases of the peak at 52.2759° have intensities of 262.85 cts, FWHM of 1.0368°, and 1.74856 Å. Ethylene, the monomer of polyethylene is expected to be confirmed with the XRD peaks. However, ethylene urea ($C_3H_6N_2O$), tri-methylene di-phosphonic acid ($C_3H_{10}O_6P_2$), and ethylenediamine hydrogen selenite ($C_2H_{10}N_2O_4Se$) see ethylene products resulted from interaction of ethylene

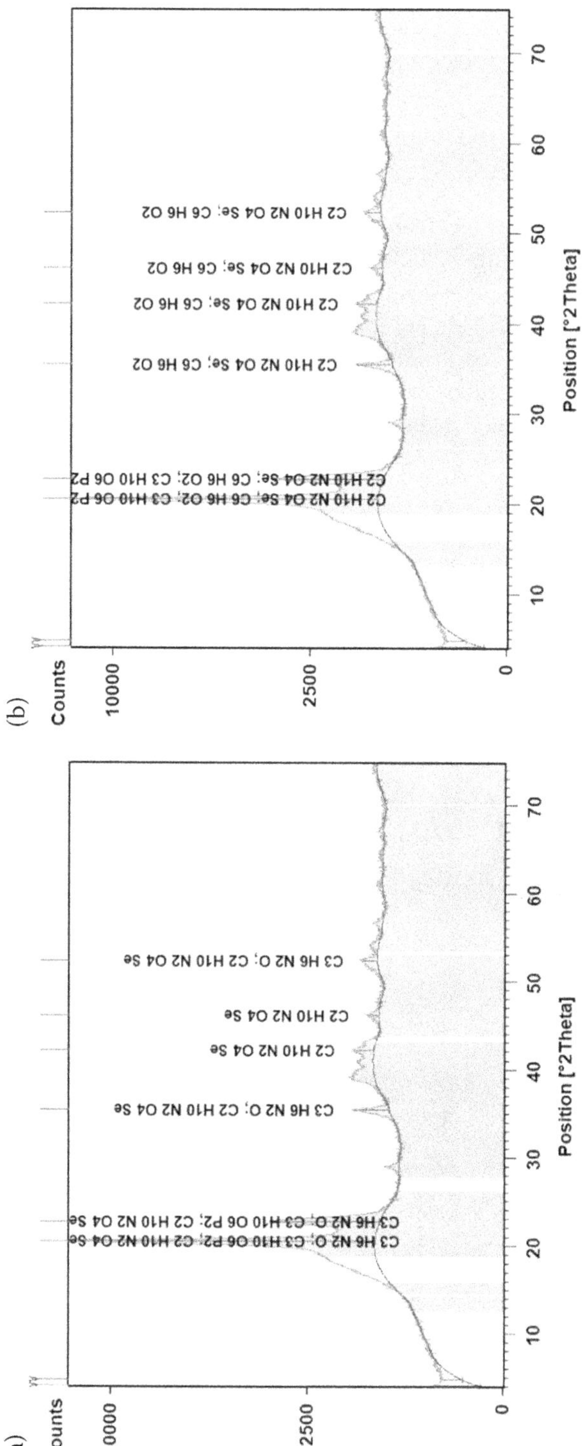

FIGURE 7.1 X-ray diffractograms of the pristine (a) recycled low-density polyethylene, (b) virgin low-density polyethylene.

with additives like plasticiser, lubricant, and flame retardants added to the LDPE while manufacturing to enhance their properties.

7.3.2 Microstructural Properties of *Delonix regia* Pod Particle-Reinforced Low-Density Polyethylene

7.3.2.1 Transmission Electron Microscopic Images and Sizes of Reinforcing Particles

Delonix regia pod particles contain many spherical individual tiny particles existing in colonies or clusters to form big spherical bodies (Figure 7.2). An existing form of reinforcement is attributed to van der Waal's attraction due to particle refinement. Literature has established an increased particle surface area [7] and energy as the particle refinement is in progress, which initiates physical fusion (van der Waal's attraction) of individual particles to form big particles with consequent reduction in energy [3].

Spherical shape of individual particles is advantageous, as it gives spaces to be filled by the LDPE melt during processing for effective bonding of particles together to give composites with high structural integrity. Moreover, particle colonies are challenges since they produce clusters forming structural inhomogeneity and

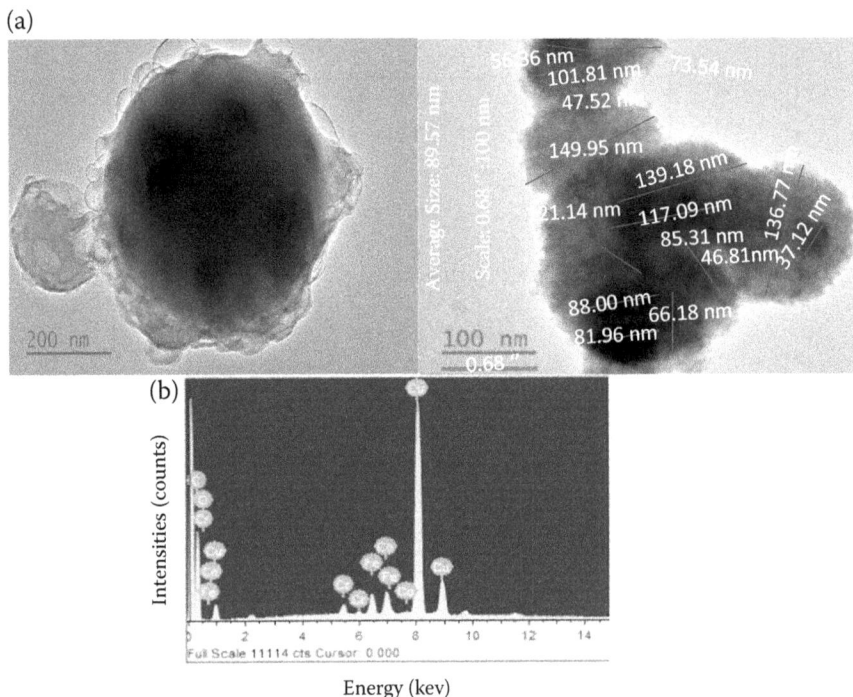

FIGURE 7.2 (a) TEM images of Delonix regia pod nanoparticles at 200 nm and 100 nm resolutions and (b) elemental composition of the Delonix regia pod nanoparticles.

discontinuity in materials that normally impair properties of the materials [8,9]. In the study reported in this chapter, *Delonix regia* pod particles were dissolved in absolute ethanol to break up agglomerates prior to mixing with the LDPE. An energy dispersive X-ray spectroscope shows that *Delonix regia* pod particles have C, O, Cr, Cu, Co, Fe, and Cu, implying the presence of minerals like $CuCO_3$, Cu_2CO_3, $FeCO_3$, $Fe_2(CO_3)_3$, $CoCO_3$, and $CrCO_3$ in the pod. However, it contains mainly Cu carbonates as minerals because of their tall peak among other metals, in addition to carbonaceous materials like cellulose, hemicellulose, and lignin being originated from woody materials. The presence of hydrogen in addition to C will affirm organic components, but its weight is too small to be detected by the EDX. The *Delonix regia* pod particles can be thermally treated to liberate volatile components or used without thermal treatment. Although thermal treatment will enhance carbon contents and ash contents according to literature [10], there is a consequent loss of large amounts of materials. More than 70% of carbonaceous material like *Delonix regia* is lost to thermal treatment [10]. For material management and conservation and to prevent the release of hazardous gases to the atmosphere, *Delonix regia* pod particles used in this study were not thermally treated. Sizes of the *Delonix regia* pod particles were determined from one of TEM images in Figure 7.2a using the Microsoft Word shape tool (line), as explained in [2,11]. Based on the blue line drawn below the resolution of 100 nm in Figure 7.2a, a scale of 0.68'' to represent 100 nm in Figure 7.2a was used for the size determination. Various *Delonix regia* pod particles in Figure 7.2a have different sizes, with an average size of 89.57 nm. Similarly, synthesis of nanoparticles from agricultural wastes using a top-down approach were reported by different researchers across the globe. Average sizes, 49.847 ± 0.48 nm of coconut shells milled for 70 hours at 10 charge ratios, were found, as reported in [3]. Average size, 106.81 nm, was disclosed in [12] for uncarbonised maize stalk milled for 70 hours. Refinement of lignin by wet ball milling with formation of a porous structure was found in [13]. Moreover, milling of poly (amide imide) for 7 hours at 3.4–4.9 ms^{-1} using attritor mill with particle size of 3 μm was confirmed as reported in [14]. Eggshell particles of an average size of 76.25 nm were revealed in [2] and *Delonix regia* pod particles obtained in the study reported in this chapter agree with findings in literature.

7.3.2.2 Scanning Electron Micrograph of *Delonix regia* Pod Particle-Reinforced Low-Density Polyethylene Composites

Pristine RLDPE appears as a continuous solid body developed from fusion of numerous polymeric grains during processing. Microstructures at different magnifications in Figure 7.3 are representatives of materials with high structural integrity. Polymeric grains of different colours characterise the pristine RLDPE, i.e., RLDPE without any filler addition, as shown in Figure 7.3a. The white regions represent some polyethylene grains/phases. They appear white due to the reflection of electron beams. This observation is affirmed by the microstructure in Figure 7.3b captured at a greater magnification than that in Figure 7.3a.

There are many grains in Figure 7.3b with fibrous, polygonal, and irregular shapes firmly connected together, affirming structural continuity of the pristine

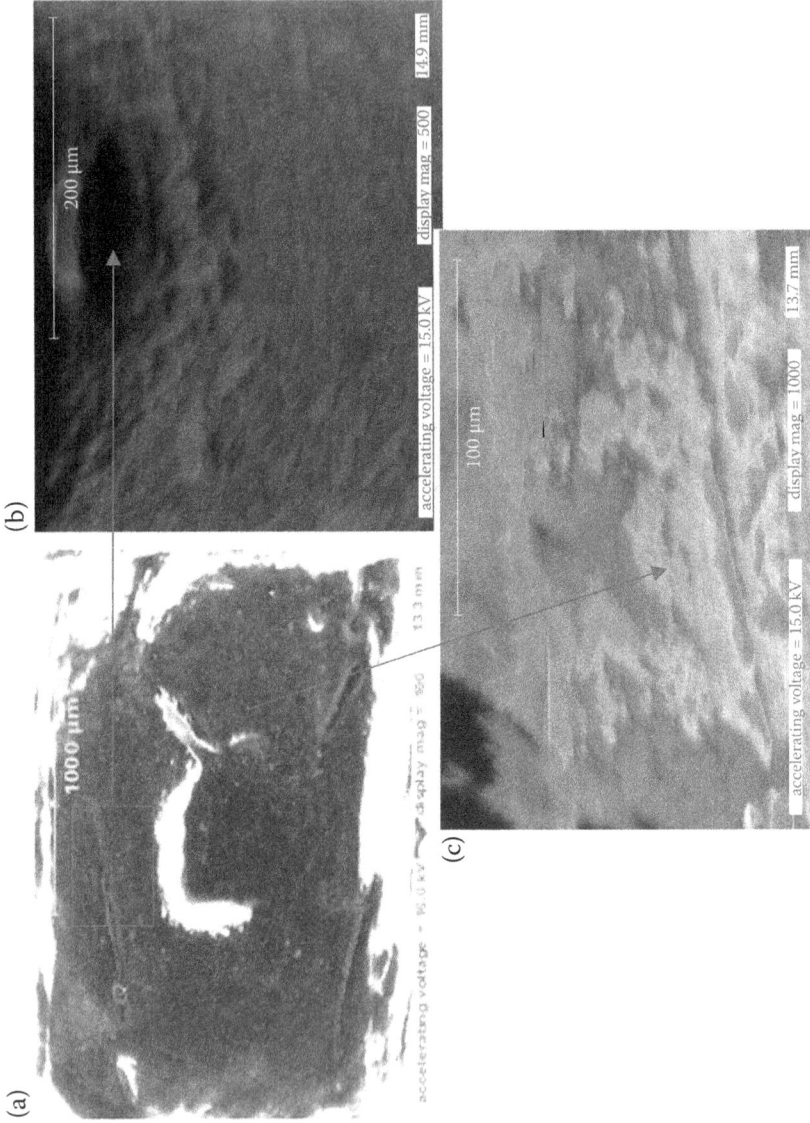

FIGURE 7.3 Microstructures of unreinforced recycled polyethylene at different magnifications (a) × 100, (b) × 500, (c) × 1000.

RLDPE. Moreover, by magnifying a portion of the central white grain in Figure 7.3a, it is observed that the portion harbours tiny polymeric grains of different geometries (Figure 7.3c) like dendrites explaining crystal growth during solidification of metals. Differences in colour can be linked with phases of varied chemical formulas confirmed with the pristine RLDPE by XRD in Figure 7.1a. In addition, pristine virgin low-density polyethylene (VLDPE) is a solid material with a continuous structure in a form like a parent rock harbouring many tiny rock debris. The structure in Figure 7.4 was produced from different layers of VLDPE lying over each other and interconnected together during fusion and compaction. Alternate white and black phases occupy the whole structure, affirming observation from the XRD result, revealing different chemical compounds belonging to the pristine VLDPE.

Moreover, differences in appearances of RLDPE and VLDPE may be linked with different configurations formed by the RLDPE and VLDPE grains during the processing. After the addition of *Delonix regia* pod particles to both RLDPE and VLDPE to produce *Delonix* regia particle-reinforced LDPE composites, the structures obtained are presented in Figures 7.5–7.8. The arrangement of grains in space in Figures 7.5–7.8 of the *Delonix regia* particle-reinforced LDPE is different from observations with the structure of pristine RLDPE and VLDPE in Figures 7.3–7.4, implying that both RLDPE and VLDPE demonstrated different behaviours when the reinforcing particles were added to each of them. At 8 wt%, *Delonix regia* pod particle addition to RLDPE, the *Delonix regia* pod particles are observed distributing within the entirety of the RLDPE matrix to form different composite layers that rest over one another, fusing together after compaction to form a continuous structure with different dendritic patterns. For thorough investigations, different points on the composite were captured at the same resolution (100 nm)/nearly the

(a)   (b)

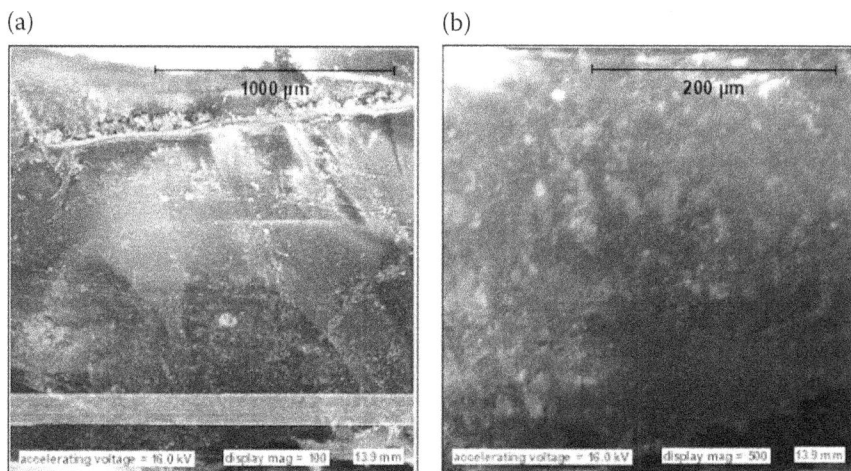

FIGURE 7.4 Microstructure of the pristine virgin low-density polyethylene at 100 and 500 magnifications.

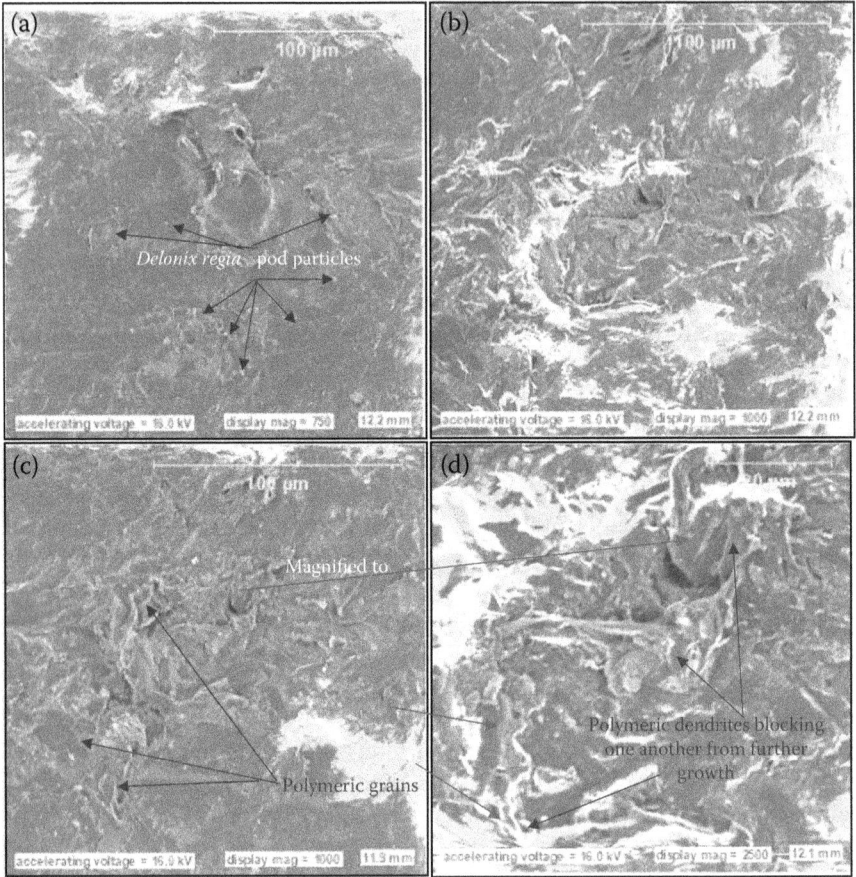

FIGURE 7.5   Microstructure of recycled low-density polyethylene 8% *Delonix regia* pod particle composites at (a) × 750, (b) × 1000, (c) × 1000 captured at another point, and (d) × 2500.

same magnifications and at smaller resolution/greater magnifications to enlarge the image. At greater magnification (Figure 7.5d), different patterns of composite grains are clearly seen. For instance, a continuous horizontal white grain appears like a skeletal spinal cord with many ribs that stays on top of the dark structure with undulating layers. Another observed pattern is a vertically grown structure with different fingers that appear like aloe vera leaves. All structures in different patterns connect to assume a strengthened structure free from any defects. Inference from this study agrees with changes in the microstructure of LDPE due to the bagasse addition as encountered in [15], breadfruit seed hull ash addition disclosed in [16], and bean pod ash addition, come across in [17]. Moreover, the structure (Figure 7.6) of 12 wt% *Delonix regia* pod particle-reinforced RLDPE composite does not resemble the structure in Figure 7.3 of the RLDPE composite at lower *Delonix regia* pod particle contents.

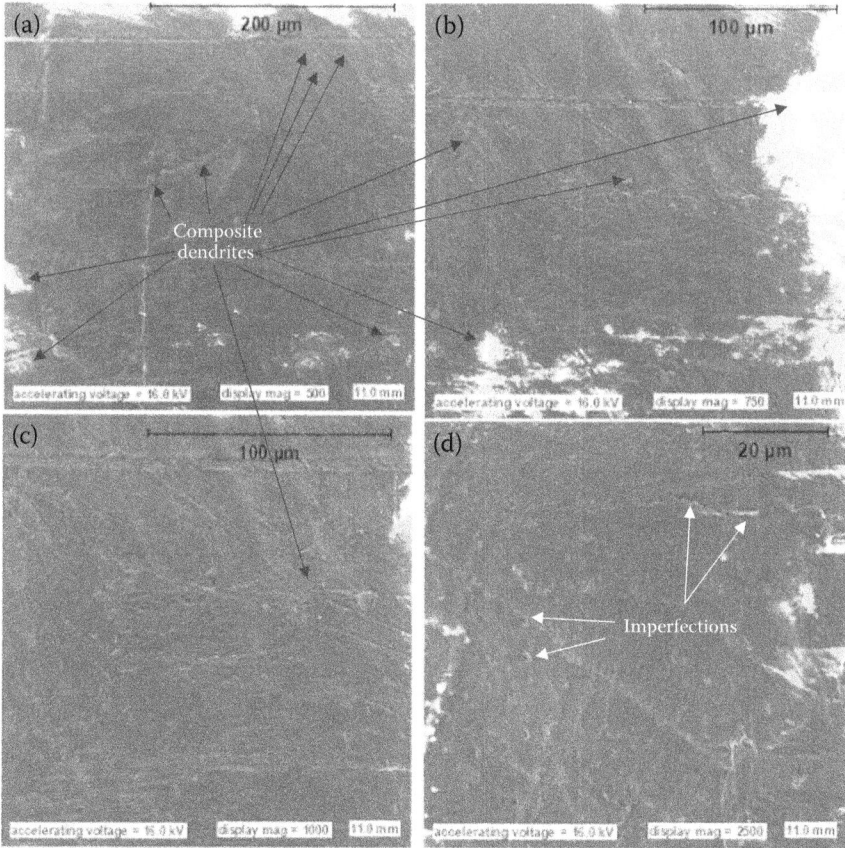

FIGURE 7.6 Microstructures of recycled low-density polyethylene 12% *Delonix regia pod* particulate composites at (a) × 500, (b) × 750, (c) × 1000, and (d) × 2500.

Twelve wt% *Delonix regia* pod particle-reinforced RLDPE composite contains numerous fine grains attributed to a high rate of solidification from composite melt. It is clear that at RLDPE fusion temperature (150°C), the *Delonix regia* pod particle still remains at a solid state. Metallic ions like Cu^{2+}, Cu^{+}, Fe^{2+}, and Fe^{3+} from mineral content of the *Delonix regia* pod particle may assist in quick heat transfer from the melt to immediate environment. This aids the atmospheric environment in chilling the composite and enhances the solidification/cooling rate. Such a relatively high rate of solidification promotes the formation of a large number of composite nuclei transforming to composite dendrites that do not have sufficient time to grow until solidification completes. A fine-grain structure, as observed in Figure 7.5, is expected to improve the strength of the composite. Moreover, microstructures of *Delonix regia* pod particle-reinforced RLDPE composites at 8 and 12 wt% *Delonix regia* pod particle addition are distinct and

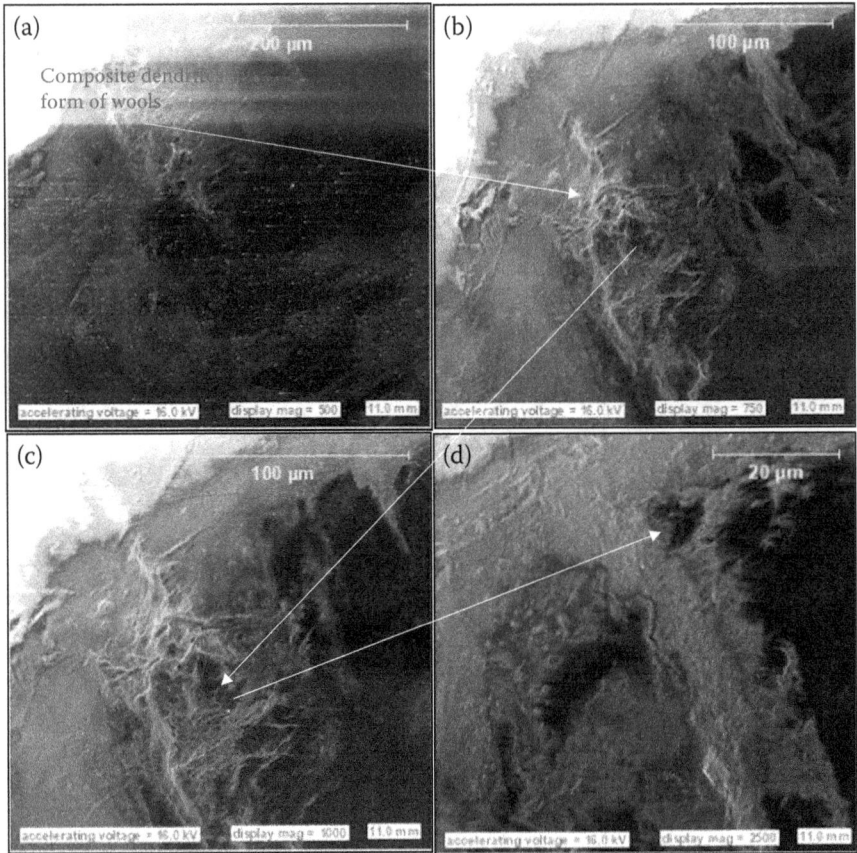

FIGURE 7.7 Microstructures of virgin low-density polyethylene 8% *Delonix regia* pod particulate composites at (a) × 500, (b) × 750, (c) × 1000, and (d) × 2500.

each of them is different from those of the pristine RLDPE polymer. Therefore, the observation agrees with literature [17].

In addition, spatial configuration (Figure 7.7) of VLDPE/8 wt% *Delonix regia* pod particle-reinforced composite grains conform to a rigid body with a last portion of composite dendrites in the form of wool on top of the continuous structural body. These structures and those in Figure 7.8 possess features different from those of the pristine VLDPE. Despite the fact that structures in Figure 7.8 belong to *Delonix regia* pod particle-reinforced VLDPE composites though at different *Delonix regia* pod particle contents, grain orientations in those figures vary from each other and 12 wt% *Delonix regia* pod particle-reinforced VLDPE composite structure has greater integrity than that of 8 wt% *Delonix regia* pod particle-reinforced VLDPE composites with some voids and it is expected to perform better under the influence of a mechanical load.

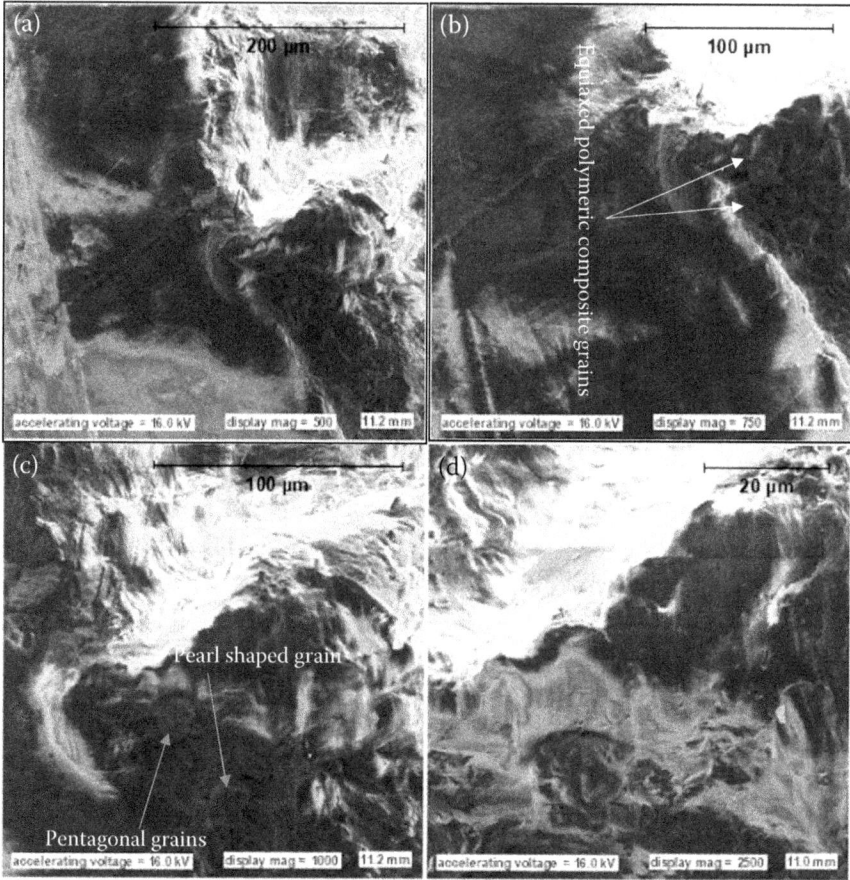

FIGURE 7.8 Microstructures of virgin low-density polyethylene 12% *Delonix regia pod* particulate composites at (a) × 500, (b) × 750, (c) × 1000, and (d) × 2500.

7.4 MECHANICAL PROPERTIES OF *DELONIX REGIA* POD PARTICLE-REINFORCED LOW-DENSITY POLYETHYLENE

7.4.1 Tensile Properties of *Delonix regia* Pod Particle-Reinforced Low-Density Polyethylene

Low-density polyethylene composites display different tensile behaviour at varied *Delonix regia* pod particle additions, as shown in Figures 7.9–7.11. For comprehensive explanation on the tensile properties of LDPE-based composites, values of various parameters for establishing tensile properties of materials were plotted against wt% of *Delonix regia* pod particle addition to the LDPE. It was observed that progressive increases in the tensile strength due to *Delonix regia* pod particle additions to both RLDPE and VLDPE were noticed except a slight reduction at 10 wt% *Delonix regia* pod particle addition to VLDPE (Figure 7.10). Moreover,

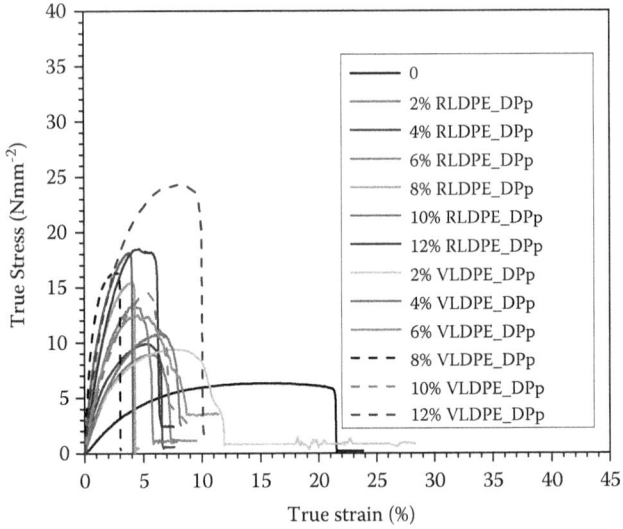

FIGURE 7.9 Flow curve of polyethylene 0–12 wt% *Delonix regia* pod particle composites.

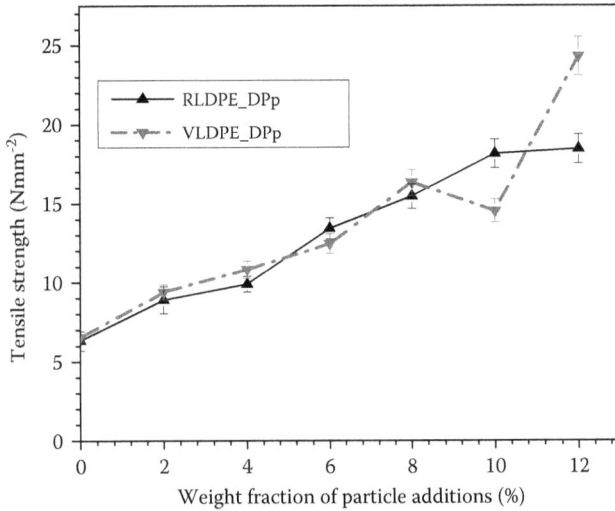

FIGURE 7.10 Tensile strength of polyethylene reinforced by *Delonix regia* pod particles.

tensile elongation of the *Delonix regia* pod particle-reinforced RLDPE decreases with an increment in the wt% of *Delonix regia* pod particles. For the *Delonix regia* pod particle-reinforced VLDPE, the tensile elongation increases significantly at 2 wt% addition of *Delonix regia* pod particle to the LDPE but above this wt%, the elongation curve appears parallel to the horizontal implying that there is not much increase in tensile elongation when the tensile elongation values (7.85, 7.88, 7.89, 8.02, and 8.05%) above 2 wt% of the *Delonix regia* pod particles addition to the

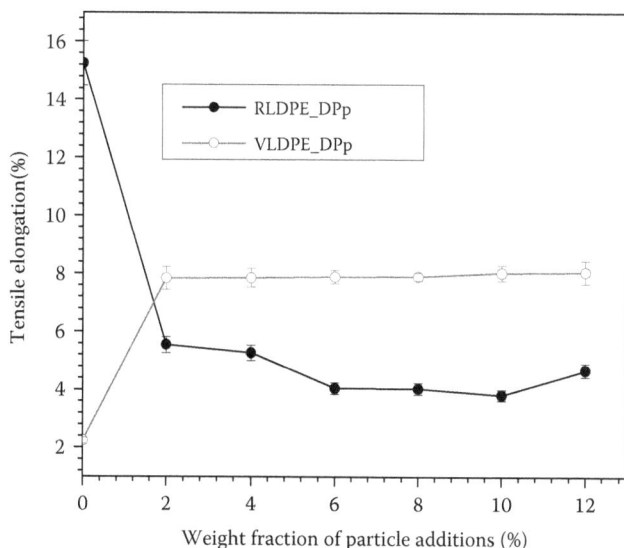

FIGURE 7.11 Tensile elongation of low-density polyethylene modified with *Delonix regia* pod.

virgin LDPE are compared with the value at 2 wt% addition. Moreover, it is noteworthy in this chapter that VLDPE performed mechanically better than RLDPE.

Generally, increases in the tensile strength of the LDPE after reinforcement connect with a degree of stiffness of the LDPE by *Delonix regia* pod particles, which in turn improves the load bearing capacity of the LDPE composite. When each of the composites was subjected to uniaxial loading, the stress from the external load was evenly distributed by the LDPE matrix to individual reinforcing *Delonix regia* pod particles. Therefore, the particles united to resist the deformation until the value of externally applied stress reached a threshold value. At this point, the particle resistance was overcome by the increasing externally applied stress either by particle breakage or particle detachment from the LDPE matrix and tearing of the matrix, resulting in formation of tiny micro voids called crazes. This is called the crazing process.

Uniaxial loading causes a pulling effect on the composite, leading to a gradual increase in deformation (increase in length) with formation of new crazes. The existing crazes propagated and grew due to elongation to fuse into one another and formed cavities (cavitation process). Since crazes' propagation and growth were resisted by the approach of reinforcing particles firmly bonded to the LDPE matrix, increasing stress was needed externally to aid the craze propagation and growth until ultimate stress level reached when fusion of voids produced large cracks causing composite fracture after necking. Therefore, the increase in the tensile strength with the increase in wt% of *Delonix regia* pod particles is justified by the increase in resistance offered by *Delonix regia* pod particles as their percentage by weight increases. Moreover, by comparing the properties of the *Delonix regia* pod particle-reinforced LDPE nanocomposite with eggshell particle-reinforced LDPE nanocomposite in [2] since the same matrix was used for both of them, greater tensile strength in respect of *Delonix regia* pod particle-reinforced LDPE than than those of eggshell particle-reinforced LDPE nanocomposite is noticed. This is attributed to various phases of the *Delonix regia* pod particle/LDPE composites with different properties from those phases occupying eggshell particle-reinforced LDPE composite structure. As indicated in Section 7.2.2.1 of this chapter, the *Delonix regia* pod particle contains mainly C and Cu. Carbon is known for high strength and rigidity. Cu is a transition metal. Its minerals are hard ceramics materials. The presence of C and Cu compounds in the *Delonix regia* pod particle-reinforced LDPE nanocomposites is expected to offer greater resistance to crack advancement and growth than Ca and K compounds in the eggshell particle reinforced LDPE. Smaller tensile strength of 10 wt% *Delonix regia* pod particle-reinforced virgin LDPE nanocomposite than expected is explained by micro cracks or voids inherently found in its structure. The voids caused discontinuities in the composite and interfered with composite flow during uniaxial loadings. Material structural discontinuities prevent even stress distribution among the reinforcing particles. Therefore, stress is concentrated to the void regions, leading to localised deformation while remaining parts of the composite bear no load. The voids acted as stress raisers, causing a stress multiplying effect on portions of the material. Since the voids are inherent to the materials, there is no craze initiation. Therefore, the crack advancement and propagation were rapid because little resistance would be offered by a portion of the composite and the failure/deformation occurred at a stress level lower than expectations if the ideal composite free from defects was involved. However, the increase in *Delonix regia* pod particle-reinforced VLDPE nanocomposite elongation is linked with higher flexibility of the VLDPE that permits firm adhesion of the reinforcing particles to the VLDPE matrix, leading to composite stretching with sluggish crack advancement and growth. Similar improvement in the strength of polyethylene due to increment in volume fraction of discontinuous carbon fibres was reported in [18], but the decrease in tensile elongations of their composite agrees with a reduction in the tensile elongation of RLDPE composites but opposed tensile elongation increase of VLDPE nanocomposites reported in this chapter. Similarly, the decrease in tensile strength of high-density polyethylene (HDPE) reinforced with hemp fibre attributed to poor adhesion between the fibre and the matrix was found in [19], but use of palm kernel oil as a lubricant or binding agent in the study reported in this chapter resulted in the strong adhesion of the reinforcing particles to the LDPE molecules accounting for improved tensile strength of the developed composites. Also, the increase in tensile strength of HDPE from 15.7 ± 1.1 to 23.2 ± 0.6 due to 40 wt% addition of sugarcane bagasse cellulose fibres was divulged in [20] and the decrease in tensile elongation due to peanut husk addition to the LDPE was read in [21]. This establishes the fact that addition incorporation of agricultural reinforcement into polyethylene does not enhance the tensile properties in respect of all agricultural resources. Enhancement or degradation of properties can take place, depending on the nature of the agricultural waste reinforcement, interaction of the agricultural reinforcement with polyethylene matrix, structure developed from reinforcement-matrix interaction, and adhesion force between the matrix and the reinforcements.

7.4.2   Flexural Properties of *Delonix regia* Pod Particle-Reinforced Low-Density Polyethylene Composites

The addition of *Delonix regia* pod particles to RLDPE produced composites having greater bending strength than itself at all levels of reinforcement except at 6 wt% of *Delonix regia* pod particles addition where the composite's bending strength was smaller than that of the pristine RLDPE (Figure 7.12). The bending strength of *Delonix regia* pod particles reinforced VLDPE increase would have been progressive if not for the bending strengths of 8% *Delonix regia* pod particles reinforced VLDPE and 10 wt% *Delonix regia* pod particles reinforced VLDPE which are smaller than that of 6 wt% *Delonix regia* pod particles reinforced VLDPE. Smaller bending strength of all indicated composites than that of 6 wt% *Delonix regia* pod particles reinforced VLDPE is linked with imperfections inherent in them. Generally, the increase in the bending strength due to *Delonix regia* pod particles addition is justified by the ability of the composite to resist the simultaneous tension and compression experienced by the examined composites, leading to their flexure prior to their failure or breakage. Rigidity of the reinforcing particles is the source of the composites' resistance to flexural deformation. As structural integrity of the composites has been established by the SEM micrographs in Figures 7–8, during the flexural investigation, stresses emanated from compressive force applied on the composite samples would equally be transferred by the LDPE matrix to all the particles embedded evenly within the matrix. Therefore, loading effects would not be felt by one particle than others. This would not give room for local deformation that may lead to the composites' failure at a lower stress. Generally, VLDPE containing *Delonix regia* pod particles has greater bending strengths than those of the RLDPE-based nanocomposites.

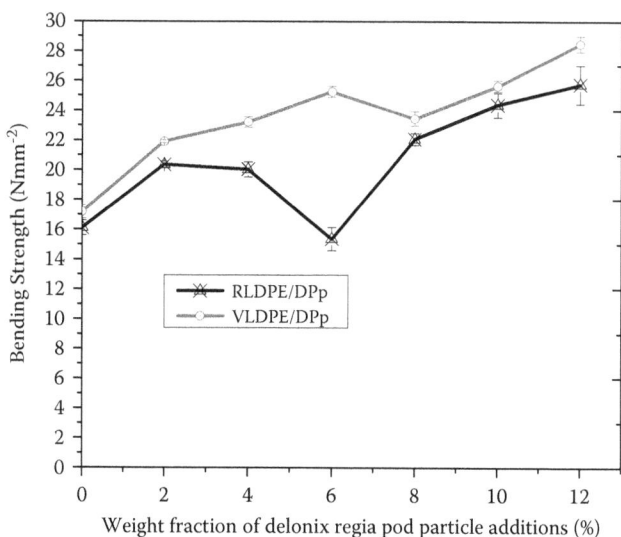

FIGURE 7.12   Bending strength of *Delonix regia* pod particle-reinforced low-density polyethylene.

7.4.3 HARDNESS VALUES OF *DELONIX REGIA* POD PARTICLE-REINFORCED LOW-DENSITY POLYETHYLENE

A gentle slope line in Figure 7.13 indicates a slightly progressive increase in the hardness values with the increment in the *Delonix regia* pod particle additions to the VLDPE. About a 13.74% improvement at 12 wt% *Delonix regia* pod particle addition is noted. A maximum hardness was confirmed at 10 wt% *Delonix regia* pod particle addition to the RLDPE, corresponding to 8.97% enhancement. This indicates a better performance of the *Delonix regia* pod particle-reinforced VLDPE than its counterpart RLDPE nanocomposite in resisting indentations due to load of equal magnitudes imposed on each nanocomposite by the indenter though the addition of the *Delonix regia* pod particle to both RLDPE and VLDPE demonstrates increases in hardness values. A decrease in hardness value above 10 wt% of *Delonix regia* pod particle addition to RLDPE indicates saturation of the RLDPE at 10 wt%. Similar behaviour is noted with RLDPE-based composite in Figure 7.10 that indicates nearly equal tensile strength at 10 and 12 wt% *Delonix regia* pod particle additions to the RLDPE. These testify to the fact that continuous addition of *Delonix regia* pod particle to the RLDPE above 10 wt% could lead to degradation in the tensile strength and hardness values but improvement in the flexural strength of the *Delonix regia* pod particle-reinforced RLDPE; meanwhile, the red curves in Figures 7.10, 7.12, and 7.13 propose increases in the tensile strength, flexural strength, and hardness values of the *Delonix regia* pod particle-reinforced VLDPE nanocomposites above 10 wt% *Delonix regia* pod particle additions. This affirms better mechanical property performances of the VLDPE than RLDPE-based composites, though greater cost of the VLDPE would make the products from *Delonix regia* pod particle-reinforced VLDPE nanocomposites more expensive than those from the RLDPE-based nanocomposites.

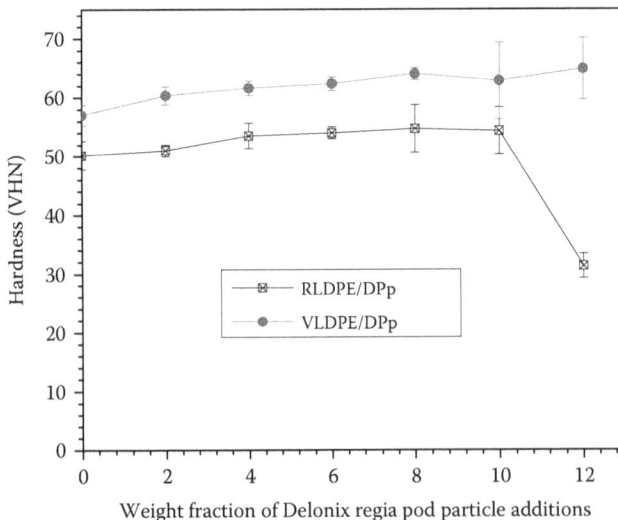

FIGURE 7.13 Hardness values of polyethylene *Delonix regia* pod composites.

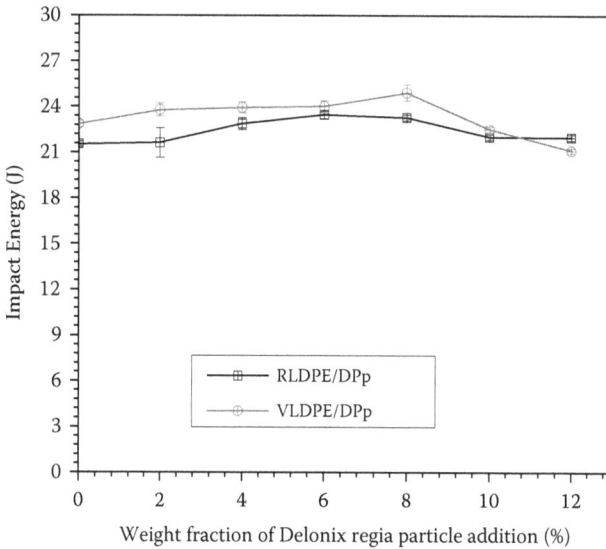

FIGURE 7.14 Impact energy of polyethylene *Delonix regia* pod composites.

7.4.4 IMPACT ENERGY OF *DELONIX REGIA* POD PARTICLE-REINFORCED LOW-DENSITY POLYETHYLENE

Incorporation of *Delonix regia* pod particles in the VLDPE does not bring profuse improvement in the impact energy of the LDPE composites. For RLDPE-based nanocomposites, the increase was observed in the impact energy up to 8 wt% *Delonix regia* pod particles with a 8.05% increase at 8 wt% addition (Figure 7.14). Moreover, addition of *Delonix regia* pod particles to VLDPE produced 8.99% maximum enhancement at 8 wt% while the impairment in impact energy occurred from 10 wt% *Delonix regia* pod particles addition to the VLDPE like those of *Delonix regia* pod particle-reinforced RLDPE nanocomposites. The impairment could either be linked with hardness values of the analysed RLDPE- and VLDPE-based nanocomposite samples. Greater improvement in impact energy of the *Delonix regia* pod particle-reinforced VLDPE than those of the RLDPE-based nanocomposites is a further confirmation that VLDPE is a better matrix to the *Delonix regia* pod particles than RLDPE. Similar observations have been found in literature in respect to eggshell nanoparticle-reinforced LDPE [2].

7.5 CONCLUSIONS AND SUMMARY

New engineering composites produced from *Delonix regia* pod nanoparticles, recycled and virgin low-density polyethylene were reported in this chapter. Properties of the nanocomposites based on recycled low-density polyethylene were compared with those of the *Delonix regia* pod nanoparticle reinforced low-density polyethylene. It was noted that tensile strength and hardness values of the *Delonix regia* pod nanoparticle reinforced low-density polyethylene indicates saturation of the

recycled low-density polyethylene at 10 wt% addition of *Delonix regia* pod nanoparticle addition; meanwhile, no saturation of the virgin low-density polyethylene
is noticed up to 12 wt% *Delonix regia* pod nanoparticle addition. This suggests that
further investigation is necessary on the virgin low-density polyethylene above
12 wt% of the *Delonix regia* pod nanoparticle addition. Based on the better mechanical properties of the *Delonix regia* pod nanoparticle-reinforced virgin low-
density polyethylene than its counterpart recycled low-density polyethylene-based
nanocomposite, it is inferred that the virgin low-density polyethylene is the better
matrix to develop new nanocomposites using the *Delonix regia* pod nanoparticle as
reinforcement than the recycled low-density polyethylene.

ACKNOWLEDGEMENT

This is to acknowledge Nigeria Academy of Engineering/ARCO Petrochemical
Engineering Company for 2019 Postdoctoral Award/Grant from which the study
reported in this chapter was sponsored.

REFERENCES

[1] Tangya, H., and Daniela, S. (2005). 10 Years of Recycling: The Good, The Bad
and The Ugly. in D. Jon, Editor, pp. 1–35.

[2] Bello, S.A., Raji, N.K., Kolawole, M.Y., Adebayo, M.K., Adebisi, J.A., Okunola,
K.A., and AbdulSalaam, M.O. (2021). Eggshell Nanoparticle Reinforced Recycled
Low-Density Polyethylene: A New Material for Automobile Application. *Journal of
King Saud University - Engineering Sciences*. 10.1016/j.jksues.2021.04.008

[3] Bello, S.A., Agunsoye, J.O., and Hassan, S.B. (2015). Synthesis of Coconut Shell
Nanoparticles Via a Top Down Approach: Assessment of Milling Duration on the
Particle Sizes and Morphologies of Coconut Shell Nanoparticles. *Materials Letters*,
159, 514–519. 10.1016/j.matlet.2015.07.063

[4] ASTM D638-14. (2014). *Standard Test Method for Tensile Properties of Plastics*.
ASTM International, West Conshohocken, PA. https://www.astm.org/Standards/D638

[5] ASTM D7264 /D7264M-15. (2015). *Standard Test Method for Flexural Properties
of Polymer Matrix Composite Materials*. ASTM International, West Conshohocken,
PA. 10.1520/D7264_D7264M-15

[6] ASTM D256-10(2018). (2018). *Standard Test Methods for Determining the Izod
Pendulum Impact Resistance of Plastics*. ASTM International, West Conshohocken, PA.

[7] Loh, Z.H., Samanta, A.K., and Sia Heng, P.W. (2015). Overview of Milling
Techniques for Improving the Solubility of Poorly Water-Soluble Drugs. *Asian
Journal of Pharmaceutical Sciences*, **10**(4), 255–274. 10.1016/j.ajps.2014.12.006

[8] Bello, S.A. (2019). Agglomerates within Ball-Milled Lignocellulosic Particles
Using Minimum Crystal Sizes. *Unilag Journal of Medicine, Science and
Technology*, **7**(1), 15–40.

[9] Bello, S.A., Agunsoye, J.O., Adebisi, J.A., and Hassan, S.B. (2018). Optimisation of
Charge Ratios for Ball Milling Synthesis: Agglomeration and Refinement of
Coconut Shells. *Engineering and Applied Science Research (EASR)*, **42**(4),
262–272. 10.14456/easr.2018.36

[10] Bello, S.A., Agunsoye, J.O., Adebisi, J.A., Kolawole, F.O., and Suleiman, B.H. (2106).
Physical Properties of Coconut Shell Nanoparticles. *Kathmandu University Journal of
Science, Engineering and Technology*, **12**(1), 63–79. 10.3126/kuset.v12i1.21566

[11] Bello, S.A. and Kolawole, M.Y. (2020). Recycled Plastics and Nanoparticles for Green Production of Nano Structural Materials, in *Handbook of Nanomaterials and Nanocomposites for Energy and Environmental Applications*, O.V. Kharissova, L.M.T. Martínez, and B.I. Kharisov, Editors, Springer International Publishing, Cham. pp. 1–33, 10.1007/978-3-030-11155-7_93-1

[12] Agunsoye, J.O., Bamigbaiye, A.A., Bello, S.A., and Hassan, S.B. (2020). Mechanical Properties of Maize Stalk Nano-Particle Reinforced Epoxy Composites. *Arabian Journal for Science and Engineering*, **45**, 5087–5097. 10.1007/s13369-02 0-04345-5

[13] Qu, Y., Luo, H., Li, H., and Xu, J. (2015). Comparison on Structural Modification of Industrial Lignin by Wet Ball Milling and Ionic Liquid Pretreatment. *Biotechnology Reports*, **6**, 1–7. 10.1016/j.btre.2014.12.011

[14] Wolff, M.F.H., Antonyuk, S., Heinrich, S., and Schneider, G.A. (2014). Attritor-Milling of Poly(Amide Imide) Suspensions. *Particuology*, **17**, 92–96. 10.1016/j.partic.2013.11.005

[15] Agunsoye, J.O. and Aigbodion, V.S. (2013). Bagasse Filled Recycled Polyethylene Bio-Composites: Morphological and Mechanical Properties Study. *Results in Physics*, **3**, 187–194. 10.1016/j.rinp.2013.09.003

[16] Atuanya, C.U., Aigbodion, V.S., and Nwigbo, S.C. (2014). Experimental Study of the Thermal and Wear Properties of Recycled Polyethylene/Breadfruit Seed Hull Ash Particulate Composites. *Materials & Design*, **53**, 65–73. 10.1016/j.matdes.2013.06.057

[17] Atuanya, C.U., Edokpia, R.O., and Aigbodion, V.S. (2014). The Physio-Mechanical Properties of Recycled Low Density Polyethylene (RLDPE)/Bean Pod Ash Particulate Composites. *Results in Physics*, **4**, 88–95. 10.1016/j.rinp.2014.05.003

[18] Brahmakumar, M., Pavithran, C., and Pillai, R. (2005). Coconut Fibre Reinforced Polyethylene Composites: Effect of Natural Waxy Surface Layer of the Fibre on Fibre/Matrix Interfacial Bonding and Strength of Composites. *Composites Science and Technology*, **65**(3-4), 563–569. 10.1016/j.compscitech.2004.09.020

[19] Dauvegis, R. and Rodrigue, D. (2015). The Mechanical Properties of Density Graded Hemp/Polyethylene *Composites*, **1664**, 060001. 10.1063/1.4918419

[20] Mulinari, D.R., Voorwald, H.J.C., Cioffi, M.O.H., and da Silva, M.L.C.P. (2016). Cellulose Fiber-Reinforced High-Density Polyethylene Composites—Mechanical and Thermal Properties. *Journal of Composite Materials*, **51**(13), 1807–1815. 10.11 77/0021998316665241

[21] Obasi, H.C. (2015). Peanut Husk Filled Polyethylene Composites: Effects of Filler Content and Compatibilizer on Properties. *Journal of Polymers*, **2015**, 1–9. 10.1155/2015/189289

Section III

Agricultural and Domestic Waste–Based Nanocomposites of Other Polymers

8 Particulate-Reinforced Polylactic Composites: Synthesis, Properties, and Applications

Collieus Lebudi, Babatunde Abiodun Obadele,
Oluseyi Philip Oladijo, and
Enoch Nifise Ogunmuyiwa
Department of Chemical, Materials and Metallurgical
Engineering, Botswana International University of Science
and Technology, Palapye, Botswana

CONTENTS

DOI: 10.1201/9781003170549-11

8.1　INTRODUCTION

Over the years, the modern industrial world has become heavily dependent on synthetic polymers derived from petroleum oil (petroleum-based polymers). Examples of these polymers include nylon, polyethylene (PE), polypropylene (PP), polytetrafluoroethylene (PTFE), and polystyrene (PS) and so many others. Petroleum-based polymers are embedded in everyday life with applications ranging from the most menial like hair combing to the most sophisticated technologies in aircraft propulsion systems. Such a great versatility in application of these polymers is attributable to their abundance in the natural environment and their highly sought-after properties. These properties include resilience, chemical and biological inertness, high strength-to-weight ratio, toughness, transparency, and lack of electrical and thermal conductivity. Moreover, they are easily processed and have relatively lower service maintenance costs compared to other classes of materials [1]. However, despite these numerous desirable aspects of petroleum-based polymers, great environmental concerns have been raised worldwide. The voluminous use of polymers derived from petroleum oil means that non-renewable fossil reserves are being depleted at a much faster rate [2]. Moreover, at every stage of their life cycle – extraction, transport, refining, production, and disposal – enormous amounts of greenhouse gasses are being emitted [2]. Disposal after usage has also become problematic due to properties like chemical and biological inertness which inhibits degradation thereby leading to accumulation of polymer waste in landfills and natural environments thus causing pollution and contamination. Petroleum-based polymers have therefore become a global environmental crisis. In efforts to remediate the environmental impact of synthetic polymers, the development of natural-based polymers (biopolymers) with a high degree of biodegradability have been at the forefront of scientific research with the aim of possibly using these polymers as replacements for synthetic polymers. Compared to other biopolymers, polylactic acid (PLA) is by far the most explored biopolymer due to its eco-friendliness (biodegradability, recyclability, and compostability), biocompatibility, ease of fabrication through different processes like injection moulding, blow moulding, thermoforming, extrusion, etc., and higher energy savings of about 25%–55% less relative to petroleum-based polymers [3,4]. PLA is derived from completely renewable resources such corn, sugarcane, and sugar beets. As a result of their properties and complete renewability, PLA polymers have found applications in many industries ranging from food packaging and cutlery, compostable yard bags, fibres and textiles, scaffolds, and many others. The production of PLA polymers and their applications are therefore expected to grow exponentially in the coming years as they gradually replace synthetic polymers. However, there are still many performance-related challenges associated with PLA polymers that limit their widespread application. These challenges include poor toughness, poor melt strength, low thermal stability, low heat deflection temperature (HDT), slow degradation rate that becomes a problem in PLA-based implants and disposal of consumer products, hydrophobicity that lowers cell affinity and may cause inflammatory responses in biomedical applications, and lack of reactive functional groups that limit surface modifications [3]. A growing interest in polymer science has therefore been to enhance the properties of PLA polymers by modification through copolymerisation and blending with elastomers, biodegradable and petroleum-based non-biodegradable polymers and introducing additives to improve flame retardation, photostability, heat stability, and impact resistance. One fascinating and relatively unexplored modification has been the development of PLA-based composites through reinforcement with nanoscale fillers and natural and synthetic fibres. These biocomposites have been reported to possess improved mechanical properties capable of competing with petroleum-based polymers thus providing renewable, biodegradable, and environmentally friendly alternatives to the market. In this chapter, an insightful review into the synthesis methods, properties, and applications of these biocomposites is presented, with interest in particulate-reinforced PLA composites.

8.2 STRENGTHENING OF POLYLACTIC ACIDS USING AGRICULTURAL WASTE NANOPARTICLES

8.2.1 POLYLACTIC ACID

8.2.1.1 Structure of Polylactic Acid

Polylactic acid (PLA) is an environmentally friendly biopolymer with high recyclability, compostability, and biodegradability. This polymer is obtained from various renewable feedstock sources that mostly constitute materials derived from the processing of plants. Conventional polymer processing techniques such as those of moulding, extrusion, and forming can be used to process PLA into various shapes and products. PLA has also been used as a feedstock material for 3D printing. In comparison to polymers derived from petroleum sources, PLA offers up to 55% energy savings during production, which results in major cost savings [4]. These are some of the reasons PLA has received great attention as an alternative sustainable polymer in the modern industries. In a technical sense, PLA is an aliphatic biodegradable polymer produced from the polymerisation of lactic acid derived. Lactic acid (2-hydroxy propionic acid) is mostly found in fermented dairy products like sour cream, cheese, and yoghurt [5]. On an industrial scale, it is produced from the fermentation simple sugars sourced from different carbohydrate feedstock like corn starch, potato starch, dairy, and in some instances agricultural waste [5]. These simple sugars are broken down through the action of fungi or bacteria, each depending on the type of feedstock used. Lactic acid exists as D (-) lactic acid or (D-lactic acid) and L (+) lactic acid or (L-lactic acid) optical isomers, as shown Figure 8.1. Lactic acid contains both hydroxyl and carbonyl groups and the monomers are joined together ester linkages to

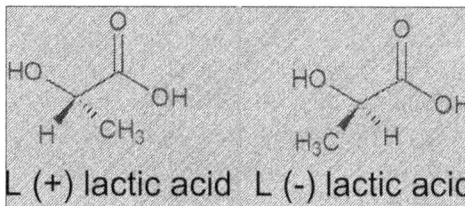

FIGURE 8.1 Structural isomers of lactic acid.

produce polylactic acid. The biopolymer can either be a homopolymer consisting only of D-lactic or L-lactic acid monomers. Alternatively, a mixture of the two isomers can be formed resulting in a heteropolymer produced from monomers known as meso-lactide or racemic lactide. The ester linkages can easily be broken down through hydrolysis and bacterial action – a property which has given polylactic acid great attention as an engineering material throughout past decades [6].

Polylactic acid can be produced by different routes. Polycondensation reaction of lactic acid monomers is the simplest and most widely known method [7]. However, the polymerisation reaction in this method quickly reaches equilibrium with the de-polymerisation reaction, which means that long reaction times at high temperatures are required for substantial yield [8]. Another major implication of this is that this method is only limited to producing low molecular weight polylactic acid which narrows the application base of PLA. Moreover, as water molecules are produced during the polycondensation reaction, they need to simultaneously be removed during the reaction to avoid hydrolysis or depolymerisation of polymerised PLA chains. In order to produce PLA of higher molecular weight, the ring opening polymerisation reaction was developed. This method first produces oligomers from the lactic acid monomers that are then catalytically converted to cyclic dimers of lactic acid known as lactide [9]. These lactides are used as monomers for polymerisation and, by controlling their purity and the synthesis conditions, a wide variety of high molecular weight polylactic acid chains can be produced with unique and desirable properties. The superior advantage of ROP over condensation methods has allowed it made the number one PLA synthesis routes for the world's leading producers.

The crystallinity of PLA plays a significant role in determining polymer properties such as hardness, elastic modulus, tensile strength, impact strength, and melting point amongst many others. Generally, crystallinity is greatly influenced by the stereochemistry of lactic acid monomer units, the molecular weight of polymer chains, the degree of branching from the backbone chain and the thermal history of the polymer. The degree of crystallinity can be adjusted by fine-tuning the compositional ratio of the D-lactic and L-lactic acid isomers in the biopolymer. A higher crystalline (semi-crystalline) PLA biopolymer can be obtained by increasing the content of L-lactic acid to 90% and above. Increasing the content of D-lactic acid to beyond 10% decreases the crystallinity and results in an amorphous biopolymer. Moreover, the glass transition temperature Tg, and melting temperature Tm of the biopolymer also decreases. The Tg is one of the most important physical properties of polymers as it directly influences other physical properties and mechanical and rheological properties. Therefore, by controlling the amount of the D-lactic acid isomer in PLA biopolymers, the final properties of PLA can be predicted from the Tg. In the case of semi-crystalline PLA biopolymers with less D-lactic acid isomer, both the Tg and the Tm are used in predicting the final properties of PLA. In addition to the composition of PLA biopolymer, the molecular weight also influences the Tg and Tm of the polymer. It has been shown that the Tg and Tm increases with molecular weight until the two temperatures reach a plateau at a molecular weight of 100 kg/mol [10].

Despite having numerous advantages over conventional petroleum-based polymers and other biopolymers, PLA has some drawbacks which limit its widespread application. These drawbacks are briefly discussed. Low ductility and toughness – The stiffness and tensile strength of PLA can compete with that of petroleum-based polymers. However, it has poor ductility and toughness, experiencing failure at elongations of less than 10% [4]. Therefore, PLA becomes unsuitable in high loading applications that require a significant degree of plastic deformation.

Slow degradation rate – The degradation rate of PLA occurs through an auto-catalytic hydrolytic breakdown of ester linkages which reduce the molecular weight of the biopolymer [3]. Further breakdown is then carried out by microorganisms to produce carbon dioxide and water. This degradation process is controlled by various factors such as crystallinity, molecular weight, reactivity of polymer chains, moisture levels, and the temperature of the environment amongst many other factors [3]. As a result, the degradation of PLA can take up to several years depending on these factors which is of great concern especially in packaging and biomedical applications.

Slow crystallization rate – The slow crystallisation rate of PLA is very slow, which results in lower degrees of crystallinity consequently affecting many phy-sical, mechanical, thermal, and rheological properties [11]. Moreover, this means that production time becomes longer with conventional processing techniques such as moulding and extrusion.

Hydrophobicity – The hydrophobic nature of PLA greatly affects the biomedical industry as it limits its compatibility with biological fluid systems that are inherently hydrophilic [3]. PLA biomedical devices can therefore cause inflammations when is introduced into living organisms.

Low dimensional stability – The Tg range of PLA is 50–60°C, which is relatively lower than that of competing petroleum-based polymers. As the Tg range determines the allowable service temperatures of most applications, this limits the use of PLA in high-temperature environments exceeding Tg. Beyond Tg, the physical and me-chanical properties of polymers begin to drastically change as they become more rubbery. Lower Tg values are often described by low heat distortion or deflection temperature (HDT).

To address the drawbacks and widen the application of PLA, various chemical and physical treatments or modifications have been undertaken in recent decades. The nature of these modifications is oftentimes governed by the intended applica-tion of the biopolymer. One interesting sustainable and environmentally friendly modification to PLA is the use of agricultural waste products as fillers to improve some of the already discussed drawbacks. Agricultural wastes have been used as PLA reinforcements to produce biocomposites; however, the focus has been on fibre composites and little attention has been given to particulate-reinforced com-posites, which might offer unique and superior composite properties.

8.2.1.2   Agricultural Waste as Reinforcement

Most of the agricultural waste products used as reinforcements in biopolymers are derived from plants. This is primarily due to the high content of cellulose in plant wastes which has countlessly been shown to significantly improve mechanical properties like tensile strength and elastic modulus. Plant agricultural wastes, like those shown in Table 8.1, have gained widespread popularity as reinforcements for polymeric materials as opposed to animal waste products [12]. However, factors

TABLE 8.1
**Different Types of Agricultural Waste That Are
Sources of Polymer Reinforcement**

Type	Examples
Wheat	Straw, husk
Sorghum	Husk, straw, cobs, stover, leaves
Rice	Husk, hull, straw, and stalk
Millet	Husk, straw, cobs, stover, leaves
Coconut	Fronds, husk, shells
Coffee	Hull, husk, ground
Cotton	Stalks
Peanut	Shells
Sugarcane	Bagasse
Groundnuts	Shells
Delonix regia	Seeds, pods, fibres
Date fruit	Seeds
Maize	Stalk, husk, cob

such as poor interfacial adhesion to the polymeric matrix, thermal instability, and high moisture absorption are detrimental to properties thereby posing major challenges to the performance and application of biocomposites produced from these types of reinforcements. Plant agricultural waste is a lignocellulosic material containing cellulose, hemicellulose, and lignin as their major components amongst many other components. The composition of these three major components greatly varies from one waste to the other depending on the conditions in which the plants were grown, the post-harvest processing of the plants, and waste. The properties of composites depend on both the properties of both the reinforcing and the matrix material. To understand these properties, the chemical structure of the reinforcement must be considered.

8.2.2 CELLULOSE

Cellulose is considerably the most ubiquitous biopolymer in the world. This polymer is mostly found in plants and is mainly responsible for the structural stability of plant cell walls which has had many implications in structural engineering applications. In plants, cellulose commonly exists in a matrix primarily made of hemicellulose and lignin with a variety of other minority components such as ash, phenolics, peptides and proteins [13]. Agricultural waste derived from plants is rich in cellulose, as reported in [13], and typical contents of cellulose in agricultural waste like algae (green), cotton, and cornstalk are 20%–40%, 80%–95%, and 39%–47%, respectively.

The chemical structure of cellulose is made from D-glucopyranose monomer units covalently bound together by an oxygen atom at the first and fourth carbon

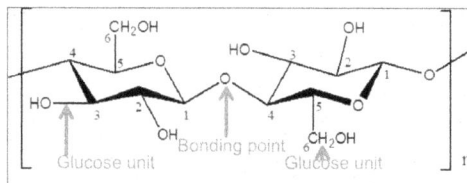

FIGURE 8.2 Chemical structure of cellulose molecule.

atoms of respective adjacent glucose rings, as seen in Figure 8.2 [14]. These bonds are the so called β-(1, 4)-glycosidic linkages which form linear chains (poly-saccharides) of repeating glucose rings ranging from 500–15,000 units [13]. Each of these glucose monomer units contains three hydroxyl functional groups of which one is a primary group while the other two are secondary hydroxyl groups. Within a single cellulose molecule, the β-(1, 4)-glycosidic bonding promotes intramolecular hydrogen interactions between a hydroxyl group of one unit and a ring oxygen atom of a neighbouring unit. These interactions promote structural stability within cellulose molecules giving them rigid structures. Moreover, these hydrogen interactions facilitate the packing of multiple parallel cellulose molecules forming the so-called fibrils with diameters ranging from 5 to 50 nm and lengths in the order of microns [15]. Within the fibrils, both crystalline and amorphous regions exist where cellulose molecules arrange themselves in a highly ordered and disorder manner, respectively. The crystalline cellulose regions which are hydrophobic can be thought of as a filler in an amorphous hydrophilic cellulose matrix. The crystallinity of cellulose can range anywhere from 50% to 80% in natural cellulose and can be modified through a series of physical and physiochemical treatments thus altering its hydrophobic and hydrophilic characters [16]. Several chemical and mechanical treatments have been done on cellulose fibrils to extract these crystalline regions at nano- and micro-scales to be used in applications ranging from food packaging, pharmaceutical, and photonics amongst many others.

8.2.3 LIGNIN

The woody tissues of plant cell wall are usually made up of 2%–40% lignin [13]; most of the lignin (60%–80%) is in the secondary wall of plant tissues [17]. Lignin is widely accepted to be an amorphous heteropolymer made primarily from the polymerisation of phenylpropane units of p-coumaryl, coniferyl, and sinapyl alcohols with structures shown in Figure 8.3. These alcohols and their derivatives are most joined through β-O-4 aryl ether linkages. Despite having hydroxyl groups, lignin is hydrophobic, a property that eases the transport of water and nutrients within the plants and acts as a protective barrier from attack by microorganisms, flame, ultra-violet (UV) radiation, and other harsh weather conditions [18,19]. Moreover, lignin is known to be covalently bound to hemicellulose side groups and other cell wall structural polymers, thus forming an amorphous matrix that covers cellulose fibrils [20], as shown in Figure 8.4. This provides the additional mechanical strength to cell walls and gives plants structural support.

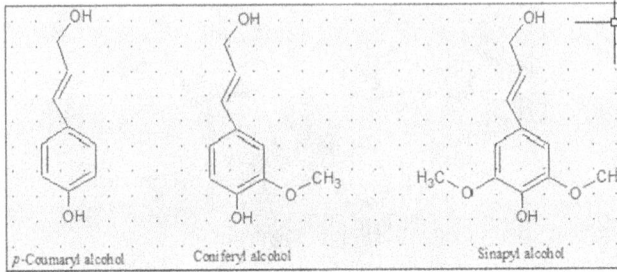

FIGURE 8.3   Building blocks of lignin.

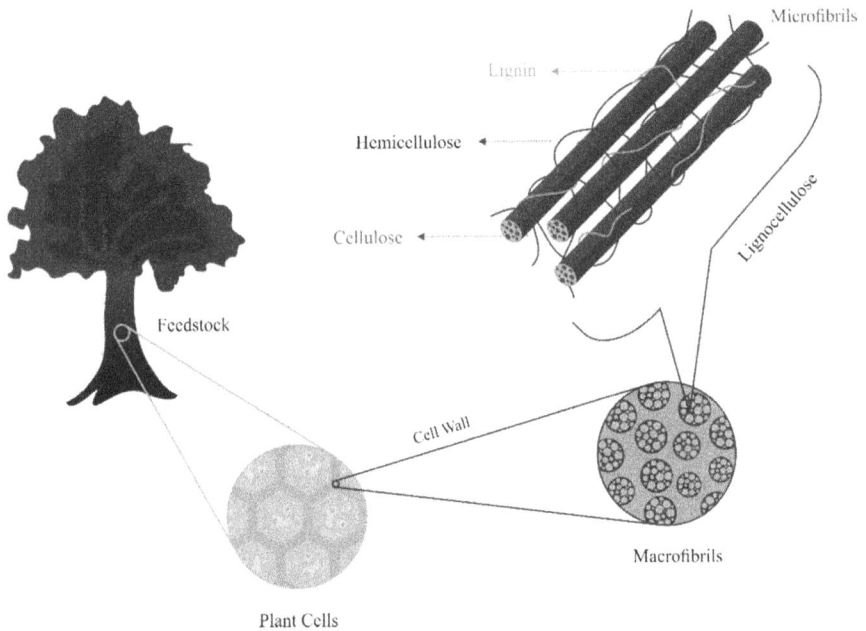

FIGURE 8.4   Schematic diagram showing the interaction between cellulose, hemicellulose, and lignin.

8.2.4   HEMICELLULOSE

Hemicellulose constitutes approximately 20–35 wt% of dry weight of plant cell walls. Unlike cellulose which is only made from glucose units, hemicellulose in addition to glucose consists of many other different monosaccharide units such as xylose, galactose, and mannose each in varying amounts depending on the species of plant, their stage of development, the environment in which they grow and the type of tissues that exist within them [21,22]. These simple sugars orient themselves randomly forming amorphous heteropolysaccharide structures of monomer units ranging from 100 to 200, which is significantly lower than those that form cellulose. Consequently, hemicellulose is shorter, highly branched and has lower strengths than cellulose. Due to its amorphous nature, hemicellulose is hydrophilic and therefore tends to facilitate water retention in plant cells, a quality that is highly desirable in plants but may be detrimental to composites engineering materials made with fillers containing hemicellulose. On the brighter side, this quality allows hemicellulose to be easily extracted from crystalline cellulose through hydrolysis with dilute bases and acids or through dissolution in high-temperature water. The nature of chemical interactions between hemicellulose, lignin, and cellulose have been the subject of debate for many decades. However, it is generally accepted that hemicellulose interacts with cellulose through hydrogen bonding and with lignin through covalent bonding [23].

8.2.5 Extractives

In addition to cellulose, hemicellulose, and lignin, there are other several organic and inorganic components that are found in lignocellulose cell walls. These components are not chemically bound and can be extracted through various chemical treatments using solvents [24]. Ash is the inorganic component of lignocellulosic materials and is mainly composed of silica with other oxides of alkali and alkali earth metals. These minerals can be introduced into plants through mineral uptake or soil contamination during harvesting, storage, and handling [24]. Organic components are composed of a mixture of phenols, proteins, ethers, alkaloids, and many other organic components.

8.3 PROPERTIES OF POLYLACTIC ACID

8.3.1 Mechanical Properties

Polylactic acid (PLA) composites reinforced with particulate agricultural waste reported in various studies have shown that factors such as the particle size of the reinforcement, the interfacial adhesion bond strength between the reinforcement and the polymer matrix, and the amount of reinforcement loaded into the polymer matrix greatly influences the mechanical properties of the composites.

One of the major drawbacks of using lignocellulose agricultural waste as a reinforcement for PLA is the poor interfacial compatibility between the two phases. These two phases may interact with each other in three different ways: (a) mechanically through interlocking, (b) physically through secondary or intermolecular attractions, and/or (c) primary through chemical bonding [25]. However, lignocellulosic materials are inherently polar and extremely hydrophilic due to the abundance of hydroxyl and carbonyl groups found in lignin, hemicellulose, amorphous cellulose, pectin, and other amorphous components, while on the other hand, the PLA matrix is non-polar and hydrophobic [26]. The presence of many hydroxyl groups in lignocellulosic agricultural waste results in secondary hydrogen interactions between particles of the reinforcing phase which leads to agglomeration. Oftentimes, this is observed microscopically as poor dispersion of particles in the polymer matrix [27]. Moreover, these hydroxyls facilitate water absorption and retention, a property that leads to poor dimensional stability because of periodic swelling of the reinforcing phase.

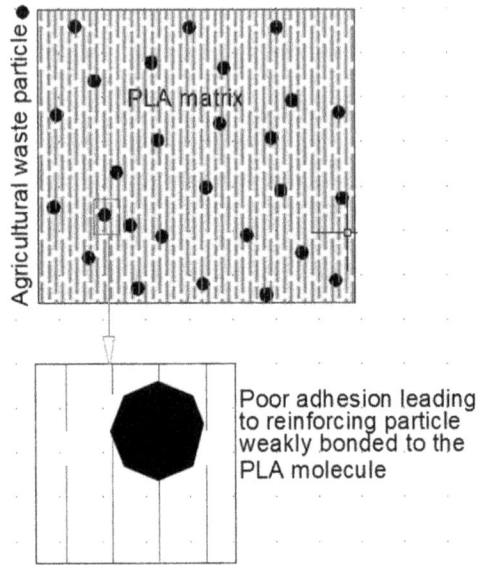

FIGURE 8.5 Two-dimensional geometric model demonstrating weak interfacial adhesion between untreated lignocellulosic particles and polylactic (PLA) molecules.

The great difference in polarity and surface energy between lignocellulosic reinforcing phases and the PLA polymer matrix results in weak interfacial adhesion bond strength and poor wettability, as seen in Figure 8.5. This has adverse effects on the mechanical properties of the resulting composites. This lowers the capability of the polymer matrix to transfer load to the reinforcing phase which becomes detrimental to any strength related properties of the resulting composite [28]. Generally, natural lignocellulosic agricultural waste is only useful as a PLA reinforcing phase at low applied loads when the effect of weak interfacial adhesion bond strength is negligible. This is seen by the great improvement in the stiffness of the composite. However, in cases where the applied load is significantly high, interfacial adhesion bond strength becomes a crucial factor to consider and introduces natural lignocellulosic agricultural waste into the PLA matrix seems futile in improving the composite's strength [28]. At higher applied loads, the weak interfacial adhesion bond strength results in debonding of the reinforcing phase from the matrix and particle pull-outs become frequent, which drastically reduces the strength of the composites.

8.3.1.1 Stiffness (Elastic Modulus)

The stiffness of lignocellulosic agricultural waste is higher than that of PLA; therefore, reinforcing PLA with lignocellulosic agricultural waste generally results in improved stiffness of the composite relative to unreinforced PLA, as already mentioned. Groundnut shell nanoparticles having an average particle size range of between 50 nm to 100 nm were used as reinforcements in the fabrication of PLA composites by Olajide et al. [29]. The elastic modulus of the composite was relatively higher with approximate values of 2.31 GPa as compared to the 2.23 GPa of the unreinforced PLA.

Barczewski et al. [30] also reinforced PLA using chestnut shell waste of approximately 10 μm as a filler material. The composites showed a significant increase in the elastic modulus at higher concentrations of the agricultural waste filler material. Composites with filler concentrations below 10 wt% had elastic modulus values like that of unreinforced PLA of approximately 3.1 GPa. However, increasing the content of the agricultural waste filler to 30 wt% raised the elastic modulus to a significant value of 3.8 GPa. Similar works which used ground agricultural waste such as wheat straw as a filler material for PLA composites also showed that concentrations above 10 wt% provide a practical improvement in the elastic modulus. The improvement of elastic modulus in agricultural waste–reinforced PLA composites is mainly attributed to the increased crystallinity of the composites. Moreover, agricultural waste products are typically more rigid than PLA, thereby consequently improving the stiffness of the polymer when introduced as reinforcements.

8.3.1.2 Tensile Strength

Tensile strength refers to the maximum amount of uniaxial tensile load per unit area that a material can withstand before failure. In all particulate-reinforced composites, this property greatly depends on the ability of the soft polymer matrix to effectively transfer the applied load to the reinforcement [28]. Wu et al. [27] varied the concentration of rice husks in PLA from 10 wt% to 40 wt%, at an interval of 10%, the tensile strength drastically decreased in a linear manner from approximately 45 MPa of the unreinforced PLA to about 20 MPa at the maximum concentration of 40 wt%. Barczewski et al. [30] also observed a decrease in tensile and impact strength as the concentration of chestnut shell waste reinforcement in PLA was varied from 2.5 wt.% to 30 wt%. At the maximum concentration of 30 wt%, the tensile and impact strengths of the composites were approximately 40.8 MPa and 2.7 J/mm^2, respectively, relative to the unreinforced PLA with tensile and impact strengths of 69.0 MPa and 6.5 J/mm^2, respectively.

A few surfaces chemical treatments on both agricultural wastes and the polymer have therefore been investigated by various scholars with the primary aim of strengthening the adhesion bond at the matrix/reinforcement interface, consequently improving the load-bearing capacity of the composites. Although some of these treatments may result in the introduction of chemical species that are not eco-friendly to the bio-polymer fabrication process, their significance is paramount in developing bio composites having properties competitive with those of petroleum-based composites. Wu et al. [27] fabricated rice husk–reinforced PLA composites (RH/PLA) without any prior surface treatments done on the raw materials and another set of composites produced from PLA grafted with acrylic acid as a coupling agent and rice husks treated in a solution of cross-linking agents containing tetraethyl orthosilicate and lactic acid catalyst in tetrahydrofuran. A comparative study on the strength of the two sets of composites was carried out and for the same content of reinforcement, the treated composites showed a significantly higher tensile strength than those without any prior treatment. With pre-surface treatment of the PLA and the rice husks, the wettability of the two is improved. The improvement could be attributable to strong adhesion of the reinforcement to the matrix which allowed easy transfer of load from the matrix to the reinforcing phase.

Other prior treatments of lignocellulosic materials may take a thermal approach at elevated temperatures to produce nanosized charcoal and ash particles that are rich in carbon and silica, respectively. These by-products have excellent mechanical properties and are oftentimes preferred as PLA reinforcements. The strength of PLA composites reinforced with thermally treated agricultural waste tends to increase with the content of the reinforcement up-to a certain threshold concentration. Ho et al. [31] investigated the tensile strength of PLA composites reinforced with bamboo charcoal particles (BC/PLA) at concentrations of 2.5 wt%, 5 wt%, 7.5 wt%, and 10 wt%. The tensile strength of the BC/PLA composites steadily increased before reaching a maximum of approximately 51 MPa at a concentration of 7.5 wt%. A further increase in the concentration to 10 wt% led to a drop in the tensile strength to about 37 MPa. In another study, nano-sized groundnut shell ash particles were introduced into the PLA matrix, the optimal concentration was achieved at 20 wt% when it was varying from 10 wt%, 20 wt%, to 30 wt%.

8.3.2 THERMAL STABILITY

The glass transition temperature (Tg) is a temperature at which the physical properties of amorphous polymers change from a glassy state to a rubbery state [1]. This temperature is characteristic of amorphous polymers or amorphous regions in semi-crystalline polymers and is very crucial in determining the allowable temperature range for any intended end-use of the polymer and its composites. At temperatures above Tg, the mobility of amorphous polymer chains is increased, and they begin to display rubbery flow, while temperatures below Tg result in stiffer and more rigid polymer chains, as shown in Figure 8.6. The addition of agricultural chestnut shell and rice husks waste to PLA has been shown to increase the Tg and

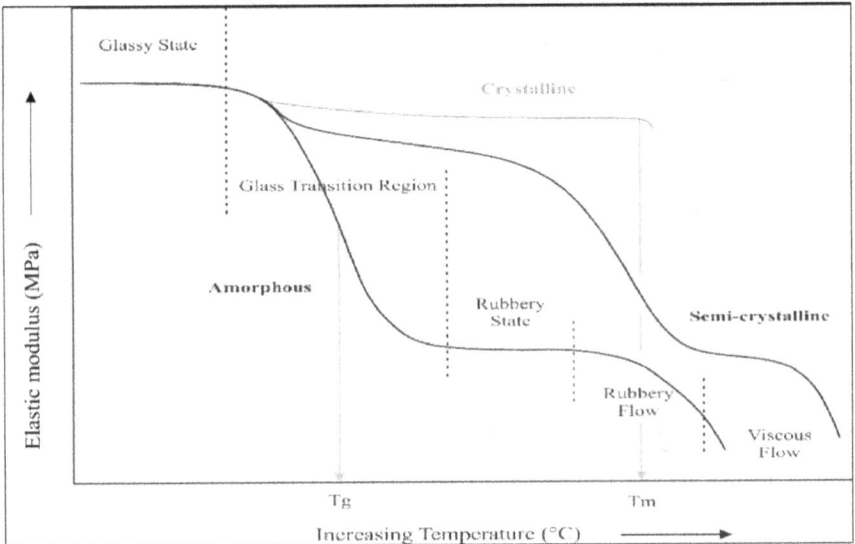

FIGURE 8.6 Thermo-mechanical behaviour of amorphous and semi-crystalline polymers.

decrease the melting point (Tm) of the resulting PLA composites measured by differential scanning calorimetry (DSC).

The high-temperature application of PLA is mainly limited by its relatively low HDT. Increasing the crystallinity of the polymer could possibly increase its HDT. However, PLA has a low rate of crystallisation, which does not only affect the HDT but also affects the production time through conventional extrusion processes. The addition of agricultural waste fillers to PLA matrix has the potential of lowering the surface free energy barrier required for nucleation [32]. This will promote the formation of ordered arrangement of polymer chains during cooling from melt, which in turn would increase crystallinity, consequently increasing the HDT. Chestnut shell waste particles were shown to increase the crystallinity of PLA with increasing concentration of the filler [14]. At a maximum filler concentration of 30 wt%, the percent degree of crystallinity (Xc) was calculated at 49.2% relative to unreinforced PLA having 41.6%. This increase in crystallinity with addition and increase of filler concentration is, however, not generic and will depend on the type of filler being used, the type of interactions that occur at the filler-matrix interface.

8.4 APPLICATIONS OF POLYLACTIC ACID

8.4.1 Food Packaging Applications

Food packaging materials require a combination of mechanical, thermal, optical, and anti-bacterial properties amongst many others. Development of biodegradable and sustainable food packaging materials has profound economic and environmental implications. The shelf life of food products can be prolonged whilst environmental footprint is kept to a minimum. Biodegradable polymers like PLA have been used for production of food films, trays, bowls, and paper bags [33]. However, poor moisture and air barrier properties are still major challenges that limit their food packaging applications. Food packaging materials should be able to regulate the moisture within food packets and maintain a certain concentration of a mixture of gasses like oxygen, nitrogen, and carbon dioxide inside the food packets. This ensures that food products remain flesh and do not spoil before the end of their shelf life. One way of improving barrier properties of biodegradable polymers like PLA is to increase their crystallinity [34]. The addition of agricultural waste fillers to PLA as nucleating agents are a potential alternative to improving the crystallinity of the resulting composite thus improving their gas and moisture barrier properties. However, such fillers still require great improvements to optimize their nucleating effects. Moreover, there are some concerns that the addition of agricultural waste fillers to PLA might affect its excellent optical transparency, which is greatly desirable for food packaging. Lignocellulosic materials are promising as they have not only been shown to improve crystallinity but to also preserve the transparency of PLA composites [35].

8.4.2 Application in Automotive Industry

Plastics amount to approximately 11% of the total construction materials in a modern sedan. It is believed that all these materials can be replaced by biopolymers

[36]. PLA composites can be used for construction of various exterior and interior components of automobiles such as air filter boxes, door panels, upholstery, carpets and floor mats, and dashboards amongst many others [36,37]. These components experience service temperatures mainly ranging from –40°C to 85°C depending on the climatic conditions of the country [37]. When the temperatures increase above Tg, the thermal stability of the polymer is lost as strength and rigidity decrease. Addition of agricultural waste fillers to PLA has been shown to increase Tg which improves the thermal stability of PLA. The nucleating agent effect of agricultural waste fillers increases the crystallinity of PLA composites, thus resulting in higher HDT. Moreover, the crystallisation rate of PLA is greatly increased, which eases processability of the biopolymer and reduces production times of automobile components. One of the most important considerations in the automotive industry is weight and energy savings. By adding agricultural waste to PLA, the rigidity of the biopolymer can be improved, which allows for fabrication of components with thinner walls. As a result, significant weight savings can be achieved, thereby reducing fuel consumption without losing the required strength of components. However, improvements in terms of ductility and flexibility of PLA-agricultural waste composites are still necessary to be able to compete with petroleum-based polymers.

8.4.3 Agricultural Mulching Materials

PLA composites reinforced with agricultural waste have great prospects as crop mulching materials. Mulching is often employed in horticulture to reduce evaporation of water and keep the soil warm thereby facilitating the development of crop rooting systems. Petroleum-based polymers such as low-density polyethylene have been at the forefront of mulching materials. However, their superb mechanical properties make them extremely durable hence allowing them to linger on in the environment even after they have exhausted their service life. A promising alternative is to use biodegradable polymers like PLA filled with lignocellulosic agricultural waste particles which will not only improve the mechanical properties of PLA but also increase the rate of its degradation. Both PLA and lignocellulosic agricultural waste fillers can break down under humid conditions through hydrolysis and/or through the action of microbes. However, lignocellulosic agricultural waste fillers are more susceptible to breakdown, which accelerates the degradation rate of PLA [38].

8.5 CONCLUSIONS AND SUMMARY

PLA is at the forefront of biodegradable polymers with the capacity to compete with conventional petroleum-sourced polymers in certain applications like industrial packaging, biomedical devices, and automotive. However, there are still many performance challenges with PLA polymers that narrow down their widespread adoption. These challenges include poor toughness, slow crystallisation rate, slow degradation rate, and low thermal stability. Improving and tailor making its chemical, physical, thermal, and biodegradable properties amongst many others through the addition of reinforcement filler materials could broaden the application base of PLA in already existing industries where petroleum polymers have dominated for long. Moreover, there are possibilities of discovering novel and innovative applications of PLA based on the performance of composites. Agricultural waste is an abundant, renewable, sustainable, and biodegradable resource that has desirable chemical and mechanical properties. These properties could be harnessed to improve the properties of PLA by using particulate agricultural waste as a reinforcing filler material. Nut shells, rice husks, and wheat straw are amongst the many examples of filler materials that have been used to improve the properties of PLA. However, since most of the agricultural waste fillers used for reinforcement are lignocellulosic materials and therefore hydrophilic in nature, prior chemical and/or physical treatments need to be carried out to improve the compatibility of these materials with the hydrophobic PLA matrix. Only through such treatments can PLA agricultural waste composites have properties capable of competing with petroleum-based polymers. Factors such as the type, particle size, composition, and amount of filler loading of agricultural waste in PLA agricultural waste composites are crucial as they greatly influence composite properties. Therefore, the challenge to optimise these factors remains in order not to only achieve the maximum performance but also to make strides towards a greener engineering material and the battle with climate change.

REFERENCES

[1] Shrivastava, A. (2018). *Introduction to Plastics Engineering*. Elsevier, Cambridge.
[2] Harding, K.G., Dennis, J.S., Blottnitz, H.v., and Harrison, S.T.L. (2007). Environmental Analysis of Plastic Production Processes: Comparing Petroleum-Based Polypropylene and Polyethylene with Biologically-Based Poly-β-Hydroxybutyric Acid Using Life Cycle Analysis. *Journal of Biotechnology*, **130**(1), 57–66.
[3] Farah, S., Anderson, D.G., and Langer, R. (2016). Physical and Mechanical Properties of PLA, and Their Functions in Widespread Applications—A Comprehensive Review. *Advanced Drug Delivery Reviews*, **107**(1), 367–392.
[4] Rasal, R.M., Janorkar, A.V., and Hirt, D.E. (2010). Poly(lactic acid) Modifications. *Progress in Polymer Science*, **35**(3), 338–356.
[5] Komesu, A., Oliveira, J.A.R.D., Martins, L.H.D.S., Maciel, M.R.W., and Filho, R.M. (2017). Lactic Acid Production to Purification: A Review. *BioResources*, **12**(2), 4364–4383.
[6] Elsawy, M.A., Kim, K.-H., Park, J.-W., and Deep, A. (2017). Hydrolytic Degradation of Polylactic Acid (PLA) and Its Composites. *Renewable and Sustainable Energy Reviews*, **79**(1), 1346–1352.
[7] Balakrishnan, R., Rajaram, S.K., and Sivaprakasam, S. (2020). Biovalorization Potential of Agro-forestry/Industry Biomass for Optically Pure Lactic Acid Fermentation: Opportunities and Challenges in *Biovalorisation of Wastes to Renewable Chemicals and Biofuels*, Elsevier, Amsterdam, pp. 261–276.
[8] Hu, Y., Daoud, W.A., Cheuk, K.K.L., and Lin, C.S.K. (2016). Newly Developed Techniques on Polycondensation, Ring-Opening Polymerization and Polymer Modification: Focus on Poly(Lactic Acid). *Materials* (Basel), **9**(3), 133.
[9] Thomas, C.M. (2010). Stereocontrolled Ring-Opening Polymerization of Cyclic Esters: Synthesis of New Polyester Microstructures. *Chemical Society Reviews*, **39**(1), 165–173.

[10] Saeidlou, S., Huneault, M.A., Li, H., and Park, C.B. (2012). Poly(lactic acid) Crystallization. *Progress in Polymer Science*, **37**(12), 1657–1677.

[11] Jiang, L., Shen, T., Xu, P., Zhao, X., Li, X., Dong, W., Ma, P., and Chen, M. (2016). Crystallization Modification of Poly(lactide) by Using Nucleating Agents and Stereocomplexation. *E-Polymers*, **16**(1), 1–13.

[12] Abba, H.A., Nur, I.Z., and Salit, S.M. (2013). Review of Agro Waste Plastic Composites Production. *Journal of Minerals and Materials Characterization and Engineering*, **1**(5), 271–279.

[13] Saini, J.K., Saini, R., and Tewari, L. (2015). Lignocellulosic Agriculture Wastes as Biomass Feedstocks for Second-Generation Bioethanol Production: Concepts and Recent Developments. *3 Biotech*, **3**(4), 337–353.

[14] Hussin, M.H., Husin, N.A., Bello, I., Othman, N., Bakar, M.A., and Haafiz, M.K.M. (2018). Isolation of Microcrystalline Cellulose (MCC) from Oil Palm Frond as Potential Natural Filler for PVA-LiClO4 Polymer Electrolyte. *International Journal of Electrochemical Science*, **13**, 3356–3371.

[15] Moon, R.J., Martini, A., Nairn, J., Simonsen, J., and Youngblood, J. (2011). Cellulose Nanomaterials Review: Structure, Properties And Nanocomposites. *Chemical Society Reviews*, **40**(7), 3941–394.

[16] Ioelovich, M. 2021. Adjustment of Hydrophobic Properties of Cellulose Materials. *Polymers*, **13**(8), 1241.

[17] Sun, R.-C. (2010). *Cereal Straw as a Resource for Sustainable Biomaterials and Biofuels*. Elsevier, Amsterdam.

[18] Zhang, Y., and Naebe, M. (2021). Lignin: A Review on Structure, Properties, and Applications as a Light-Colored UV Absorber. *ACS Sustainable Chemistry & Engineering*, **9**(4), 1427–1442.

[19] Shikinaka, K., Nakamura, M., and Otsuka, Y. (2020). Strong UV Absorption by Nanoparticulated Lignin in Polymer Films with Reinforcement of Mechanical Properties. *Polymer*, **190**, 122254.

[20] Geies, A., Abdelazim, M., Sayed, A.M., and Ibrahim, S. (2020). Thermal, Morphological and Cytotoxicity Characterization of Hardwood Lignins Isolated by In-Situ Sodium Hydroxide-Sodium Bisulfate Method. *Natural Resources*, **11**(10), 427–438.

[21] Brunner, G. (2014). *Hydrothermal and Supercritical Water Processes*. Elsevier, Amsterdam.

[22] Machmudah, S., Wahyudiono, Kanda, H., and Goto, M. (2017). Water Extraction of Bioactive Compounds – From Plants to Drug Development. in *Hydrolysis of Biopolymers in Near-Critical and Subcritical Water*, Elsevier, Amsterdam, pp. 69–107.

[23] Naidjonoka, P., Hernandez, M.A., Pálsson, G.K., Heinrich, F., Stålbrand, H., and Nylander, T. (2020). On the Interaction of Softwood Hemicellulose with Cellulose Surfaces in Relation to Molecular Structure and Physicochemical Properties of Hemicellulose. *Soft Matter*, **16**(30), 7063–7076.

[24] Hörhammer, H., Dou, C., Gustafson, R., Suko, A., and Bura, R. (2018). Removal of Non-structural Components from Poplar Whole-Tree Chips to Enhance Hydrolysis and Fermentation Performance. *Biotechnology for Biofuels*, **11**(1), 1–12.

[25] Orue, A., Eceiza, A., and Arbelaiz, A. (2018). Pretreatments of Natural Fibers for Polymer Composite Materials in *Lignocellulosic Composite Materials*, S. Kalia, Editor, Springer, pp. 137–175.

[26] Zhai, S., Liu, Q., Zhao, Y., Sun, H., Yang, B., and Weng, Y. (2021). Research Progress in Modification of Poly (Lactic Acid) by Lignin and Cellulose. *Polymers*, **13**(5), 776.

[27] Wu, C.-S. and Tsou, C.-H. (2019). Fabrication, Characterization, and Application of Biocomposites from Poly(lactic acid) with Renewable Rice Husk as Reinforcement. *Journal of Polymer Research volume*, **26**(2), 1–9.

[28] Fu, S.-Y., Feng, X.-Q., Lauke, B., and Mai, Y.-W. (2008). Effects of Particle Size, Particle/Matrix Interface Adhesion and Particle Loading on Mechanical Properties of Particulate–Polymer Composites. *Composites Part B: Engineering*, **39**(6), 933–961.

[29] Olajide, O.S., Yaro, S.A., Asuke, F., and Aponbiede, O. (2017). Experimental Correlation between Process Parameters and Tensile Strength of Polylactic Acid/ Groundnut Shell Nanoparticle Biocomposites. *The International Journal of Advanced Manufacturing Technology*, **93**(1), 717–726.

[30] Barczewski, M., Matykiewicz, D., Krygier, A., Andrzejewski, J., and Skórczewska, K. (2018). Characterisation of Poly(lactic acid) Biocomposites Filled with Chestnut Shell Waste. *Journal of Material Cycles and Waste Management*, **20**(1), 914–924.

[31] Ho, M.P., Lau, K.T., Wang, H., and Hui, D. (2015). Improvement on the Properties of Polylactic Acid (PLA) Using Bamboo Charcoal Particles. *Composites Part B: Engineering*, **81**, 14–25.

[32] Beltrán, F.R., Gaspar, G., Chomachayi, M.D., Jalali-Arani, A., Lozano-Pérez, A.A., Cenis, J.L., Orden, M. U. d. l., Pérez, E., and Urreaga, J.M.M. (2021). Influence of Addition of Organic Fillers on the Properties of Mechanically Recycled PLA. *Waste and Biomass Management & Valorization*, **28**(19), 24291–24304.

[33] Jamshidian, M., Tehrany, E.A., Imran, M., Jacquot, M., and Desobry, S. (2010). Poly-Lactic Acid: Production, Applications, Nanocomposites, and Release Studies. *Comprehensive Reviews in Food Science and Food Safety*, **9**(5), 552–571.

[34] Wu, F., Misra, M., and Mohanty, A.K. (2021). Challenges and New Opportunities on Barrier Performance of Biodegradable Polymers for Sustainable Packaging. *Progress in Polymer Science*, **117**, 101395.

[35] Ncube, L.K., Ude, A.U., Ogunmuyiwa, E.N., Zulkifli, R., and Beas, I.N. (2020). Environmental Impact of Food Packaging Materials: A Review of Contemporary Development from Conventional Plastics to Polylactic Acid Based Materials. *Materials*, **13**(21), 4994.

[36] Barillari, F., and Chini, F. (2020). Biopolymers - Sustainability for the Automotive Value-added Chain. *ATZ worldwide*, **122**(11), 36–39.

[37] Arjmandi, R., Hassan, A., and Zakaria, Z. (2017). Polylactic Acid Green Nanocomposites for Automotive Applications. in *Green Biocomposites*, M. Jawaid, S. Mohd Sapuan, and O.Y. Alothman, Editors, Springer, Cham, pp. 193–208.

[38] Hemmati, F., Farizeh, T., and Mohammadi-Roshandeh, J. (2021). Lignocellulosic Fiber-Reinforced PLA Green Composites: Effects of Chemical Fiber Treatment. in *Biocomposite Materials*, M.T. Hameed Sultan, M.S. Abdul Majid, M.R. Mohd Jamir, M.I. Azmi, and N. Saba, Editors, Springer, pp. 97–204.

9 Parquetina nigrescens: Date Seed Pod Particle Polymethylmethacrylate Nanocomposites for Biomedical Applications

Sefiu Adekunle Bello and Sunday Wilson Balogun
Department of Materials Science and Engineering, Kwara State University, Malete, Kwara State, Nigeria

Raphael Gboyega Adeyemo
Gateway (ICT) Polytechnic, Saapade, Ogun State, Nigeria

Timothy Adewale Adeyi
Department of Mechanical Engineering, Lead City University, Ibadan, Oyo State, Nigeria

Kemi Audu and Boluwatife Olukunle
Department of Materials Science and Engineering, Kwara State University, Malete, Kwara State, Nigeria

Kazeem Koledoye Olatoye
Department of Food Science and Technology, Kwara State University, Malete, Kwara State, Nigeria

CONTENTS

DOI: 10.1201/9781003170549-12

9.1 INTRODUCTION

Poly (methyl methacrylate) (PMMA) cement has been a popular and promising biomaterial for decades that finds its noble applications in clinical and biomedical fields like bone replacement, in orthopaedic surgery. However, there are some drawbacks and flaws associated with this biomaterial, such as thermal necrosis to surrounding tissue due to heat produced when MMA polymerises, infection of the prosthesis, aseptic loosening, mechanical and biological shortcomings, and its radiolucency as an orthopaedic material. Efforts to overcome this flaw and draw-back by incorporating additives to PMMA bone cement, such as magnesium oxide, hydroxyapatite, chitosan, barium sulfate, and silica have been intensified by re-searchers to target those problems. Bioceramic, filler, and antibacterial; porogens; biological agents; and mixed additives that are synthetic are incorporated to PMMA to change or alter the properties to make it suitable for use in clinical and biome-dical fields. Besides the high cost of PMMA, synthetic additives contribute to the high cost of artificial implants produced from PMMA-based composites. To address the high cost of the PMMA-based implants and make them more affordable, this chapter proposes uses of agricultural wastes/residue like *Parquetina nigrescens* pods and date seed nanoparticles as potential additives for developing PMMA-based composites for biomedical applications.

9.2 POLYMETHYLMETHACRYLATE (PMMA) AND ITS PROPERTIES

Polymethyl methacrylate $(C_5O_2H_8)_n$ is a tough and rigid transparent thermoplastic organic polymer material with a density of 1.8 g/cm^3, melting point at 160°C, and refractive index of 1.4905 at 589.3 nm [1,2]. PMMA is light and colourless. It has a high optical light transmittance properties, cellular compatibility [3], resistance to UV radiation, and impact mechanical property that makes it suitable for the repair of craniofacial deformities [4,5]. PMMA has wide applications such as antibacterial adhesives for bone and tooth repair [6], used as optical pneumatic actuator/micro sensors [7–9], drug delivery [10], bone cement in clinical applications (orthopaedic surgery [11,12], total joint replacement surgery [13,14], vertebroplasty [15], bone defect reconstruction [15–17], and infectious lesion treatment [18–23], denture base to hold teeth during mastication [24–26], packaging material for the implantable medical device in the bladder [27,28], and coating for corrosion inhibition [29]. PMMA is used in nanotechnology to prepare carbon nanotube polymer composites [30]. The unique mechanical property of this polymer makes it the hardest thermoplastic with a high resistance against scratch or other chemical assaults. However, its resistivity is relatively low against a few chemical groups such as chlorinated and aromatic hydrocarbons, esters, or ketones [31]. Evaluation of PMMA for cytotoxicity for biomedical application and bioactive bone cements were carried out by [32–36]. PMMA is also employed in the moulding process for microstructure fabrication. Oh et al. (2016) [37] employed PMMA in the treatment of osteomyelitis. The effect of nanoparticles and alternative monomer on the exothermic temperature of PMMA bone cement was studied by [38]. Some researchers also synthesised and modified PMMA as biomaterials in orthopaedic surgery and for medical applications [39–46].

9.3   SYNTHETIC ADDITIVES TO POLYMETHYLMETHACRYLATE (PMMA) AND PROPERTIES

Compositions of PMMA bone cements are polymer powder (PMMA, benzoyl peroxide, and/or methacrylate copolymers) and a liquid monomer (MMA). Benzoyl peroxide (BPO) enhances a free radical polymerization reaction [13]. Conventional PMMA bone cements have some flaws and drawbacks such as mechanical, biological shortcomings, and strong exothermic properties from its monomer methyl methacrylate (MMA) because of the amount of heat being produced when MMA undergoes the polymerisation process, and its radiolucency as an orthopaedic material [47,48]. As a result, researchers have intensified their efforts to overcome these flaws and drawbacks by incorporating additives to PMMA bone cement, such as magnesium oxide, hydroxyapatite, chitosan, barium sulfate, and silica. When additives are incorporated to PMMA, it alters the properties and makes it more suitable for the function it is meant/designed for [41]. Haas *et al.* (1975) [49] increased the PMMA ratio to the liquid monomer (MMA); the resultant effect shows reduction in exothermic temperature without change in the strength of the material. Conversely, a decrease in the mechanical strength of the cement is obtained when the PMMA ratio to the liquid monomer is reduced [50,51]. Commercial PMMA cement was doped with MPn [21], which decreases the effect of the strong exothermic reactions during a PMMA setting and decreases the mechanical properties of the PMMA cement, which enhances good clinical practice/requirements. Kim *et al.* (2015) [27] did intravesical implantation of PMMA-coated ball-shaped metal into the bladders of rats to evaluate the biocompatibility of PMMA with a macrophage migratory inhibitory factor (MIF), and inflammatory cytokines in the bladder decrease in inflammatory cytokines such as IL-1, IL-6, and TNF- were observed/noticed which indicate/suggest PMMA is biocompatible with MIF for the implanted medical devices. Stefan Beck and Andreas Boger (2009) [52] investigated different preparation methods for a PMMA/MMA mixture. The flaws or the drawbacks remain unresolved because a suitable level comparable to regular vertebroplasty cement/cancellous bone could not be achieved. Tang *et al.* (2019) [53] carried out research to reduce the elastic modulus of the bone cement to the relatively close level of the cancellous bone by administering dosage of gelatin as the porogen as an additive. It was reported that mechanical and thermal properties were successfully reduced by the porous structure compared to PMMA bone cement. Gelatin as an additive altered the mechanical properties of PMMA as it affected the porosity. Feuser *et al.* (2014) [32] synthesised PMMA nanoparticles to deduce the cytotoxicity for biomedical applications in acute monocytic leukemia cells (THP1) and in human lung adenocarcinoma cells (A549). The result shows it can be used as encapsulate antitumor agents for biomedical application. Harper 1998 [33] published a review paper in which he concluded that bioactive additive can improve PMMA cement. Zhai *et al.* (2018) [54] incorporated a different percentage of Mg microspheres into PMMA to prepare degradable and bioactive PMMA–Mg composite bone cements. The presence of the additive Mg microspheres caused porosity that resulted in a decrease in temperature and an increase in the compressive strength of the composite bone cement.

Research to improve the fracture toughness of an implant–cement interface by evaluating the effects of fibres and loading angles on the interface fracture toughness of implant–cement specimens with and without fibres at the interface implant for orthopaedic applications was carried out using titanium as the implant, poly methyl methacrylate (PMMA) as cement, and polycaprolactone (PCL) as fibre materials [55]. They arrived at a conclusion that fracture toughness of Ti–PMMA samples with fibres was higher compared to those Ti–PMMA samples without fibres for all loading angles and that loading has a significant effect on the fracture toughness of Ti–PMMA samples, which means load-invariant implant–cement interfaces can be created by the application of fibres at the interface. Silica was incorporated into conventional PMMA bone cement which caused an increase in the flexural strength and fracture toughness for orthopaedic and orthodontic applications as disseminated in [56]. Ricker *et al.* (2008) [57] in their study found PMMA cement samples incorporated with nanoparticles of magnesium oxide (MgO) or $BaSO_4$ dramatically experienced a drop in temperature during PMMA solidification which improves the properties of PMMA for orthopaedic applications. One of the PMMA cement flaws is its very high exothermic temperature which causes thermal necrosis to the surrounding tissues. Composite PMMA bone cement with microcapsules material (paraffin) (PCMc) were mixed. PCMc was found to absorb generated heat leading to reduction in thermal necrosis and increased setting time, lower compressive strength, significantly lower compressive modulus, comparable biocompatibility, and significantly smaller thermal necrosis zone. Conclusively, composite cement containing 20% PCMc may be suitable for use in percutaneous vertebroplasty and balloon kyphoplasty [58]. Comparison of the clinical effect of biomimetic mineralised collagen (MC) modified PMMA bone cement, and tradi-tional PMMA bone cement was carried out as reported in [11] for the treatment of osteoporotic vertebral compression fractures (OVCF). It was asserted that PMMA bone cement modified with mineralised collagen MC) showed good clinical out-comes and better mechanical properties than the PMMA traditional bone cements. An *in vitro* and *in vivo* studies were performed using hydrophilised poly methyl-methacrylate (PMMA) bone cements containing Pluronic F68 as a hydrophilic additive and vancomycin (F68-VA cements). Observation shows that the use of Pluronic F68 (7 wt%)-VA cement as a hydrophilic additive for antibiotic-eluting PMMA bone cement was effective for the treatment of osteomyelitis as well as prophylaxis of bone infection without any critical effect [37].

Modification of bioactive PMMA-based bone cement was done with γ -methacryloxypropyltrimethoxysilane (MPS) and calcium acetate calcined to in-vestigate setting time, apatite formation and compressive strength in a simulated body fluid (Kokubo solution) by molecular weight ratio. PMMA powder of higher molecular weight showed a smaller decrease in compressive strength than that of smaller molecular weight [39]. A bioactive bone cement (BABC) that consists of apatite and wollastonite containing glass ceramic (AW-GC) powder and bisphenol-A-glycidyl dimethacrylate (Bis-GMA) based resin was developed for repair of segmental bone defects under load-bearing conditions to study the effectiveness of the bioactive bone cement. Results indicated that the bioactive bone cement was useful as a bone substitute under load-bearing conditions [36]. An *in vivo* bio-compatibility of linoleic acid (LA)–modified acrylic bone cement was studied using a huge animal. LA modified cement exhibited a reduction in the elastic modulus of up to 65% with lower mechanical properties which showed good biocompatibility as promising material for use in treatment of osteoporotic vertebral fractures [34].

PMMA cement was reinforced with strontium-containing borate bioactive glass (SrBG) to improve their bioactivity and osseointegration. SrBG in the composite cement acted as the reinforcement phase and bioactive filler of PMMA cement. SrBG/PMMA composite cements presented better osseointegration than PMMA bone cement and there was appreciable decreased in polymerization temperature when compared with PMMA. SrBG/PMMA composite cements also retained properties of appropriate setting time and high mechanical strengthas abetter al-ternative to PMMA cement. SrBG/PMMA composite cements presented better osseointegration than PMMA bone cement. SrBG/PMMA composite cement can serve as a better alternative to PMMA cement in clinical applications and has promising orthopaedic applications [15].

Evaluation of bone behaviour to porous PMMA cement and PMMA loaded with hydroxyapatite (HA) particles was studied to determine their tolerance to the sur-rounding tissue of a bone defect created in rabbit mandible. Additive particles of HA that were incorporated to PMMA improves the osteoconductive behaviour and support bone repair by enhancing bone ingrowth [16]. Moreover, based on various papers reviewed on PMMA-based composites, many synthetic materials have been established as PMMA additives for biomedical applications. Various properties of those synthetic materials have been part of the previous discussion in this chapter. On the other hands, there are many agricultural by-products/wastes that possess some of those features, making synthetic materials suitable as additives for bio-medical applications. Future directions of agricultural by-products as sources of materials for biomedical applications can only be achieved through experimental investigations. There are gaps to be filled by interested researchers across the globe.

9.4 POTENTIALS OF *PARQUETINA NIGRESCENS* PODS AS SOURCES OF NANOPARTICLES FOR BIOMEDICAL APPLICATIONS

9.4.1 *PARQUETINA NIGRESCENS*

Parquetina nigrescens (Figure 9.1) are found in many African countries. They have many uses including treatments of infections like hypertension, genuine kidney is-sues, extreme obstruction, venereal illnesses, stomach ulcers, looseness of the bowels, gonorrhea, the runs, cardio poisonousness, GIT issues, diabetes, exhaustion, feminine issue, measles, intestinal worms, diarrheal, loose bowels, diabetes, feminine pro-blems, respiratory illnesses, and venereal infections. Their leaves are the regular fixing in drug to treat craziness. *Parquetina nigrescens* are broadly employed in conventional medication and in customary medication practice as a restorative spice as disclosed in [59]. *Parquetina nigrescens* produce flowers and fruits in the rainy season [60–62]. They survive in savanna, secondary forest, vegetation bordering roads, gallery forest, and on anthills. It grows on various types of soil including marshy areas and the average 1000-seed weight is 11.2 g, as read in [61]. The plant parts are harvested throughout the year. The leaves can be used fresh or dried for later use. The plant is not prone or threatened by genetic erosion [63]. *Parquetina ni-grescens* (Periplocaceae), a shrub has been utilised in customary medication practice for quite a long time and is guaranteed by some society medication experts. The leaves, roots, and latex are the pieces of the plant utilised for customary medication [59]. *Parquetina nigrescens* (Periplocaceae) has rough fibre and is broadly utilised in conventional medication as a restorative spice. Normally, modest quantities of *Parquetina nigrescens* are utilised as it is poisonous, particularly the latex, and many crucial mishaps have been recorded. A plant or leave decoction is taken as a bowel purge to treat genuine kidney issues, a few stoppages, and to prompt early termina-tion. Occasionally, new squashed leaves are taken as an emetic to treat extreme stoppage. A plant, leave decoction, imbuement alone or with part of other plant species is smashed to treat measles, intestinal worms, diarrheal, loose bowels, dia-betes, feminine problems, and venereal infections. In very little amounts, it is given to youngsters to treat respiratory illnesses. A leave decoction with nectar added cures diminish weakness, jaundice, stomach ulcer, and sickliness as a tonic. It is likewise taken to treat hypertension and to ease a youngster birth. The body is washed with a leave decoction to treat general weakness. The leaves are the regular fixing in drugs to treat craziness. It was reported that phytochemical screening of the organic product bark shows the presence of alkaloids, tannins, and flavonoids and general examination shows dampness content, crude proteins, crude fats, and carbohydrates [59,62–66].

The seeds (Figure 9.1c) of *Parquetina nigrescens* are attached to light fibre (Figure 9.1d) like materials like tridax. The seeds are dispersed by the wind. The

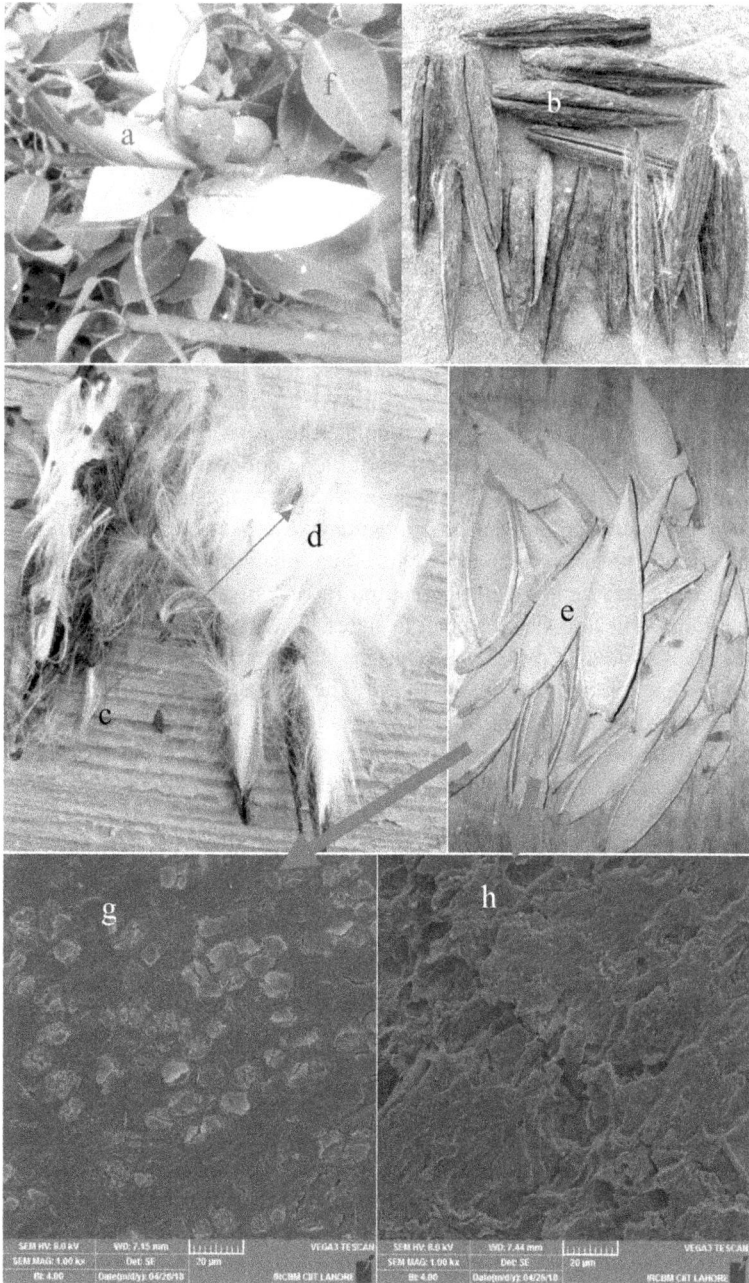

FIGURE 9.1 *Parquetina nigrescens* (a) fresh pod, (b) dried pod showing outer layer, (c) seeds, (d) fibres, (e) inner layer, (f) leaf, (g) microstructure of the inner surface of the pod, and (h) microstructure of the outer surface of the pod.

fibre-like material is very light and is easily carried by the wind. The fruit has double layers (Figure 9.1b, e) and very thin fibres in between the layers. It appears green (Figure 9.1a) at first and then turn dark brown (Figure 9.1b) on drying. The fibres in the pods appear offwhite [67]. The outer covering of the pod forms a bark-like structure when dried. Few or little research work have been carried out on the pod.

9.4.2 PROPERTIES AND USES OF *PARQUETINA NIGRESCENS*

Parquetina nigrescens is wealthy in heart glycosides (cardenolides). These glycosides, altogether called strophanthins, are generally plentiful in the latex. The watery part of an ethanolic root extricate improved the morphology of sickled cells in vitro, and furthermore hindered and switched the sickling cycle in a moderate manner. A fluid leaf extricates orally regulated to iron-deficient rodents showed a critical and portion subordinate increment of blood boundaries. An examination was done to decide the impact of a home-grown readiness, containing *Parquetina nigrescens*, sorghum bicolour, and *Harungana madagascariensis*, on blood boundaries in pallid bunnies, brought about by a contamination with *Trypanosoma brucei*. The weak hares not treated with the natural readiness died before the finish of the investigation, while for the treated frail hares the blood boundaries were reestablished to ordinary levels. A fluid leaf separate showed moderate antimicrobial action against a scope of pathogenic microbes. The ethanol extricate was viable just against *Pseudomonas aeruginosa* and *Salmonella typhi*. Unrefined hot and cold water and ethanolic concentrates of the leaves showed moderate antibacterial action against *Salmonella typhi*. A fluid leaf extricates showed moderate anthelmintic action against parasitic (*Onchocerca ochengi*) and free-living (*Caenorhabditis elegans*) nematodes [62]. *Parquetina nigrescens* is generally utilised in conventional medication. Generally modest quantities are utilised, as the plant is exceptionally poisonous, particularly the latex. Numerous deadly mishaps have been recorded. A plant or leaf decoction is taken as a purification to treat genuine kidney issues, extreme stoppage, and to prompt fetus removal. Occasionally, new squashed leaves are taken as an emetic to treat extreme stoppage. A plant or leaf decoction or implantation, occasionally with parts of other plant species, is plastered to treat measles, intestinal worms, loose bowels, diarrhea, diabetes, feminine issues, and venereal illnesses. In minuscule amounts, it is given to youngsters to treat respiratory sicknesses. A leaf decoction with nectar added is tipsy to treat weakness, jaundice, stomach ulcers and pallor, as a tonic. It is additionally taken to treat hypotension and to ease labor. The body is washed with a leaf decoction to treat general weariness. The leaves are a typical fixing in drugs to treat madness. Antioxidant properties are present in *Parquetina nigrescens* and these oxidants can help in protection against free radical induced ulcer [68,69]. The latex and leaf sap create a consuming uproar on the skin and are remotely applied to tumors, abscesses, wounds, and consumes; they likewise darken scars and are applied to thistles in the skin to remove them. Leaves are applied as a dressing for wounds; squashed leaves are applied to skin sicknesses and head lice. A leaf maceration is applied to the legs of youngsters with rachitis, to the head to treat migraine and to the side to treat torment in the side. Pounded bark is applied to skin cuts to treat stiffness. Leaf sap, without latex, is utilised as eye drops to treat conjunctivitis and jaundice. The latex is considered to cause visual impairment. Pounded root bark, scoured on the body, is viewed as a powerful love potion. Squashed roots with Capsicum peppers are utilised as a purification to treat venereal infections. Squashed roots blended in with shea spread (*Vitellaria paradoxa* C.F.Gaertn.) is scoured on all fours as a snake repellent. Root glue is additionally applied to snakebites and a root decoction is known for a similar reason. Root glue is likewise applied to scarification on the joints to treat spasms in youngsters. The fluid of bubbled roots with lemon juice is taken to treat nicotine harming; it is a tobacco revulsive. The latex is exceptionally harmful and generally utilised as an element of bolt poison, particularly in Central Africa, to chase hedge meat. The latex is utilised alone, or blended in with different plants, contingent upon the customs of the toxin creator and accessibility of these plants. It is utilised likewise as Strophanthus and Acokanthera bolt harms. The fibre, root, sap, and latex also have their relevant applications. Some of its advantages are as follows:

The Leaf
For medical applications, cutting across:

   a. General healing.
   b. As painkillers.
   c. Stomach troubles.
   d. Cutaneous, subcutaneous parasitic infections.
   e. Malnutrition and debility.

The Sap
For medical applications, including:

   a. Arthritis and rheumatism treatment.
   b. In addition, with the leaf, it is used for eye treatments and laxatives.
   c. Vermifuge, diuretics, abortifacients, and ecbolics.
   d. Also dyes, stains, inks, tattoos, and mordants.

The Root
It is used for the following:

   a. Venital stimulants/depressants and reptile-repellents.
   b. Antidote for venomous bites, stings, etc.
   c. Treating paralysis, epilepsy, convulsions, and spasms.

The Stem

   a. They are used in processing exudations-gums and resins.

The Fibre

a. Used in the production of farming, forestry, hunting, and fishing apparatus.

The Latex
It is used for treating:

a. Tumours and cancer.
b. Arrow-poisons and fish poisons.
c. Miscellaneously poisonous or repellent [62].

9.4.3   SYNTHESIS, CHEMISTRY, AND TOXICITY OF *PARQUETINA NIGRESCENS* POD NANOPARTICLE

Uses of most parts of the *Parquetina nigrescens* have been established in literature [59,60,62–65,69]. Moreover, no scientific documentation has been found on *Parquetina nigrescens* pods. During the dry season, pods falling from the *Parquetina nigrescens* plant/shrubs burst to release the fibres and seeds remain as waste littering the plant environment. Being a component of the *Parquetina nigrescens* plant, it is likely to contain phytochemicals and other beneficial materials for biomedical applications. Being an effort to add to its value, the pods, after washing and drying, were converted to nanoparticles through a top-down approach using a disc grinder and 2 kg capacity tumbler ball mill. Powders obtained from disc grinding were milling for 40 hours at 10 charge ratios and drum speed of 150 revolution per minutes. Chemistry of the synthesised nanoparticles were studied using gas chromatography/mass spectrometer (GC/MS). The GCMS analysis describes the analytical procedure, which involves the combination of features of both gas chromatography and mass spectrometry for identification of different substances within a specimen component matrix (parent material). One of the most accurate analyses available is the GC/MS analysis, which is useful for materials with requirements of low detection limits, and it can be used to examine samples in varying chemical state, even when the quantity of the sample is limited.

Table 9.1 highlights the various peaks attained by each compound detected in *Parquetina nigrescens* pod nanoparticles, as shown in Figure 9.2. Various compounds (Table 9.1) detected in the *Parquetina nigrescens* pod nanoparticles were used in evaluating the level of toxicity of the *Parquetina nigrescens* pod nanoparticles. Confirmed compounds were identified using a Java software, "Toxtree version 3.1.0.1851". The software was employed in identifying the various level of toxicity of each compound present in the *Parquetina nigrescens* pod nanoparticles. It categorises the toxicity level of each compound into three different classes which are classes I, II, and III. Class I indicates low toxicity, class II, indicates intermediate level of toxicity and class III shows a high level of toxicity. Toxtree was able to identify majority of the compounds, leaving only a few. The analysis from the Toxtree shows that there are various toxic substances present in *Parquetina nigrescens* pod nanoparticles. Moreover, GCMS result identifies some compounds detected in the *Parquetina nigrescens* pod nanoparticles, which have a certain usefulness. An example is the Oleic

TABLE 9.1

Compounds Present in *Parquetina nigrescens* Pod Nanoparticles Detected from Gas Chromatography/Mass Spectroscopy

Peak	Retention Time	Area (%)	Compound	Qual	Toxicity Class
1	2.493	0.14	L-Methioniol	36	III
			1-Cyclopentyl-1-diisopropylsilyloxyethane	23	III
			4-Fluoro-1-ribofuranosylimidazole-5-carboxamide	22	III
2	2–600	0.06	Vinyl ethyl sulfoxide	64	III
			Sulfoxide	43	III
			2-chloroethyl vinyl Mecysteine	43	–
3	3.726	13.78	2- Dimethylsilyloxytridecane	12	III
			D-Fucose	12	I
			Methyl 3-hydroxytetradecanoate	10	II
4	3.888	23.30	Aziridine	9	III
			2-methyl-2-Propenal	9	II
5	4.508	31.75	1-Octanol,	42	I
			3,7-dimethyl-Cyclopropane	40	II
			1-methyl-2-(3-methyl pentyl)-	38	–
6	5.127	0.99	Guadine	7	–
			methyl-Methane	7	I
			isothiocyanato-2-[2-[2-Methoxyethoxy]ethoxy-1,3-dioxalane	5	III
7	5.415	3.07	Peroxide	9	–
			DimethylUrethane	5	I
8	5.721	8.85	Propanenitrile	43	III
			2-hydroxy-2-Propen-1-amine, 2-methyl-	9	III
9	6.459	1.13	Pentane	9	I
			1- (1-ethoxyethoxy)-2,4-Dimethylpentan-3-yl ethyl carbonate.	9	I
			1,3-Dioxolane	7	III
10	7.091	0.47	Hexanoic acid	87	I
			ethyl ester	64	I
			Pentanoic acid, 4-methyl-, ethyl ester	64	I
11	9.468	0.15	Thiophosphoric acid	64	III
			S-[[2-(trimethylsilylmethylamino)ethylamino] ethyl ester	64	–
			Phenylpropanolamine	59	II
			bis(trimethylsilyl) l-Alanine	59	III
			N-(trimethylsilyl)-, trimethylsilyl ester	53	–
12	12.458	15.35	9-Octadecenoic acid	90	I
			(E) – Oleic acid	90	I
			6-Octadecenoic acid	89	I

(*Continued*)

TABLE 9.1 (Continued)

Compounds Present in *Parquetina nigrescens* Pod Nanoparticles Detected from Gas Chromatography/Mass Spectroscopy

Peak	Retention Time	Area (%)	Compound	Qual	Toxicity Class
13	13.928	0.75	Oleic acid	50	I
			n-Hexadecanoic acid	38	I
14	14.091	0.22	1,2,3-Triazolo[4,5-f] benzotriazole-4,8 (4H,8H) – dione	32	III
			Furfurylidenimionosulpurpentafluoride	9	–
			Cyclohex-2-enone, 3-(2H-tetrazol-5-ylamino)-	9	III

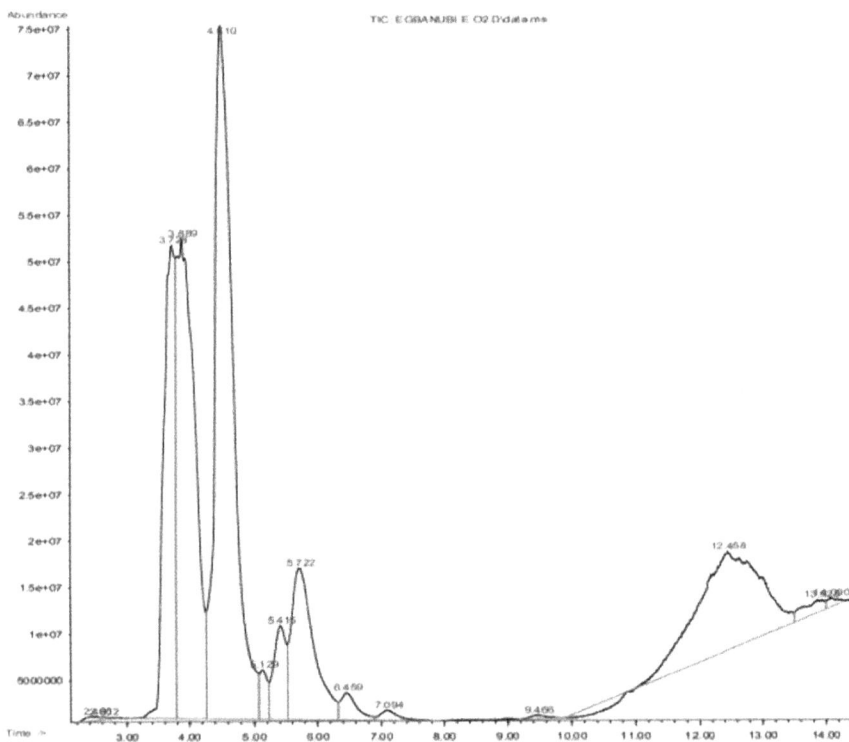

FIGURE 9.2 GCMS spectrograph showing peaks attained by various compounds detected in *Parquetina nigrescens* pod nanoparticles.

acid, which has an open chain, aliphatic structure with some functional groups. It is useful in food additives as flavouring agents and used in the production of agrochemicals such as herbicides, insecticides, etc. Oleic acid has been verified to be of low concern and it has a toxicity level of "class I", indicating it has a low toxicity level. The presence of a toxic substance in the *Parquetina nigrescens* pod nanoparticles does not prevent its use as a PMMA additive for developing nanocomposite for biomedical applications. Its presence in PMMA can act as an antibacterial additive, improving antibacterial properties of the PMMA in adhesives for bone and tooth repair, in addition to probable improvement in mechanical properties of the PMMA. This implies that experimental investigations are imperative to determine an amount of *Parquetina nigrescens* pod nanoparticles to be incorporated in the PMMA to develop nanocomposites implants saved to human and to ascertain the proposed applications of *Parquetina nigrescens* pod nanoparticles as additives in the PMMA for biomedical applications. On the other hand, research can be focused on determining saved chemicals that can dissolve toxic components of the *Parquetina nigrescens* pod nanoparticles. This approach can lead to extraction of toxic components from the *Parquetina nigrescens* pod nanoparticles to leave remnants saved to the body as additives in the PMMA for improving the mechanical properties, though other benefits like an antibacterial property may be lost by the *Parquetina nigrescens* pod nanoparticles in this technique.

9.4.4 OPTICAL PROPERTIES OF *PARQUETINA NIGRESCENS* POD NANOPARTICLES

Ultraviolet visible spectroscopic analysis was performed on *Parquetina nigrescens* pod nanoparticle to determine its optical properties (which includes its ability to absorb ultraviolet ray) using photons of known wavelengths. This analysis is useful to ascertain its possible application in an area demanding absorption of the ultraviolet rays like in sunblock cream to prevent skin from cellular damage. Ultraviolet-visible spectroscopy or ultraviolet-visible (UV-Vis or UV/Vis) spectrophotometry refers to absorption spectroscopy or reflection spectroscopy in part of the ultraviolet and all adjacent visible regions of the electromagnetic spectrum. It is a quantitative technique used to determine the optical properties of materials under investigations [70–73]. It studies the way electromagnetic radiation interacts with the characterised material in the ultraviolet visible region of the electromagnetic spectrum. In the ultraviolet visible region, radiation can be transmitted, reflected, absorbed, or emitted [73]. This means that it uses light in the visible and adjacent areas. Absorption or reflection in the visible range directly influences the perceived colour of the chemicals involved. In this part of the spectrum, atoms, and molecules go through electronic transitions. Absorption spectroscopy is complementary to fluorescence spectroscopy in that fluorescence deals with the transitions of electrons from the excited state to the ground state, while absorption measures the transitions from the ground state to the excited state [74,75]. Moreover, terms used in the UV-Vis spectroscopy is explained in Sections 9.3.4.1–9.3.4.3 of this chapter.

9.4.4.1 Transmittance

Transmittance (T) is defined as ratio of transmitted light (output), I to incident light (input), I_o (Equation 9.1). Percentage transmittance (%T) is the transmittance expressed in percentage (Equation 9.2) [75].

$$T = I/I_o \tag{9.1}$$

$$\left(\frac{I}{I_o}\right) \times 100 \tag{9.2}$$

If T = 0.5, then %T = 50%. 50% transmittance would mean that 50% of light has passed through the material and gone through to the other side.

9.4.4.2 Absorbance and Absorbance Coefficient

Absorbance is related to the transmittance as expressed in Equation 9.3, while the absorbance coefficient (α) is presented in Equation 9.4[75], t = thickness of the solution on the substrate.

$$Absorbance = 2 - \log 10(\%T) \tag{9.3}$$

$$\alpha = 2.303 \left(\frac{A}{t}\right) \tag{9.4}$$

9.4.4.3 Extinction Coefficient

Extinction coefficient (k) is a measure of the rate of diminution of transmitted light via scattering and absorption for a medium [73]. It is also defined as the ratio of maximum to minimum transmission of a beam of light that passes through a polarization optical train [74]. The formula for extinction coefficient is shown in Equation 9.4, α = absorbance coefficient, λ = wavelength and π = pi.

$$k = \frac{\alpha\lambda}{4\pi} \tag{9.5}$$

9.4.4.4 Analysis of Parquetina Nigrescens Pod Nanoparticle

In examining *Parquetina nigrescens* pod nanoparticles, chloroform, methanol, and ethanol were used as solvents for dissolving the material under investigation. 250 mg of the *Parquetina nigrescens* pod nanoparticles were mixed with 10 ml each of the reagent (solvent) in different three beakers using magnetic stirrer operated for 2 hours to ensure homogeneity of each solution. Then, the solution from each beaker was deposited on different substrates in a spin coater using a syringe. The spin coater was set to operate at 2000 revolutions per minute (rpm) for 30 seconds to produce a thickness of 98 nm of the solution on the substrate. After labelling the substrates to avoid a mix-up error, each of the substrates (solid samples) were exposed to the visible light of the UV-Visible spectrophotometer (model: Avalight-DH-5-BAL) and the transmittance were recorded. The analysis was carried out at the Materials Laboratory, Department of Materials Science and Engineering, Kwara State University, Malete, Nigeria. Results of the UV-Vis examination on the *Parquetina nigrescens* pod nanoparticles are presented in Figures 9.3–9.6.

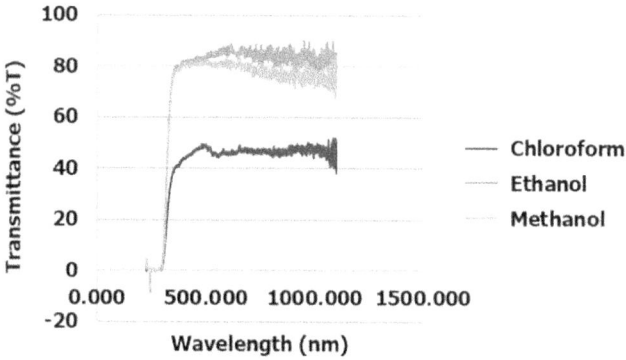

FIGURE 9.3 Ultraviolet visible spectrographs showing relationship between light transmitted through *Parquetina nigrescens* pod nanoparticles and photon wavelength at different reagents used for dissolving the *Parquetina nigrescens* pod nanoparticles.

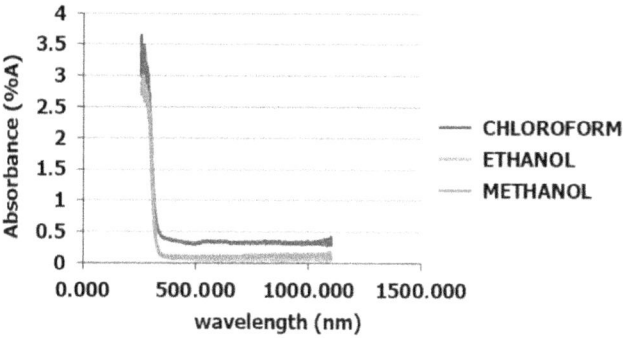

FIGURE 9.4 Ultraviolet visible spectrographs showing relationship between amount of light absorbed by *Parquetina nigrescens* pod nanoparticles and photon wavelength at different reagents used for dissolving the *Parquetina nigrescens* pod nanoparticles.

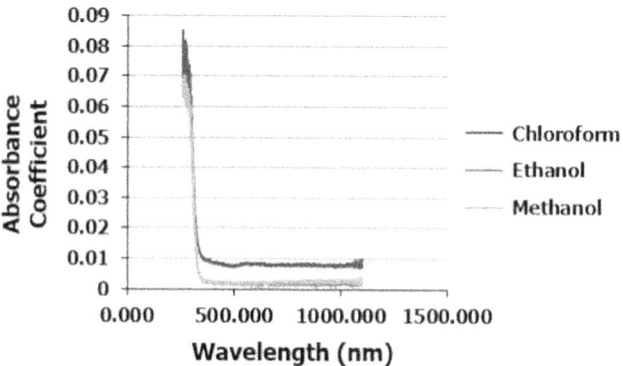

FIGURE 9.5 Ultraviolet visible spectrographs showing relationship between absorbance coefficient (cm^{-1}) of *Parquetina nigrescens* pod nanoparticles and photon wavelength at different reagents used for dissolving the *Parquetina nigrescens* pod nanoparticles.

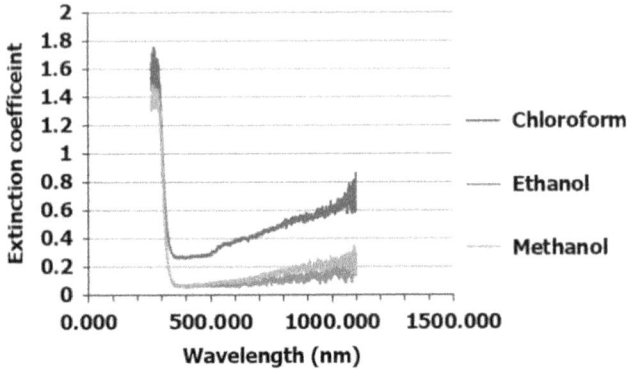

FIGURE 9.6 Ultraviolet visible spectrographs showing relationship between extinction coefficient of *Parquetina nigrescens* pod nanoparticles and photon wavelength at different reagents used for dissolving the *Parquetina nigrescens* pod nanoparticles.

It is observed from Figures 9.3–9.6 that *Parquetina nigrescens* pod nano-particles dissolved in a chloroform-based solution shows the least transmittance but greatest absorbance while those dissolved in ethanol and methanol-based solutions have a close range of transmittances and absorbances. This implies that a chloroform-based reagent is the best solvent for dissolving *Parquetina nigrescens* nanoparticles. Absorption of radiant energy (ultraviolet rays) ascertains its potential usage as an additive in the sunblock cream to control cellular skin damages. However, further analysis like animal study is recommended to determine an amount of the *Perquetina nigrescens* nanoparticles to be added in the sunblock cream preparation to ensure safety of the human body because of its toxic components (Table 9.1).

9.5 DATE SEED NANOPARTICLES: PROPERTIES AND APPLICATIONS

9.5.1 DATE SEED

Date palm refers to both date palm tree and fruit produced from the date palm tree. The date seeds are also known as stones, kernels, or pits [76]. It constitutes between 10% and 15% of the weight of the date fruit [77]. The date palm trees are a dioecious species [78,79]. The botanical name of date or date palm is *Phoenix dactylifera L.* [77,80], and it is commonly called *Dabino* by the Hausa-speaking tribe in Nigeria [81]. The origin of date palm in Nigeria dates to the early eighth century, where it was traded by Arab traders from North Africa in exchange for dry leaves of the Henna plant (*Lawsonia inermins*) used by women mostly for body decoration. The date fruits are valued among communities in Nigeria and are a surplus during the festive seasons and fasting period [81]. The date palm fruit is composed of a fleshy pericarp, mesocarp, endocarp, and the date seed [76]. The date palms are dioecious species [77,78].

9.5.2 Compositions and Benefits of Date Seeds

The date seeds are considered agricultural waste because 50% of propagated date seeds develop to produce male date palm trees that do not bear fruits [77,79]. Since the date seeds from date fruits have proven to be futile during propagation, other usefulness has been found with the goal that it is not productive, but it is discarded as waste. The composition of the date seed constituents such as moisture, fat, protein, ash, and carbohydrate depend on the date varieties. The date seeds are a good source of dietary fibres and dissolvable sugars. The seeds are high in potassium, magnesium, phosphorus, calcium, and a modest degree of iron. They have higher amounts of phenolic, antioxidants, and dietary fibre than the fleshy part of the date, as reported in [82]. The date seeds can be utilised as an alternative for coffee (espresso). It is essential to utilise an ideal processing conditions when preparing the date seed based espresso powder which is free of caffeine and has low amount of total phenolic compounds than coffee [76], such as taking time to remove the white skin from the dried date seeds before they are roasted, crushed, and grinded to powder since the white skin is a natural foaming substance that could cause foam to be in the espresso [82]. The oil extracted from the date seeds with organic solvents such as hexane and diethyl ether [83] are edible and are used in pharmaceuticals and cosmetics [77,83]. Since the date seed oil can absorb high UV, it serves as a useful ingredient for producing sun block creams, thereby protecting the skin from cellular damage. However, due to the low oil yield from the seeds when compared to customary oil crops, the date seed oil is expensive. The level of unsaturation and the amount of linoleic acid in the date seed oil is lower when compared with frequently ingested vegetable oils. The oil from the date seeds was found to have oxidative stability that is equivalent to that of olive oil and higher than that of the most consumed vegetable oils. Date seed oil could as well be regarded as a possible source of natural phenolic compounds, besides their commitment to protection from oxidative rancidity, and their interest in presenting a particular flavour to the oil [77]. The evidence of dietary fibre in the date seed makes them appropriate for the preparation of fibre-based food varieties like bread, biscuits, and cakes; and dietary supplements [83]. The addition of milled date seed powder in bread increases the dietary fibre contents. Research shows that bread containing 10% coarse date seed powder had higher dietary fibre content than the wheat grain control, and comparable sensory properties, whereas bread containing fine date seed powder is found to reduce in colour, flavour, odour, chewiness, uniformity, and overall acceptability sensory scores apart from having higher dietary fibre contents [76,82,83]. With the increasing advancement in the field of technology and the revolution of nanoparticles, agricultural waste has proven to be useful in the creation of new materials while modifying existing ones for different applications.

9.5.3 Cultivation of Date Palm

Date palm flourishes in arid climate, in areas that are very dry and hot, with temperature between 25°C and 32°C and an adequate supply of water [78,79]. Irrigation is necessary to feed the date palm, to ensure the growth and development of the berries. High humidity and rainfall adversely affect the quality of the date, causing damage and inhibiting pollination. Date palm can be planted on various soil types possessing varying amounts of organic and mineral nutrients but yields more on sandy loam soil. It is known to have a higher salt tolerance than any other fruit crop [79]. It can be propagated sexually using the date seeds, asexually through vegetative propagation using the offshoots from the female date palm, and by the recently developed tissue culture techniques (in vitro propagation), [77,81]. Since 50% of date seeds developed to produce male date palms that do not bear fruits and offshoots, vegetative propagation is the most used method for propagation. The offshoots are cut from female date palms (mother) and transplanted into a nursery that has good growing conditions to support the development of the roots. The young date palms are transplanted to their permanent site after one year depending on how well the root system develops and the number of palm leaves on them (usually about 10 to 12 leaves are recommended). Offshoots with developed roots already from the mother are planted directly to the permanent site [79].

9.5.4 HARVESTING STAGE AND VARIETIES OF DATE FRUITS

Visible changes such as the colour of the date fruits, the degree of ripeness, and invisible changes such as the activities of various enzymes and an amount of sugar and water in the dates determine the right time for harvesting the date fruit [84]. There are five different fruiting stages of the date palm: hanabauk, kimri, khalal, rutab, and tamar [76]. Dates (Figure 9.7) are harvested at the last three developmental stages [84].

The first dates to be harvested are the dates that have reached the khalal stage, the first development stage. At this stage, the harvested dates are hard and crisp, with a moisture content between 50% and 85% [84]. These dates are partially ripe,

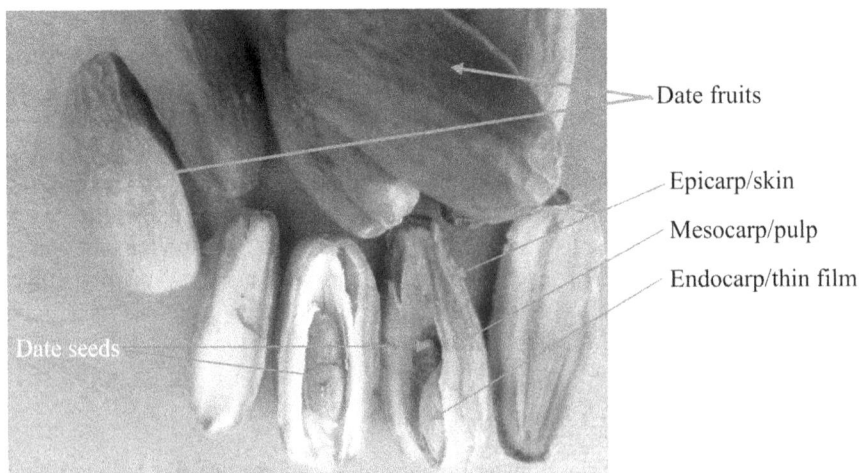

FIGURE 9.7   Image of date fruit revealing stages of development and components of the date fruit.

so they are either bright yellow or red in colour, depending on the variety. Since the dates are not fully ripe before harvesting and the moisture content at this stage is high, the amount of tannin is also high. Date varieties with a low tannin content at the khalal stage of development are the only dates best suited for consumption. The date varieties produced at the khalal stage are Barhee, Zaghlool, Hayany, and Khalas [76]. Rutab is the second stage in which the dates become edible and are ready to be harvested. The fruits colour turns partially brown, the moisture content is reduced to between 30% and 45%, and the fibres are softened at this stage. The date varieties harvested at the rutab stage are called Medjool and Deglet Nour. The final stage for harvesting dates is the tamar stage. The dates are fully ripe at this stage and changes occur when their colour turns from amber to dark brown. There is a further reduction in the moisture content from 25% to 10% and less, and the texture turns from soft to firm to hard. Dayri, Halawy, Khadrawy, Thoori, Zahidi, Sayer, and Aliig are the date varieties harvested at this stage. The dates harvested at the khalal and rutab development stage are perishable, while those produced at the tamar stage are non-perishable [76,84]. Experienced workers are required for harvesting since the process needs to be faultless and clean so as not to damage the dates. The harvested dates are kept in a container during transportation to prevent damages.

9.5.5 Properties and Uses of Date Fruits

Dates, the product of the date palm, can be considered an ideal food that gives a wide scope of fundamental supplements with numerous potential medical advantages [85,86]. The fruit contains about 70% carbohydrate, particularly the sugar, sucrose, fructose, and glucose. The sugars in the date are effectively processed; after consumption, they are quickly moved to the blood and are rapidly digested to deliver energy for different cell exercises. Dates are likewise a rich source of valuable fibres, and they contain many significant nutrients and minerals, including critical measures of some other elements [85,87]. Research shows that a date contains properties such as antioxidant, anti-inflammatory, anti-cancer, anti-microbial, and anti-diabetic activity among others [87]. Dates are generally eaten like that or taken with coffee, milk, or yoghurt. The dates are also taken as paste, syrup, jam, jellies, prickles, and are utilised in many confectionary items along with coconut, vinegar, honey, chocolates, after they have been refined [76,85–87]. Since the date seed from the date has proven to be futile during propagation, other usefulness has been found with the goal that it does not wind up as waste. Moreover, different species of date seeds have similar contents but with different proportions [77]. For instance, Khalas has 7.5% moisture, 10.5% fat, 5.7% protein, 1.05% ash and 78.3% carbohydrate; Fardh has 9.5% moisture, 8.2% fat, 5.8% protein, 1.2% ash and 78% carbohydrate; Lulu has 10.9% moisture, 10.5% fat, 5.2% protein, 0.91% ash, and 74.7% carbohydrate; Deglet Nour has 11.2% moisture, 10.1% fat, 5.6% protein, 1.1% ash, and 83.1% carbohydrate; and Allig has 10.3% moisture, 12.5% fat, 5.2% protein, 1.10% ash, and 81% carbohydrate. Similarly, Mebselli, Um-shella, Shalal, Barhe, Shikatalkahlas, Sokkery, and Booman have their respective proportions of moisture, fat, protein, ash, and carbohydrate per 100 g of fruit. The composition of the date seed constituents such as moisture, fat, protein, ash, and carbohydrate relies on the variety of the date

[77]. The date seed is also a source of dietary fibres and dissolvable sugars that include average values of glucose, raffinose, stachyose, sucrose, fructose, and galactose [88]. The seeds are high in potassium, magnesium, phosphorus, calcium, and a modest degree of iron [88]. Moreover, it has been reported that date seed has a higher amount of phenolic, antioxidants, and dietary fibre, than the fleshy part of the date [85]. The phenolics, antioxidants, and dietary fibre contents of date seed are different from one variety of date to another. The date seed can be utilised as an alternative for coffee (espresso) [76,88]. It is essential to utilise an ideal processing condition when preparing the date seed–based espresso powder, which is free of caffeine and has a low amount of total phenolic compounds than coffee [75], such as taking time to remove the white skin from the dried date seeds before they are roasted, crushed, and grinded to powder since the white skin is a natural foaming substance that could cause foam to be in the espresso [76,88]. The oil extracted from the date seeds with organic solvents such as hexane and diethyl ether [83] are edible and can also be used in pharmaceuticals and cosmetics [83]. Since the date seed oil can absorb high UV, it serves as a useful ingredient to produce sunblock creams, thereby protecting the skin from cellular damage [77,83]. However, due to the low oil yield from the seeds when compared to customary oil crops, the date seed oil is expensive. The level of unsaturation and the amount of linoleic acid in the date seed oil is lower when compared with frequently ingested vegetable oils [77]. The oil from the date seeds has oxidative stability that is equivalent to that of olive oil and higher than that of the most consumed vegetable oil. Date seed oil could as well be regarded as a possible source of natural phenolic compounds, besides their commitment to protection from oxidative rancidity, and their interest in presenting a particular flavour to the oil [76–78,80–82,84–86]. The evidence of dietary fibre in the date seed makes them appropriate for the preparation of fibre-based food varieties like bread, biscuits, and cakes; and dietary supplements [83]. The addition of milled date seed powder in bread increases the dietary fibre contents. Research shows that bread containing 10% coarse date seed powder had higher dietary fibre content than the wheat grain control, and comparable sensory properties, whereas bread containing fine date seed powder is found to reduce in colour, flavor, odor, chewiness, uniformity, and overall acceptability sensory scores apart from having higher dietary fibre contents [76,83,84]. With the increasing development in the field of technology and the revolution of nanoparticles, the seeds of the date palm have proven to be useful in the creation of new materials while modifying existing ones for different applications. There are reasons to be worried as to why there is a high amount of selenium present in some date types. Some studies have also recorded abnormal reaction (hypersensitivity) of the immune system to the date. Nevertheless, no record is found on side effects of date seed [77]. Moreover, uses of date seed powders for improving male reproductivity has been confirmed by traditional/local medicinal experts without any body side effects though this needs scientific confirmation through research. Therefore, this remains as a gap to be filled by interested researcher across the globe. Besides the benefits of the date seed as disclosed based on information from various articles reviewed for the purpose of this chapter, there are other benefits from date seeds. Date seeds are hard materials that make their nanoparticles suitable for reinforcing polymer like PMMA or metals like aluminium. This implies that addition of date seed nanoparticles in PMMA to

develop biomedical nanocomposites could serve many purposes including imparting antioxidant properties in the PMMA due to presence of antioxidants and phenolics in the date seeds. Antioxidant properties will make date seed particle-reinforced PMMA nanocomposites suitable to avert cell damages that are caused by free radical oxidation especially when used as a body absorbable artificial stent implant. Additionally, date seed particle reinforced PMMA nanocomposites will also serve as a source of beneficial minerals like potassium, magnesium, phosphorus, calcium, and iron to the body. The main importance is that its potential for improving mechanical properties of the PMMA to meet up with the required mechanical properties for targeted biomedical applications remain intact.

9.5.6 UV-Visible Optical Properties of Date Seed Nanoparticles

The date seed nanoparticles examined in this chapter were produced using the top-down approach which involves hammer crushing of the date seeds to reduce the size before milling to produce the nanoparticles. The date seeds were thoroughly rinsed in water to remove white skin and dirt and dried (Figure 9.8a) using Uniscope oven set at 120°C for four days at 8 hours per day. A hammer was used to reduce the size of the date seeds before they were grinded using a heavy-duty disc grinder to produce the date seed powders (Figure 9.8b) that were refined at 10 charge ratios for 40 hours using a 2 kg-capacity tumbler ball mill to produce date seed nanoparticles (Figure 9.8c). The particles obtained after 40 hours were analysed using transmission electron microscope and UV-Vis spectrophotometer. Transmission electron microscopic images of date seed nanoparticles are presented and discussed in Section 9.5 of this chapter.

Hydrochloric acid (HCl), chloroform (CHCl$_3$), and methanol (CH$_3$OH) were used to prepare the date seed nanoparticles solutions for the UV characterisation. The same technique described in Section 9.3.4.4 for the *Parquetina nigrescens* pod nanoparticles was used for determining the optical properties of the date seed nanoparticles. Transmittance obtained from the UV-Vis investigation on date seed nanoparticles were

(a)   (b)   (c)

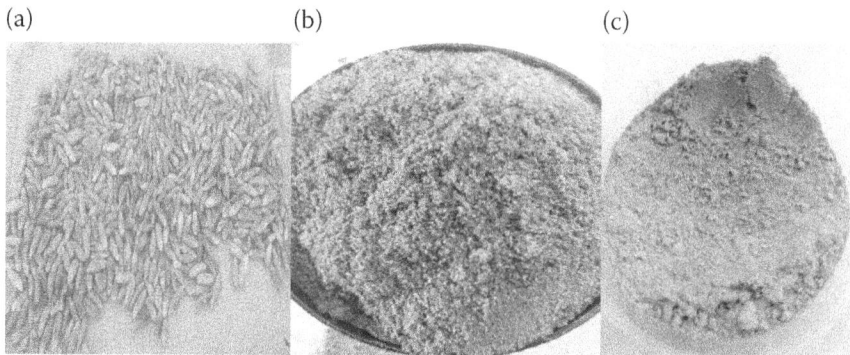

FIGURE 9.8 Date seed (a) after washing and drying, (b) powders obtained after disc grinding, (c) nanoparticle obtained after milling for 40 hours at 10 charge ratios (mass of balls/mass of particles).

used to calculate the absorbance, absorbance coefficient, and the extinction coefficient using formula in Equations 9.3–9.5, as presented in Figures 9.9–9.12. The result shows that the solution containing methanol has the highest transmittance while the HCl-based solution recorded the least transmittance. The highest amount of absorbance is

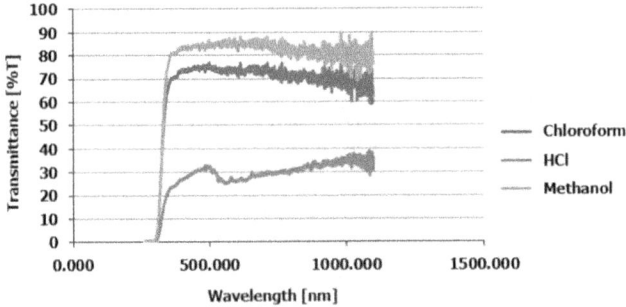

FIGURE 9.9 Ultraviolet visible spectrographs showing relationship between light transmitted through date seed nanoparticles and photon wavelength at different reagents used for dissolving the date seed pod nanoparticles.

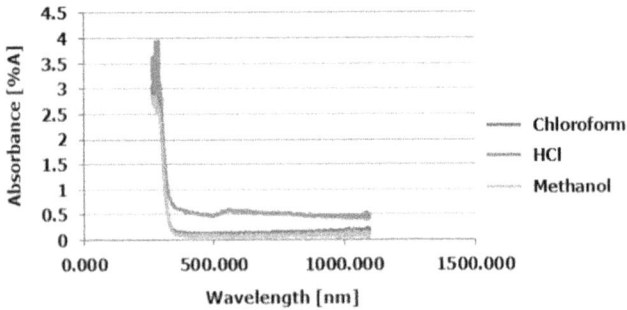

FIGURE 9.10 Ultraviolet visible spectrographs showing relationship between light absorbed by date seed nanoparticles and photon wavelength at different reagents used for dissolving the date seed pod nanoparticles.

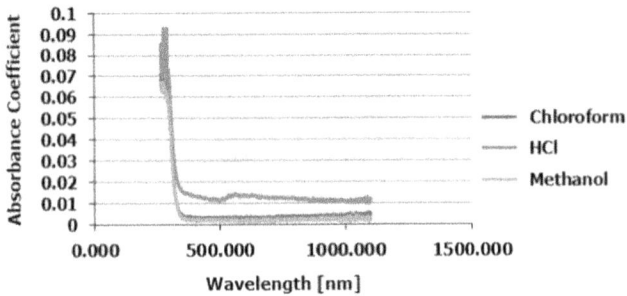

FIGURE 9.11 Ultraviolet visible spectrographs showing relationship between absorbance coefficient of date seed nanoparticles and photon wavelength at different reagents used for dissolving the date seed pod nanoparticles.

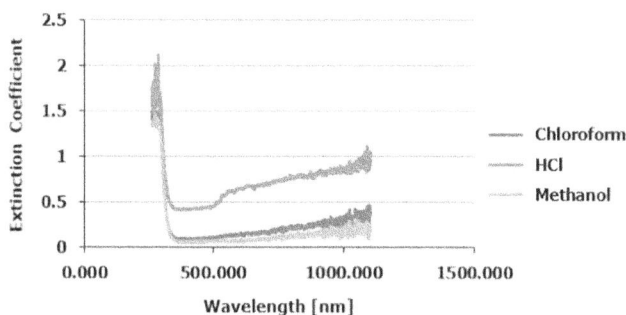

FIGURE 9.12 Ultraviolet visible spectrographs showing relationship between extinction coefficient of date seed nanoparticles and photon wavelength at different reagents used for dissolving the date seed pod nanoparticles.

observed in the HCl-based solution, having the least transmittance (Figures 9.8–9.11). It implies that the best solution for dissolving date seed nanoparticles among others is HCl. Absorbance of ultraviolet ray by date seed nanoparticles as confirmed in this chapter is in line with that of date seed oil, reported in [77,83].

9.5.7 CHEMICAL COMPOSITIONS AND TOXICITY OF DATE SEED NANOPARTICLES

The gas chromatography mass spectroscopy (GCMS) was carried out on the date seed nanoparticles treated with 5% phenyl methyl silox to determine its components and the detected compounds were identified by comparing their retention time and mass spectra with those of the standard mass spectra from NIST 05 spectra database. A gas chromatography from Agilent USA hyphenated to a mass spectrometer (Model: 7890A GC system, 5675C Inert MSD) with triple-axis detector equipped with an auto injector (10 µl syringe) was employed for date seed nanoparticle analysis and carrier gas was helium. The analysis was carried out at the Chemical Engineering laboratory, University of Ilorin, Nigeria. All chromatographic separations were carried out on a capillary column with the following specifications: length, 30 m; internal diameter, 0.2 µm; thickness, 250 µm; ion source temperature, 250°C; interphase temperature, 300°C; pressure 16.2 psia; out time, 1.8 mm; 1 µl injector in split mode with a split ratio of 0f 1:50; injection temperature, 300°C; onset column temperature at 35°C was maintained for 5 minutes and reduced to 150°C at a rate of 4°C/minute. During analysis, the temperature was raised to 250°C at a rate of 20°C/minute and held for 5 minutes at 250°C with a total elution of 47.5 minutes. The system was controlled, and data was acquired by Ms Solution Software.

The gas chromatography mass spectroscopy (GC-MS) result in Figure 9.13 is interpreted as shown in Table 9.2, which reveals compounds present in the date seed nanoparticles obtained after milling for 40 hours. In total, 14 compounds were identified. The Toxtree software (version 3.1.0-1851-1525442531402) was used to estimate levels of toxicity of the detected compounds in the date seed nanoparticles, as explained in Section 9.3.3 of this chapter.

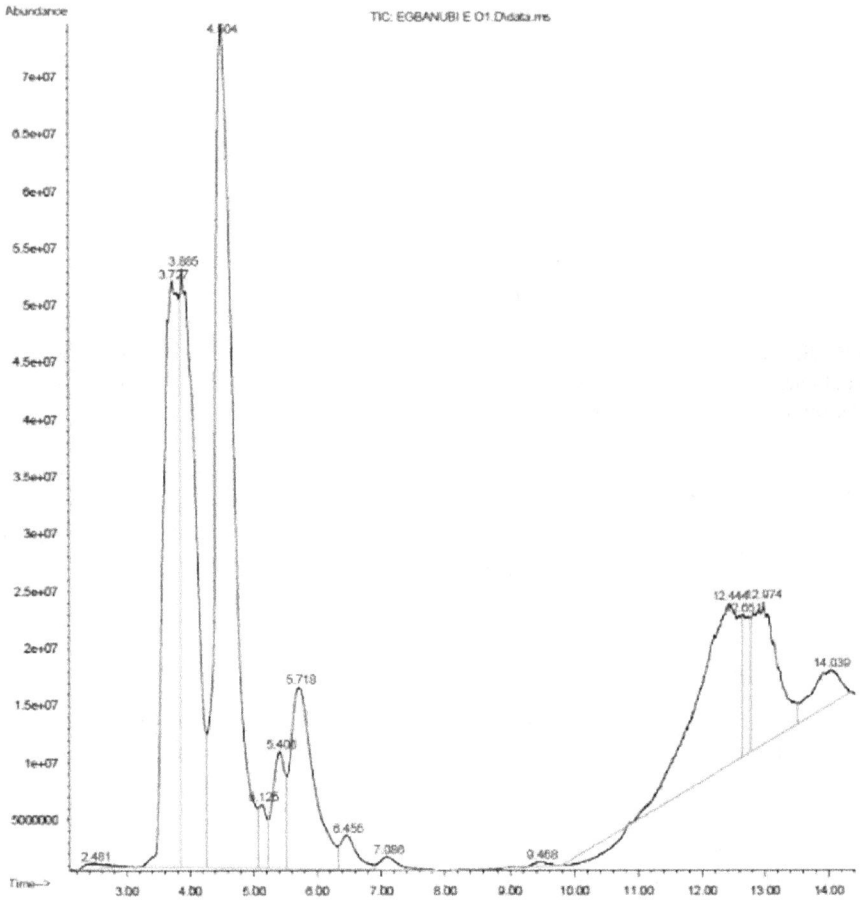

FIGURE 9.13   GCMS spectrograph showing peaks attained by various compounds detected in date seed nanoparticles.

According to Table 9.2, the predominant compounds that were identified according to the order of the peak area percent are Octyl chloroformate Cyclopropane, 1-methyl-2-(3-methylpentyl)–Aziridine, 2-ethyl-; 1-Hexene, 4-methyl-2-Propenal Aziridine, 2-methyl-; 2-Dimethylsilyloxytetradecane Methyl 3-hydroxytetradecanoate; Cis-11-Eicosenoic acid Oleic Acid cis-9-Hexadecenoic acid; and Propanenitrile, 2-hydroxy-2-Propen-1-amine, 2-methyl-2-Butene. Among the compounds identified, Octyl chloroformate Cyclopropane, 1-methyl-2-(3-methylpentyl)–Aziridine, 2-ethyl- has the highest percentage peak area of 28.99, followed by 1-Hexene, 4-methyl-2-Propenal Aziridine, 2-methyl with percentage peak area of 18.21 and 2-Dimethylsilyloxytetradecane Methyl 3-hydroxytetradecanoate with percentage peak area of 16.14. The presence of an aziridine ring in the compounds with the most predominant percentage peak area reveals that the date seed nanoparticles possess anti-tumor and anti-bacterial properties among others [16], implying that date seed addition to PMMA can improve antibacterial properties of the PMMA to treat osteomyelitis in additions to

TABLE 9.2
Chemical Composition of Date Seed Nanoparticles

Peak	Retention Time	Name	Peak Area %	Toxicity Class
1	2.481	L-Methioninol	0.27	III
		1-Cyclopentyl-1 diisopropylsilyloxyethane		III
		1,6-Dideoxy-1-mannitol		–
2	3.727	2-Dimethylsilyloxytetradecane	16.14	–
		Methyl 3-hydroxytetradecanoate		II
3	3.885	1-Hexene, 4-methyl-	18.21	I
		2-Propenal		II
		Aziridine, 2-methyl-		III
4	4.504	Octyl chloroformate	28.99	III
		Cyclopropane, 1-methyl-2-(3-methylpentyl) –		II
		Aziridine, 2- ethyl-		III
5	5.125	Methyl d-lyxofuranoside	0.94	I
		3,6,9-Trioxa-2-silaundecane, 2,2-dimethyl-		III
		Methane, dipropoxy-		I
6	5.406	Peroxide, dimethyl	2.78	III
		Urethane		I
		Peroxide, dimethyl		III
7	5.718	Propanenitrile, 2-hydroxy-	8.06	III
		2-Propen-1-amine, 2-methyl-		I
		2-Butene		I
8	6.455	2,4-Dimethylpentan-3-yl ethyl carbonate	1.08	I
		Pentane, 1-(1-ethoxyethoxy) –		I
		1,3-Dioxolane, 2-(1-methylethyl) -		III
9	7.086	Hexanoic acid, ethyl ester	0.48	I
		Pentanoic acid, 4-methyl-, ethyl ester		I
		Hexanoic acid, ethyl ester		I
10	9.468	Urea,N,N'-diethyl-	0.08	III
		2(1H)–Pyrimidinethione, tetrahydro		–
		Silylamine, 1,1,1-trimethyl-N- (.alpha.-methylphenethyl) -		–
11	12.444	Cis-11-Eicosenoic acid	12.30	I
		Oleic Acid		I
		cis-9-Hexadecenoic acid		I
12	12.651	3-Dimethylaminoacrylonitrile	2.09	III
		2-Cyclohexen-1-one, 4-(1-methylethyl) –		II
		1H- Pyrazole, 1,5-dimethyl-		–
13	12.974	Eicosanoic acid, methyl ester	6.48	I
		Octacosanoic acid, methyl ester		I
14	14.039	Oleic Acid	2.09	I
		12-Methyl-E, E-2, 13-octadecadien-1-ol		I
		6-Octadecenoic acid		I

its benefits discussed in Section 9.4.5 of this chapter. Based on results obtained from Toxtree software analysis, date seed nanoparticles contain some toxic components though no problem has been confirmed from its uses as local medication in south-western part of Nigeria for improving male reproductivity. Moreover, its use for ex-presso as a coffee alternative and usage of its oil extract in pharmacy and cosmetics have been reported in [75,76,82] with no cases attach to it. It can be inferred that the toxic components of the date seed nanoparticle still have some benefits offer to the body. Further studies on the date seed nanoparticles for certainty of its uses as additives in food, drug, and biomedical products are suggested.

9.6   SYNTHESIS, STRUCTURES, PROPERTIES, AND APPLICATIONS OF PARTICULATE-REINFORCED POLYMETHYLMETHACRYLATE NANOCOMPOSITES

Synthesis of particulate-based nanocomposites involves two stages: (i) preparation of the additive particles and (ii) nanocomposite productions. Preparation of additive particles depend on nature of particles. Generally, particles are produced either by bottom-up or top-down approaches [89,90]. Both approaches cover many processes which have been explained in many texts [90,91]. Since particle additives to PMMA are to address thermal necrosis due to heat released from further polymerisation of MMA, infections from damaged cell or tissues from exposure to elevated temperature or rubbing of PMMA components (cement/implants) against cells or tissues between the cement and bone, cement and implant and PMMA implants and bones (depending on the applications), variations in mechanical properties of the PMMA cements or implants and those of the natural bones, different particles have been added to the PMMA. To be specific, gentamycin sulfate, vancomycin, tobramycin, and clin-damycin because of their antibacterial properties are added to the PMMA to address cell or tissue infections [6]. Bioactive glasses like silicate glasses, borate glasses, and strontium-containing borate glasses [15] and ceramics like MgO, hydroxyapatite (bone material), chitosan, $BaSO_4$, and SiO_2 are incorporated in the PMMA to reduce the thermal necrosis [38] with enhancement in compressive strength, modulus of elasticity, and reduction in flexural strength [56]. All the additive particles are syn-thetic; report of their synthesis is scarce in literature. However, both bottom up and top-down approach already mentioned in this chapter remain as technique for particle synthesis. Particle purity and mass production differ from a method to another. Crushing, disc grinding, and optimised ball milling techniques are established for date seed and *Parquetina nigrescens* pod nanoparticle production. Samples of date seed particles obtained after 40 hours of ball milling are rounded in shapes with nearly equal diameters and their scale bar measured sizes indicate 10.25 nm (Figure 9.14a). In addition, *Parquetina nigrescens* pod particles are rounded but angular in shape, with varied sizes (Figure 9.14b).

According to [92], maximum temperature for polymerisation of MMA is 88.7°C ± 2.23°C. Since both date seeds and *Parquetina nigrescens* pod were dried at tem-perature 100°C without any decomposition after rinsing in water before pulverisation. This suggests that both date seed and *Parquetina nigrescens* nanoparticles could be used as additive in PMMA to control the exothermic reaction leading to thermal

(a)   (b)

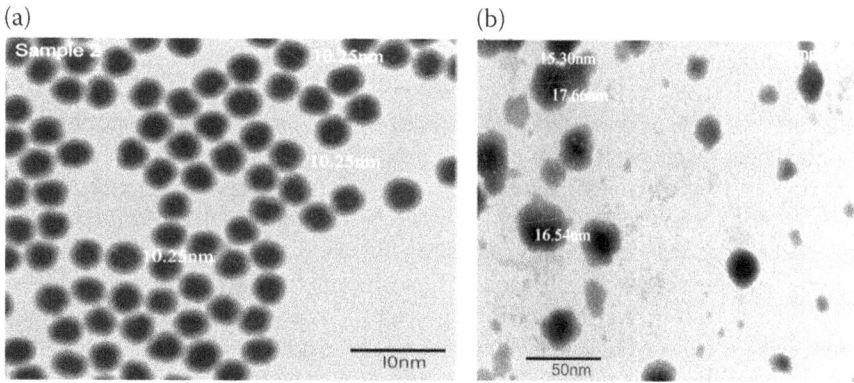

FIGURE 9.14 Transmission electron microscopic images affirming production of (a) date seed, (b) *Parquetina nigrescens* nanoparticles from aggregated techniques involving crushing, disc grinding, and 40-hour ball milling.

necrosis in addition to their probable potentials to enhance mechanical properties of the PMMA.

Particle-reinforced polymethyl methacrylate composites are produced by in situ/emulsion polymerisation and solution casting. The technique involves thorough mixing of solid additives in actual proportion. Solid additives contain bone cements, biological/antibacterial additives, and other additives to prevent a premature MMA polymerisation/exothermic reaction leading to necrosis. Once a thorough mixing of solid additives is achieved, liquid components are added and stirred. Then the solution is poured in a mould until curing takes place. The moulds have cavities which are replica of standard shapes for various analysis to perform on samples of the PMMA-based composites produced. Liquid components commonly used in the PMMA-based composites are MMA, accelerator/activator, and stabiliser. MMA undergoes polymerisation reaction to form PMMA. Accelerator/activator acts like hardener or cross-linking agents that enables MMA polymerisation while stabiliser averts premature curing of MMA. In the development of antibiotic-releasing gelatin PMMA bone cement reported in [6], 2 g of solid component was mixed with 1 ml of liquid components. Similarly, proportion of solid to liquid components in 2:1 was reported by many authors [13,38,56]. Liquid component contains 3% by weight of N, N-dimethyl-p-toluidine used as the activator, 97% by weight of MMA and 50 part per minute (ppm) of liquid hydroquinone used as stabiliser; 5% gentamicin sulfate, power of PMMA bone cement (10%–50% by weight) and 0–600 μm gelatin subject to 24 hours of continuous mixing are solid components as indicated in the study [6]. Moreover, the liquid components described in [15] has MMA monomer (3 mL) and N, N-dimethyl-p-toluidine (0.14 mL) without the addition of hydroquinone while the solid components have SrBG (10–50 mm) and PMMA (10–80 mm).

Usages of particulate particles reinforced PMMA as artificial implants and bone cements have been reported in many articles, though with identified challenges [13]. Technically, comprehensive and flexural properties (strength and modulus) are prerequisites of any materials for the target biomedical applications, forming a basis of many research dedicated to materials for artificial implants or bone cements for concentrating majorly on compressive and flexural strength and modulus. Once a particle reinforced PMMA nanocomposite is installed as an artificial implant (whether as a tooth or skeletal part in the body) or as a bone cement (connecting two different both together), it forms a network with bones for skeletal body framework operations. A lot of clinical issues [6,11,13,17] (e.g., osteoporotic vertebral compression fractures, weakness of the interface between artificial and natural bones) have been reported on property mismatch between artificial and natural bones causing inconveniences to the patient. Matching of the PMMA-based implant or bone cement with natural bones are important for safety of the patient. Therefore, knowledge of required mechanical properties of the natural and artificial bones are important before their connection to form a part of the body skeletal system. Natural bones appear in two forms depending on their density and zones: (i) cortical/impact bone; (ii) cancellous/trabecular/sponge bones. Cortical/impact bones have 90% hard/dense tissues while the remnants are soft tissues. In cortical bones, tissues appear in an organised, regularly arranged layer or sheet called lamellae. They are the external layers of all bones. They occupy 80%–90% of all bones forming the body skeletal system [92–95]. The remnant accounts for the second form of bone, which is called cancellous/trabecular/sponge bone and its surface is largely exposed to bone marrow and flow. Most metabolic activity/role like mineral exchange, production of red and white blood cells take place in the cancellous/trabecular/sponge bone. Cancellous/trabecular/sponge bone is found in both short or long bone especially in medullary cavity, metaphysis, and epiphysis at the end of long bones [96] and it transfers mechanical load to the cortical bone from the articular surface [92]. Mechanically, cortical bones are stronger, more rigid but less ductile than cancellous/trabecular/sponge bone. Their greater strength and rigidity could be due to their high calcium contents (in form of calcium phosphate or hydroxyapatite) and less porosity than cancellous bones which are approximately 70%–90% porous [97–99]. Reports on determination of bone strength shows that both cortical and trabecular bones have mechanical properties depended on directions implying that they are anisotropic in nature [98]. For instance, human femoral (cortical) bone has ultimate compressive strength of 205 Nmm^{-2} along its length (longitudinal section) and 131 Nmm^{-2} along transverse directions. Since in practice, femur carries the body weight along longitudinal direction, the longitudinal strength is meant to bear the body weight while the transverse strength resists collision impact along transverse direction. Besides directions, compositions and locations, human age and sex also affect bone's mechanical properties. As reported in [100], compressive strength of the cortical bones ranges from 130–200 MPa (1 MPa = 1 Nmm^{-2}), their compressive modulus varies from 11.5–17 GPa, and their fracture toughness are from 2–12 $MPam^{1/2}$. Moreover, compressive strength of cortical/compact bone of 85.5–329.57 Nmm^{-2} along the longitudinal direction and 31.02–323.36 Nmm^{-2} along the transverse direction were disclosed in [101]. Strength and compressive (elastic) modulus of the cortical bone are much greater than those of the trabecular bone. According to Ohman-Magi et al. [102], yield strength of trabecular bone based on the study carried out at thoracic-lumber region varies from 0.2–14 Nmm^{-2}, and compressive elastic modulus are from 0.004–0.976 GPa depending on age and sex (male or female). In addition, 0.1–16 Nmm^{-2} compressive strength and 0.12–1.1 GPa compressive (elastic) modulus of the trabecular bone were reported in [100]. Moreover, compressive strength and modulus of a PMMA polymer are 84–115 Nmm^{-2} and 1.7–3.7 GPa [6,15], respectively, and its flexural strength and modulus are 98–125 Nmm^{-2} and 2.96–3.30 GPa, respectively [103].

Since particle-reinforced PMMA targets bone cement and artificial implant applications and according to ISO 5833:2002 [104], strength requirements of bone cement materials are 70 Nmm^{-2} for compressive strength and 50 Nmm^{-2} for flexural strength. The technical requirements of the materials for bone cement are smaller than those of the PMMA polymer. This implies that pristine PMMA for bone cement applications is mechanically suitable, which implies that the PMMA for the bone cement does not need any reinforcing additives. Rather, it requires some additives that can cause reduction in compressive and flexural strength of the PMMA. In fact, uses of PMMA for bone cement underutilises its potential. Other polymers that are safe to the body and less strong than the technical requirements can replace the PMMA as the matrix for developing the material for the artificial implants. Instead of developing a porous PMMA-based composite to reduce its strength to the requirement, polymers like high-density polyethene (HDPE) can be reinforced with additive particles to improve their strength to meet up with the requirements as indicated in the standard [104]. Therefore, research on other polymer-based composites for medical applications is a gap to be filled by interested researchers across the globe. For artificial implants like hip bone replacements, further enhancement is necessary; therefore, there is a need for reinforcing fillers incorporated into the PMMA in addition to other additives.

REFERENCES

[1] Smith, W. and Hashemi, J. (2010). *Foundations of Materials Science and Engineering*. 5th ed., MC Graw-Hill.

[2] Zhou, T. and Yan, J. (2017). Glass Molding Process for Microstructures, in *Microfabrication and Precision Engineering*, J.P. Davim, Editor, Woodhead Publishing. pp. 213–262. 10.1016/B978-0-85709-485-8.00008-5

[3] Harper, C.A. and Petrie, E.M. (2003). Plastics Materials and Processes, in *Concise Encyclopedia*. Wiley. pp. 42–44.

[4] Demir, M.M., Memesa, M., Castignolles, P., and Wegner, G. (2006). Pmma/Zinc Oxide Nanocomposites Prepared by in-Situ Bulk Polymerization. *Macromolecular Rapid Communications*, **27**(10), 763–770. 10.1002/marc.200500870

[5] Ali, U., Karim, K.J.B.A., and Buang, N.A. (2015). A Review of the Properties and Applications of Poly (Methyl Methacrylate) (Pmma). *Polymer Reviews*, **55**(4), 678–705. 10.1080/15583724.2015.1031377

[6] Chen, L., Tang, Y., Zhao, K., Zha, X., Liu, J., Bai, H., and Wu, Z. (2019). Fabrication of the Antibiotic-Releasing Gelatin/Pmma Bone Cement. *Colloids Surf B Biointerfaces*, **183**, 110448. 10.1016/j.colsurfb.2019.110448

[7] Majerus, S.J., Fletter, P.C., Damaser, M.S., and Garverick, S.L. (2011). Low-Power Wireless Micromanometer System for Acute and Chronic Bladder-Pressure Monitoring. *IEEE Trans Biomed Eng*, **58**(3), 763–767. 10.1109/TBME.2010. 2085002

[8] Melgaard, J. and Rijkhoff, N.J. (2011). Detecting the Onset of Urinary Bladder Contractions Using an Implantable Pressure Sensor. *IEEE Trans Neural Syst Rehabil Eng*, **19**(6), 700–708. 10.1109/TNSRE.2011.2171368

[9] Wang, J., Hou, C., Zheng, X., Zhang, W., Chen, A., and Xu, Z. (2009). Design and Evaluation of a New Bladder Volume Monitor. *Arch Phys Med Rehabil*, **90**(11), 1944–1947. 10.1016/j.apmr.2009.06.013

[10] Dunne, N., Clements, J., and Wang, J.S. (2014). 8 – Acrylic Cements for Bone Fixation in Joint Replacement*Note: This Chapter Is an Updated Version of Chapter 10, from the First Edition of Joint Replacement Technology, Edited by P. A. Revell and Published by Woodhead Publishing, 2008*, in *Joint Replacement Technology*, P.A. Revell, Editor, Woodhead Publishing. pp. 212–256. 10.1533/9780857098474.2.212

[11] Wang, X., Kou, J.M., Yue, Y., Weng, X.S., Qiu, Z.Y., and Zhang, X.F. (2018). Clinical Outcome Comparison of Polymethylmethacrylate Bone Cement with and without Mineralized Collagen Modification for Osteoporotic Vertebral Compression Fractures. *Medicine (Baltimore)*, **97**(37), e12204. 10.1097/md.0000000000012204

[12] Ginebra, M.P. (2009). 10 – Cements as Bone Repair Materials, in *Bone Repair Biomaterials*, J.A. Planell, S.M. Best, D. Lacroix, and A. Merolli, Editors, Woodhead Publishing. pp. 271–308, 10.1533/9781845696610.2.271

[13] Sayeed, Z., Padela, M.T., El-Othmani, M.M., and Saleh, K.J. (2017). Acrylic Bone Cements for Joint Replacement, in *Biomedical Composites (Second Edition)*, L. Ambrosio, Editor, Woodhead Publishing. pp. 199–214. 10.1016/B978-0-08-100752-5.00009-3

[14] Downes, S. (1991). Methods for Improving Drug Release from Poly(Methyl) Methacrylate Bone Cement. *Clin Mater*, **7**(3), 227–231. 10.1016/0267-6605(91)90063-I

[15] Cui, X., Huang, C., Zhang, M., Ruan, C., Peng, S., Li, L., Liu, W., Wang, T., Li, B., Huang, W., Rahaman, M.N., Lu, W.W., and Pan, H. (2017). Enhanced Osteointegration of Poly(Methylmethacrylate) Bone Cements by Incorporating Strontium-Containing Borate Bioactive Glass. *J R Soc Interface*, **14**(131). 10.1098/rsif.2016.1057

[16] Sa, Y., Yu, N., Wolke, J.G.C., Chanchareonsook, N., Goh, B.T., Wang, Y., Yang, F., and Jansen, J.A. (2017). Bone Response to Porous Poly(Methyl Methacrylate) Cement Loaded with Hydroxyapatite Particles in a Rabbit Mandibular Model. *Tissue Eng Part C Methods*, **23**(5), 262–273. 10.1089/ten.TEC.2016.0521

[17] Merolli, A. (2019). 11 – Bone Repair Biomaterials in Orthopedic Surgery, in *Bone Repair Biomaterials (Second Edition)*, K.M. Pawelec and J.A. Planell, Editors, Woodhead Publishing. pp. 301–327, 10.1016/B978-0-08-102451-5.00011-1

[18] Kost, J. and Langer, R. (2001). Responsive Polymeric Delivery Systems. *Advanced Drug Delivery Reviews*, **46**(1), 125–148. 10.1016/S0169-409X(00)00136-8

[19] Shi, M., Kretlow, J.D., Spicer, P.P., Tabata, Y., Demian, N., Wong, M.E., Kasper, F.K., and Mikos, A.G. (2011). Antibiotic-Releasing Porous Polymethylmethacrylate/Gelatin/Antibiotic Constructs for Craniofacial Tissue Engineering. *Journal of Controlled Release*, **152**(1), 196–205. 10.1016/j.jconrel.2011.01.029

[20] Beruto, D.T., Botter, R., and Fini, M. (2002). The Effect of Water in Inorganic Microsponges of Calcium Phosphates on the Porosity and Permeability of Composites Made with Polymethylmethacrylate. *Biomaterials*, **23**(12), 2509–2517. 10.1016/S0142-9612(01)00385-4

[21] Xin, L., Bungartz, M., Maenz, S., Horbert, V., Hennig, M., Illerhaus, B., Günster, J., Bossert, J., Bischoff, S., Borowski, J., Schubert, H., Jandt, K.D., Kunisch, E., Kinne, R.W., and Brinkmann, O. (2016). Decreased Extrusion of Calcium Phosphate Cement Versus High Viscosity Pmma Cement into Spongious Bone

Marrow-an Ex Vivo and in Vivo Study in Sheep Vertebrae. *Spine J*, **16**(12), 1468–1477. 10.1016/j.spinee.2016.07.529

[22] Auer, J.A., Rechenberg, B.v., Bohner, M., and Hofmann-Amtenbrink, M. (2012). Chapter 77 – Bone Grafts and Bone Replacements, in *Equine Surgery (Fourth Edition)*, J.A. Auer and J.A. Stick, Editors, W.B. Saunders. pp. 1081–1096, 10.1016/B978-1-4377-0867-7.00077-6

[23] McGirt, M.J., Parker, S.L., Wolinsky, J.P., Witham, T.F., Bydon, A., and Gokaslan, Z.L. (2009). Vertebroplasty and Kyphoplasty for the Treatment of Vertebral Compression Fractures: An Evidenced-Based Review of the Literature. *Spine J*, **9**(6), 501–508. 10.1016/j.spinee.2009.01.003

[24] Joung, Y.H. (2013). Development of Implantable Medical Devices: From an Engineering Perspective. *Int Neurourol J*, **17**(3), 98–106. 10.5213/inj.2013.17.3.98

[25] Pillay, V., Seedat, A., Choonara, Y.E., du Toit, L.C., Kumar, P., and Ndesendo, V.M. (2013). A Review of Polymeric Refabrication Techniques to Modify Polymer Properties for Biomedical and Drug Delivery Applications. *AAPS PharmSciTech*, **14**(2), 692–711. 10.1208/s12249-013-9955-z

[26] Shalabi, M.M., Wolke, J.G., Cuijpers, V.M., and Jansen, J.A. (2007). Evaluation of Bone Response to Titanium-Coated Polymethyl Methacrylate Resin (Pmma) Implants by X-Ray Tomography. *J Mater Sci Mater Med*, **18**(10), 2033–2039. 10.1007/s10856-007-3160-0

[27] Kim, S.J., Choi, B., Kim, K.S., Bae, W.J., Hong, S.H., Lee, J.Y., Hwang, T.K., and Kim, S.W. (2015). The Potential Role of Polymethyl Methacrylate as a New Packaging Material for the Implantable Medical Device in the Bladder. *Biomed Res Int*, **2015**, 852456. 10.1155/2015/852456

[28] Lee, D.S., Kim, S.J., Sohn, D.W., Choi, B., Lee, M.K., Lee, S.J., and Kim, S.W. (2011). Real-Time Bladder Volume Monitoring by the Application of a New Implantable Bladder Volume Sensor for a Small Animal Model. *The Kaohsiung Journal of Medical Sciences*, **27**(4), 132–137. 10.1016/j.kjms.2010.12.006

[29] Harb, S.V., Trentin, A., de Souza, T.A.C., Magnani, M., Pulcinelli, S.H., Santilli, C.V., and Hammer, P. (2020). Effective Corrosion Protection by Eco-Friendly Self-Healing Pmma-Cerium Oxide Coatings. *Chemical Engineering Journal*, **383**, 123219. 10.1016/j.cej.2019.123219

[30] Wang, M., Pramoda, K.P., and Goh, S.H. (2005). Enhancement of the Mechanical Properties of Poly(Styrene-Co-Acrylonitrile) with Poly(Methyl Methacrylate)-Grafted Multiwalled Carbon Nanotubes. *Polymer*, **46**(25), 11510–11516. 10.1016/j.polymer.2005.10.007

[31] Nair, L.S. and Laurencin, C.T. (2006). Polymers as Biomaterials for Tissue Engineering and Controlled Drug Delivery. *Adv Biochem Eng Biotechnol*, **102**, 47–90. 10.1007/b137240

[32] Feuser, P.E., Gaspar, P.C., Ricci-Júnior, E., Silva, M.C.S.d., Nele, M., Sayer, C., and de Araújo, P.H.H. (2014). Synthesis and Characterization of Poly(Methyl Methacrylate) Pmma and Evaluation of Cytotoxicity for Biomedical Application. *Macromolecular Symposia*, **343**(1), 65–69. 10.1002/masy.201300194

[33] Harper, E.J. (1998). Bioactive Bone Cements. *Proc Inst Mech Eng H*, **212**(2), 113–120. 10.1243/0954411981533881

[34] Robo, C., Hulsart-Billström, G., Nilsson, M., and Persson, C. (2018). In Vivo Response to a Low-Modulus Pmma Bone Cement in an Ovine Model. *Acta Biomater*, **72**, 362–370. 10.1016/j.actbio.2018.03.014

[35] Revell, P.A., Braden, M., and Freeman, M.A. (1998). Review of the Biological Response to a Novel Bone Cement Containing Poly(Ethyl Methacrylate) and N-Butyl Methacrylate. *Biomaterials*, **19**(17), 1579–1586. 10.1016/s0142-9612(97)00118-x

[36] Okada, Y., Kawanabe, K., Fujita, H., Nishio, K., and Nakamura, T. (1999). Repair of Segmental Bone Defects Using Bioactive Bone Cement: Comparison with Pmma Bone Cement. *J Biomed Mater Res*, **47**(3), 353–359. 10.1002/(sici)1097-4636 (19991205)47:3<353::aid-jbm9>3.0.co;2-p

[37] Oh, E.J., Oh, S.H., Lee, I.S., Kwon, O.S., and Lee, J.H. (2016). Antibiotic-Eluting Hydrophilized Pmma Bone Cement with Prolonged Bactericidal Effect for the Treatment of Osteomyelitis. *J Biomater Appl*, **30**(10), 1534–1544. 10.1177/088532 8216629823

[38] Khandaker, M. and Meng, Z. (2015). The Effect of Nanoparticles and Alternative Monomer on the Exothermic Temperature of Pmma Bone Cement. *Procedia Eng*, **105**, 946–952. 10.1016/j.proeng.2015.05.120

[39] Mori, A., Ohtsuki, C., Miyazaki, T., Sugino, A., Tanihara, M., Kuramoto, K., and Osaka, A. (2005). Synthesis of Bioactive Pmma Bone Cement Via Modification with Methacryloxypropyltri-Methoxysilane and Calcium Acetate. *J Mater Sci Mater Med*, **16**(8), 713–718. 10.1007/s10856-005-2607-4

[40] Fuchs, R.K., Thompson, W.R., and Warden, S.J. (2019). Bone Biology, in *Bone Repair Biomaterials (Second Edition)*, K.M. Pawelec and J.A. Planell, Editors, Woodhead Publishing. pp. 15–52. 10.1016/B978-0-08-102451-5.00002-0

[41] González-Carrasco, J.L., Cifuentes Cuellar, S.C., and Lieblich Rodríguez, M. (2019). Metals, in *Bone Repair Biomaterials (Second Edition)*, K.M. Pawelec and J.A. Planell, Editors, Woodhead Publishing. pp. 103–140, 10.1016/B978-0-08-102451-5.00005-6

[42] Lacroix, D. (2019). Biomechanical Aspects of Bone Repair, in *Bone Repair Biomaterials (Second Edition)*, K.M. Pawelec and J.A. Planell, Editors, Woodhead Publishing. pp. 53–64, 10.1016/B978-0-08-102451-5.00003-2

[43] Pawelec, K.M. (2019). Introduction to the Challenges of Bone Repair, in *Bone Repair Biomaterials (Second Edition)*, K.M. Pawelec and J.A. Planell, Editors, Woodhead Publishing. pp. 1–13, 10.1016/B978-0-08-102451-5.00001-9

[44] Pawelec, K.M., White, A.A., and Best, S.M. (2019). Properties and Characterization of Bone Repair Materials, in *Bone Repair Biomaterials (Second Edition)*, K.M. Pawelec and J.A. Planell, Editors, Woodhead Publishing. pp. 65–102, 10.1016/B978-0-08-102451-5.00004-4

[45] Harish Prashanth, K.V., Lakshman, K., Shamala, T.R., and Tharanathan, R.N. (2005). Biodegradation of Chitosan-Graft-Polymethylmethacrylate Films. *International Biodeterioration & Biodegradation*, **56**(2), 115–120. 10.1016/j.ibiod.2005.06.007

[46] Krishnakumar, S. and Senthilvelan, T. (2021). Polymer Composites in Dentistry and Orthopedic Applications-a Review. *Materials Today: Proceedings*, **46**, 9707–9713. 10.1016/j.matpr.2020.08.463

[47] Bhambri, S.K. and Gilbertson, L.N. (1995). Micromechanisms of Fatigue Crack Initiation and Propagation in Bone Cements. *Journal of Biomedical Materials Research*, **29**(2), 233–237. 10.1002/jbm.820290214

[48] Ginebra, M.P., Albuixech, L., Fernández-Barragán, E., Aparicio, C., Gil, F.J., San, R.J., Vázquez, B., and Planell, J.A. (2002). Mechanical Performance of Acrylic Bone Cements Containing Different Radiopacifying Agents. *Biomaterials*, **23**(8), 1873–1882. 10.1016/s0142-9612(01)00314-3

[49] Haas, S.S., Brauer, G.M., and Dickson, G. (1975). A Characterization of Polymethylmethacrylate Bone Cement. *J Bone Joint Surg Am*, **57**(3), 380–391.

[50] Belkoff, S.M., Sanders, J.C., and Jasper, L.E. (2002). The Effect of the Monomer-to-Powder Ratio on the Material Properties of Acrylic Bone Cement. *J Biomed Mater Res*, **63**(4), 396–399. 10.1002/jbm.10258

[51] Klaus-Dieter, K. (2000). Bone Cements, up-to-Date Comparison of Physical and Chemical Properties of Commercial Materials. Springer.

[52] Beck, S. and Boger, A. (2009). Evaluation of the Particle Release of Porous Pmma Cements During Curing. *Acta Biomater*, **5**(7), 2503–2507. 10.1016/j.actbio. 2009.04.002

[53] Chen, L., Tang, Y., Zhao, K., Zha, X., Liu, J., Bai, H., and Wu, Z. (2019). Fabrication of the Antibiotic-Releasing Gelatin/Pmma Bone Cement. *Colloids and Surfaces B: Biointerfaces*, **183**, 110448. 10.1016/j.colsurfb.2019.110448

[54] Zhai, Q., Han, F., He, Z., Shi, C., Zhou, P., Zhu, C., Guo, Q., Zhu, X., Yang, H., and Li, B. (2018). The "Magnesium Sacrifice" Strategy Enables Pmma Bone Cement Partial Biodegradability and Osseointegration Potential. *Int J Mol Sci*, **19**(6). 10.3390/ijms19061746

[55] Khandaker, M., Utsaha, K.C., and Morris, T. (2014). Fracture Toughness of Titanium-Cement Interfaces: Effects of Fibers and Loading Angles. *Int J Nanomedicine*, **9**, 1689–1697. 10.2147/IJN.S59253

[56] Khandaker, M., Vaughan, M.B., Morris, T.L., White, J.J., and Meng, Z. (2014). Effect of Additive Particles on Mechanical, Thermal, and Cell Functioning Properties of Poly(Methyl Methacrylate) Cement. *Int J Nanomedicine*, **9**, 2699–2712. 10.2147/ ijn.s61964

[57] Ricker, A., Liu-Snyder, P., and Webster, T. (2008). The Influence of Nano Mgo and Baso4 Particle Size Additives on Properties of Pmma Bone Cement. *International Journal of Nanomedicine*, **3**, 125–132.

[58] Lv, Y., Li, A., Zhou, F., Pan, X., Liang, F., Qu, X., Qiu, D., and Yang, Z. (2015). A Novel Composite Pmma-Based Bone Cement with Reduced Potential for Thermal Necrosis. *ACS Applied Materials & Interfaces*, **7**(21), 11280–11285. 10.1021/ acsami.5b01447

[59] Adoga, S., Ekle, D., Kyenge, B., Aondo, T., and Ikese, C. (2019). Phytochemical Screening, Thin-Layer Chromatography and Antimicrobial Activity Study of Parquetina Nigrescens Leaf Extracts. *Ovidius University Annals of Chemistry*, **30**(2), 88–94. 10.2478/auoc-2019-0016

[60] Marks, W.H., Fong, H.H.S., Tin-Wa, M., and Farnsworth, N.R. (1975). Cytotoxic Principles of Parquetina Nigrescens (Afzel.) Bullock (Asclepiadaceae). *Journal of Pharmaceutical Sciences*, **64**(10), 1674–1676. 10.1002/jps.2600641019

[61] Schmelzer, G.H., and Gurib-Fakim A., (2008). *Medicinal plants* (2). PROTA Foundation.

[62] Alvarez Cruz, N.S. (2012). Parquetina Nigrescens (Afzel) Bullock, in *Prota 11(2): Medicinal Plants/Plantes Médicinales*, G.H. Schmelzer and A. Gurib-Fakim, Editors, Prota 11(2): Medicinal plants/Plantes médicinales 2. PROTA.

[63] Bukunmi Aborisade, A. (2017). Phytochemical and Proximate Analysis of Some Medicinal Leaves. *Clinical Medicine Research*, **6**(6), 209. 10.11648/j.cmr. 20170606.16

[64] Kayode, A.A.A., Kayode, O.T., and Odetola, A.A. (2009). Antiulcerogenic Activity of Two Extracts of Parquetina Nigrescens and Their Effects on Mucosal Antioxidants Defense System on Ethanol-Induced Ulcer in Rats. *Research Journal of Medicinal Plant*, **3**, 102–108.

[65] Imaga, N.O.A., Adenekan, S.O., Yussuph, G.A., Nwoyimi, T.I., Balogun, O.O., and Eguntola, T.A. (2010). Assessment of Antioxidation Potential of Selected Plants with Antisickling Property. *Journal of Medicinal Plants Research*, **4**(21), 2217–2221. 10.5897/JMPR10.175

[66] Sofowora, A. (1993). Medicinal Plants and Traditional Medicine in Africa. Spectrum Books. 9789782462190

[67] Bello, S.A. and Kolawole, M.Y. (2020). Recycled Plastics and Nanoparticles for Green Production of Nano Structural Materials, in *Handbook of Nanomaterials and Nanocomposites for Energy and Environmental Applications*, O.V. Kharissova,

L.M.T. Martínez, and B.I. Kharisov, Editors, Springer International Publishing. pp. 1–33. 10.1007/978-3-030-11155-7_93-1

[68] Kola-Mustapha, A.T., Ghazali, Y.O., Ayotunde, H.T., Atunwa, S.A., and Usman, S.O. (2019). Evaluation of the Antidiarrheal Activity of the Leaf Extract of Parquetina Nigrescens and Formulation into Oral Suspensions. *J Exp Pharmacol*, **11**, 65–72. 10.2147/JEP.S214417

[69] Ozaslan, M. (2011). Parquetina Nigrescens Checks the Ulceration and Oxidation. *Pakistan Journal of Biological Sciences*, **14**, 1124–1125.

[70] Dyamenahalli, K., Famili, A., and Shandas, R. (2015). Characterization of Shape-Memory Polymers for Biomedical Applications, in *Shape Memory Polymers for Biomedical Applications*, L.H. Yahia, Editor, Woodhead Publishing. pp. 35–63. 10.1016/B978-0-85709-698-2.00003-9

[71] Passos, M.L.C., Sarraguça, M.C., Saraiva, M.L.M.F.S., Prasada Rao, T., and Biju, V.M. (2018). Organic Compounds. 10.1016/b978-0-12-409547-2.14465-8

[72] Faraji, M., Yamini, Y., and Salehi, N. (2021). 3 - Characterization of Magnetic Nanomaterials, in *Magnetic Nanomaterials in Analytical Chemistry*, M. Ahmadi, A. Afkhami, and T. Madrakian, Editors, Elsevier. pp. 39–60. 10.1016/B978-0-12-822131-0.00014-5

[73] Picollo, M., Aceto, M., and Vitorino, T. (2019). Uv-Vis Spectroscopy. *Physical Sciences Reviews*, **4**(4). 10.1515/psr-2018-0008

[74] Berk, Z. (2009). Chapter 25 – Ionizing Irradiation and Other Non-Thermal Preservation Processes, in *Food Process Engineering and Technology*, Z. Berk, Editor, Academic Press. pp. 533–544. 10.1016/B978-0-12-373660-4.00025-9

[75] Contributors, W. (2021). Ultraviolet–Visible Spectroscopy, in *Wikipedia, The Free Encyclopedia*. https://en.wikipedia.org/w/index.php?title=Ultraviolet%E2%80%93visible_spectroscopy&oldid=1059796368

[76] Ghnimi, S., Umer, S., Karim, A., and Kamal-Eldin, A. (2017). Date Fruit (Phoenix Dactylifera L.): An Underutilized Food Seeking Industrial Valorization. *NFS Journal*, **6**, 1–10. 10.1016/j.nfs.2016.12.001

[77] Al-Farsi, M.A. and Lee, C.Y. (2011). Chapter 53 – Usage of Date (Phoenix Dactylifera L.) Seeds in Human Health and Animal Feed, in *Nuts and Seeds in Health and Disease Prevention*, V.R. Preedy, R.R. Watson, and V.B. Patel, Editors, Academic Press. pp. 447–452. 10.1016/B978-0-12-375688-6.10053-2

[78] Al-Yahyai, R. and Manickavasagan, A. (2012). An Overview of Date Palm Production, in *Dates Production, Processing, Food, and Medicinal Values*, A. Manickavasagan, E.M. Mohamed, and E. Sukuma, Editors, CRC Press.

[79] Buki, B., Botes, G., Zaid, A., and Emmens, J. (2002). Date Harvesting, Packing House Management and Marketing Aspects, in *Date Palm Cultivation*. FAO. https://www.fao.org/3/y4360e/y4360e0d.htm

[80] Al-Harrasi, A., Rehman, N.U., Hussain, J., Khan, A.L., Al-Rawahi, A., Gilani, S.A., Al-Broumi, M., and Ali, L. (2014). Nutritional Assessment and Antioxidant Analysis of 22 Date Palm (Phoenix Dactylifera) Varieties Growing in Sultanate of Oman. *Asian Pacific Journal of Tropical Medicine*, **7**, S591–S598. 10.1016/s1995-7645(14)60294-7

[81] Sani, L.A., Aliyu, M.D., Hamza, A., Adetunji, O.A., Gidado, R.M., and Solomon, B.O. (2010). Exploring the Nigerian Date Palm (Phoenix Dactylifera L.) Germplasm for in Vitro Callogenesis, in *IV International Date Palm Conference*, A. Zaid and G.A. Alhadrami, Editors, International Society for Horticultural Science, pp. 117–184.

[82] Gramza-Michałowska, A. and Kmiecik, D. (2016). Functional Aspects of Antioxidants in Traditional Food, Springer, pp. 3–7. 10.1007/978-1-4899-7662-8_1

[83] Ali, A., Waly, M., Essa, M.M., and Devarajan, S. (2012). Nutritional and Medicinal Value of Date Fruit, in *Dates Production, Processing, Food, and Medicinal Values*, A. Manickavasagan, M.E. Mohamed, and E. Sukumar, Editors, CRC Press, pp. 361–375.

[84] Vayalil, P.K. (2012). Date Fruits (Phoenix Dactylifera Linn): An Emerging Medicinal Food. *Crit Rev Food Sci Nutr*, **52**(3), 249–271. 10.1080/10408398.2010. 499824

[85] Hossain, M., Waly, M., Singh, V., Sequeira, V., and Rahman, M. (2014). Chemical Compositions of Date-Pits and Its Potential for Developing Value-Added Product – a Review. *Polish Journal of Food and Nutrition Sciences*, **64**(4), 215–226. 10.2478/ pjfns-2013-0018

[86] Al-Shahib, W. and Marshall, R.J. (2003). The Fruit of the Date Palm: Its Possible Use as the Best Food for the Future? *Int J Food Sci Nutr*, **54**(4), 247–259. 10.1080/ 09637480120091982

[87] Rahmani, A.H., Aly, S.M., Ali, H., Babiker, A.Y., Srikar, S., and Khan, A.A. (2014). Therapeutic Effects of Date Fruits (Phoenix Dactylifera) in the Prevention of Diseases Via Modulation of Anti-Inflammatory, Anti-Oxidant and Anti-Tumour Activity. *International journal of clinical and experimental medicine*, **7**(3), 483–491.

[88] Brouk, M. and Fishman, A. (2016). Antioxidant Properties and Health Benefits of Date Seeds, in *Functional Properties of Traditional Foods. Integrating Food Science and Engineering Knowledge Into the Food Chain*, K. Kristbergsson , and S. Ötles, Editors, Springer, vol. 12. 10.1007/978-1-4899-7662-8_16

[89] Bello, S.A., Agunsoye, J.O., and Hassan, S.B. (2015). Synthesis of Coconut Shell Nanoparticles Via a Top Down Approach: Assessment of Milling Duration on the Particle Sizes and Morphologies of Coconut Shell Nanoparticles. *Materials Letters*, **159**, 514–519. 10.1016/j.matlet.2015.07.063

[90] Pokropivny, V., Lohmus, R., Hussainova, I., Pokropivny, A., and Vlassov, S.,(2007). Introduction to Nanomaterials and Nanotechnology, pp. 192, Tartu University Press. 978–9949–11–741–3.

[91] Prasad Yadav, T., Manohar Yadav, R., and Pratap Singh, D. (2012). Mechanical Milling: A Top Down Approach for the Synthesis of Nanomaterials and Nanocomposites. *Nanoscience and Nanotechnology*, **2**(3), 22–48. 10.5923/j.nn. 20120203.01

[92] Ott, S.M. (2018). Cortical or Trabecular Bone: What's the Difference? *Am J Nephrol*, **47**(6), 373–375. 10.1159/000489672

[93] Monier-Faugere, M.-C., Chris Langub, M., and Malluche, H.H. (1998). Chapter 8 - Bone Biopsies: A Modern Approach, in *Metabolic Bone Disease and Clinically Related Disorders (Third Edition)*, L.V. Avioli and S.M. Krane, Editors, Academic Press. pp. 237–280e. 10.1016/B978-012068700-8/50009-8

[94] Compston, J. (1999). Chapter 23 – Histomorphometric Manifestations of Age-Related Bone Loss, in *The Aging Skeleton*, C.J. Rosen, J. Glowacki, and J.P. Bilezikian, Editors, Academic Press. pp. 251–261. 10.1016/B978-012098655-2/ 50025-9

[95] Morgan, E.F., Barnes, G.L., and Einhorn, T.A. (2013). Chapter 1 – the Bone Organ System: Form and Function, in *Osteoporosis (Fourth Edition)*, R. Marcus, D. Feldman, D.W. Dempster, M. Luckey, and J.A. Cauley, Editors, Academic Press. pp. 3–20. 10.1016/B978-0-12-415853-5.00001-7

[96] Bilgiç, E., Boyacıoğlu, Ö., Gizer, M., Korkusuz, P., and Korkusuz, F. (2020). Chapter 6 – Architecture of Bone Tissue and Its Adaptation to Pathological Conditions, in *Comparative Kinesiology of the Human Body*, S. Angin and I.E. Şimşek, Editors, Academic Press. pp. 71–90. 10.1016/B978-0-12-812162-7.00006-0

[97] Morgan, E.F., Barnes, G.L., and Einhorn, T.A. (2013). The Bone Organ System. pp. 3–20. 10.1016/b978-0-12-415853-5.00001-7

[98] Bevill, G., Farhamand, F., and Keaveny, T.M. (2009). Heterogeneity of Yield Strain in Low-Density Versus High-Density Human Trabecular Bone. *J Biomech*, **42**(13), 2165–2170. 10.1016/j.jbiomech.2009.05.023

[99] McNamara, L.M. (2017). 2.10 Bone as a Material ☆, in *Comprehensive Biomaterials II*, P. Ducheyne, Editor, Elsevier, pp. 202–227. 10.1016/b978-0-12-803581-8.10127-4

[100] Gerhardt, L.C. and Boccaccini, A.R. (2010). Bioactive Glass and Glass-Ceramic Scaffolds for Bone Tissue Engineering. *Materials (Basel)*, **3**(7), 3867–3910. 10.3390/ma3073867

[101] Evans, F.G. and Herbert, R.L. (1957). Tensile and Compressive Strength of Human Parietal Bone. *Journal of Applied Physiology*, **10**(3), 493–497. 10.1152/jappl.1957.10.3.493

[102] Ohman-Magi, C., Holub, O., Wu, D., Hall, R.M., and Persson, C. (2021). Density and Mechanical Properties of Vertebral Trabecular Bone-a Review. *JOR Spine*, **4**(4), e1176. 10.1002/jsp2.1176

[103] Data, M.P., (1996–2002). Overview of Materials for Acrylic, Cast *MATWEB*. https://www.matweb.com/search/datasheet.aspx?bassnum=O1303&ckck=1

[104] ISO (2002). Iso 5833. 2002 Implants for Surgery—Acrylic Resin Cements. *ISO*.

Section IV

Industrial Applications

10 Particulate Hybrid Epoxy Nanocomposite: Structures, Tensile Properties, Regression Analysis, and Applications

Sefiu Adekunle Bello
Department of Materials Science and Engineering, Kwara State University, Malete, Kwara State, Nigeria

CONTENTS

DOI: 10.1201/9781003170549-14

10.1   INTRODUCTION

Materials and designs are dependent terms which are inevitable in creating new artefacts for improving world/society we live in. Design proffers steps/guides to create new artefacts that offer solutions to societal problems; however, the artefacts constitute appropriate materials to enable performances of its responsibilities. For every design to create a new product, qualified materials are specified to offer required properties expected of the product [1]. Therefore, advances in materials developments/processes for materials evolution to meet design demands are expected to be continuous practise across the globe. Different materials are identified including metals, polymers, composites, and ceramics with peculiar properties enable their applications in agriculture, mining, oil, medicine, building, and transportation. Heavy steels/alloys are used in agriculture because of involvement of huge machineries for excavating and bulldozing. Ceramics like diamond, tungsten, and titanium carbides, being hard and resistant to wear are used in making tools (e.g., drill bits) profusely used in mining and in well operations; ytralox for optical application; mullite and spinel as refractory materials for furnace lining and partially stabilised zirconia for die making used in metal extrusion. Biomaterials are employed to replace, support, or repair part (tissue, organ, or functions) that has been malfunctioned due to ageing, diseases, or damaged/lost in accidents. Gold and seashells as dental implants form part of historical biomaterials [2]. Ultrahigh molecular weight polyethylene is employed in artificial hip arthroplasty system, polymethylmethacrylate for intraocular lenses, desk-sized pacemaker, synthetic heart valve, Mg-based alloys like Mg-Zn, Mg-Zn-Si, Mg-Zn-Sr, and Mg-Zn-Y for biomedical stent applications, and Ti-based alloy and bamboo for artificial tibia implants [3]. Materials evolutions are witnessed in space, ships, and automobiles to improve fuel efficiency, which is a global requirement for decreasing gas emissions. Initially, heavy steels were used for automobile outer parts to enhance its crashworthiness in accidents. Outer panel of recent lightweight cars are made of high strength steel, advanced high strength steel, twining induced plastic steels, and light nonferrous metal alloys like Mg and Al alloys (AA6111 and AA5754) [4,5]. Presently, further reduction in body in weight has been achieved via applications of polymeric composites in automobiles cutting across interior panels, some exterior panels such as bumper cover and engine bays. Matrix for the polymeric automobile composites range from thermoplastics (polypropylene, nylon) and thermoset of which epoxy resin is prominent because of its superior properties to those of other thermosets justifying its wide applications as the matrix for developing composites for various components forming parts of automobile, aircraft, spacecraft, and marine equipment.

Epoxy resin is a thermoset belonging to the oxirane group. It reacts with hardener to produce intermediate compounds cross-linking with one another to give rise to a relatively rigid but brittle structure susceptible to cracking. Being resistant to chemical attacks, its properties are improved through modification with graphene, rubbers, thermoplastics, reinforcing fibres, nanotubes, and particles resulting in reinforced plastics of superior properties [6,7]. Application of epoxy composites is limited to luxury applications such as in military aircraft and luxurious racing/sport cars where their high- ost emanating from expensive epoxy resin and synthetic fibres can be justified. Moreover, usages of epoxy composites can be extended to non-luxurious areas like in commercial vehicles and airplanes if a cheap approach to epoxy composites via source of reinforcement from wastes are established without compromising the established properties.

Domestic and agricultural wastes are potential risks to human health and contribute immensely to environmental problems. Domestic wastes are perfume, cream, food and drink packages like metallic and polymeric cans, bottles and films which are discarded after consuming/using their contents. Agriculture produces wide food varieties across the globe to meet food demands for increasing human populations. Remnants like shells, seeds, pods, and chaff left, after removal of edible parts, create environmental nuisances and foul smells from them when rotten threaten human health. Both domestics, if attended to, are sources of raw materials for green material productions thereby creating wealth rather environmental problems. Reports are found on aluminium cans as major material for developing metal matrix composites [8–11]. Recently, nanoparticles for polymer reinforcement was reported to have been synthesised from coconut shells [12], periwinkle shells [13], maize stalk [14] and aluminium can [15,16] using top down approach. Furthermore, investigation on mechanical, thermal, and physical properties of jute, coconut shell, kenaf, curaua, and flax have been reported [17,18] to establish their performance as reinforcements to replace expensive glass and carbon fibres as epoxy reinforcement. This approach has many benefits, including value addition to wastes, environmental hygiene, and production of eco-friendly epoxy composites. Being a contribution towards establishing eco-friendly epoxy hybrid composites and to address a high-cost barrier to widespread applications of epoxy composites in passenger vehicles and airplanes, this chapter reports properties of an epoxy nanocomposite containing mixed aluminium-coconut shell nanoparticles as a hybrid reinforcement in epoxy.

10.2 MANUFACTURING OF HYBRID PARTICLE-REINFORCED EPOXY

10.2.1 HYBRID REINFORCEMENTS

Reinforcements are load bearers that are connected by a matrix. They are fibres or particles. Hybrid reinforcement refers to at least two different materials used as the reinforcements in the matrix. The hybrid reinforcement can be classified as particle-particle, particle-fibre, fibre-fibre, agricultural-agricultural, agricultural-synthetic, and synthetic-synthetic reinforcements. Particle-particle hybrid reinforcements contain particles of different materials used as reinforcements in the matrix, like glass and graphite particles reinforced epoxy reported in [19]. Particle-fibre re-inforcements are hybrid reinforcement obtained from mixture of fibres and particles of the same or different materials. For instance, a mixture of mullite-like oxide crystal (2–8 μm diameter, 80–200 μm length) fibres and powders of boron carbides (100 μm), silicon carbide (80 μm), zirconium diboride (100 μm), basalt (100 μm), and diluvium powder (60 μm) used as hybrid reinforcement in an epoxy-Diane matrix for fabricating wind turbine blades [20] falls within the particle-fibre group. Another example on particle-fibre hybrid reinforcement are silica nanoparticle (20 nm) modified halloysite (length: 100–500 nm and diameter: 10–50 nm) in epoxy [21] and molybdenum disulfide powders and carbon nanofibers in epoxy found in [22]. Fibre-fibre hybrids have fibres of different materials used as re-inforcement in the matrix. Barley husk and coconut shell fibres in polypropylene [23], carbon and glass fibre in epoxy [24], and jute and banana fibres in epoxy [25] exemplify this class of the hybrid reinforcement. Both reinforcements can be sourced naturally from agricultural wastes (agricultural-agricultural hybrid) like jute and banana fibres in polypropylene. Similarly, both reinforcements can be synthetic materials (synthetic-synthetic hybrid) such as carbon and glass fibre in epoxy. Also, report from literature indicates hybrid reinforcement from agricultural and synthetic materials like glass-kenaf fibre in epoxy [26].

Concept of reinforcement hybridization was introduced to complement proper-ties of a mono-filler composites to give rise to high-performance composites for special applications or to cut the cost of production down. Both reinforcements forming hybrid act synergically to offer composites exceptional properties. Hybrids are chosen such that required properties that a reinforcement lacks, are provided by another reinforcement. Therefore, cost reduction and performance enhancements are reasons for reinforcement hybridisation in composites. Partial replacement of glass fibre with kenaf fibre in developing glass fibre reinforced epoxy is an attempt to add values to the agricultural wastes (kenaf) [26], reducing the amount of glass fibre used to cut the cost of production of the glass fibre reinforced epoxy. Moreover, addition of boron nitride nanoparticle to carbon nanotube reinforced epoxy [27] is meant for further improvement in epoxy composite properties. However, the addition of aluminum-coconut shell nanoparticles to epoxy resin reported in this chapter targets the improvement in properties of the epoxy. Aluminium nanoparticles are added to epoxy to improve the toughness of the epoxy composite and the coconut shell nanoparticle addition is to increase strength.

10.2.2 Synthesis of Aluminium Nanoparticles

Aluminium nanoparticles are promising materials, and their applications are found in aircrafts, automobiles, and biomedical artefacts. Also, they are used as catalysts for rocket fuel reaction and for hydrogen storage applications. They are synthesised using a chemical method, pulse laser ablation in liquid, cryogenic, wet, and dry milling. A chemical approach to aluminium nanoparticle synthesis involves a re-action of different species to eventually precipitate aluminium particles whose diameter is less 100 nm. The process called bottom-up involves a combination of atoms to produce ultrafine aluminium particles. Ball milling and pulse laser ablation techniques are top-down approaches that involve breakage of large aluminium body into ultrafine particles. Pulse laser ablation in liquid is described in [28] and uses an aluminium target of 1.2 cm submerged in vinegar of a pH 2. Laser beam of wavelength 532 nm, 2 W power at a repetition rate of 10 Hz was focused on the surface of aluminium target via a screen of 5 cm lens; 3.8 mg of aluminium particles of 12 ± 9 nm average size determined from a transmission electron microscope supported with ImageJ software was reported after 15 minutes. A small amount of alumina (Al_2O_3) or carbon having an average size of 50 nm was detected as impurities using a high resolution transmission electron microscope and Fourier infrared spectroscopy. White vinegar medium acts as an inert environment, reducing interaction of aluminium nanoparticles and environmental oxygen, limiting Al_2O_3 formation [28]. The impact of the laser beam on the target is manifested in the transformation of electromagnetic energy into electronic excitation and thermal energy, which raises the target temperature. The white vinegar cools the target and improves its brittleness, giving rise to particle attrition from the target as the exposure to the laser beam increases.

Ball milling has been successfully used for nanoparticle synthesis with a very large turn-out, but production of nanoparticle production from aluminium being very ductile requires some conditions that causes reduction its ductility. Its high ductility limits its breakage during the ball milling process. Instead of fracture/breakage due to milling ball impacts, they are flattened. This challenge is addressed in different ways, including milling in a controlled (cryogenic) environment much cooler than room temperature, milling in a liquid (water) medium known as wet milling, and dry milling using a solid lubricant. Cryogenic milling using liquid nitrogen was reported in [29] for nanoparticle synthesis from Nc-6061 Al powder coated with 0.2% stearic acid acting process control agent. Milling of 50 μm sized Nc-6061 Al powder takes place in attritor ball mill using 6.4 mm grinding medium steel balls immersed in liquid nitrogen inside stainless steel cylinder/vial rotated at 180 revolution per minute (RPM) and 30 charge ratios (weight of the balls/weight of the powder). Sizes of nanoparticles obtained at 15 hours of milling determined from TEM and XRD are 23 and 20 nm. It is worth noted that attrition milling introduced carbon, nitrogen, oxygen, and iron as impurities in the Nc-6061 nanoparticles [29]. In dry ball milling of 51 μm sized Al(1XXX) powders using carbon as the process control agent/solid lubricant, aluminium nanoparticles of 201.6, 122.6 and 107.1 nm average size were reported at 16, 46 and 70 hours of milling at 8.5 charge ratios using planetary ball mill with its vial rotates at 195 RPM. The lubricant prevents sticking of Al(1XXX) particles to the milling balls during milling and prevents their flatness.

Coarse grain aluminium powders are miniaturised by impacts, collision, and attrition of milling balls in a drum called vial rotating at variable speeds. Oxidation of aluminium powders is controlled using inert gas like liquid nitrogen or argon which also creates a cryogenic environment to aid breakage of the aluminium powders through reduction of ductility of the powders. Ultrafine aluminium particles are obtained by repeated interparticle, particle-ball and particle-vial's internal wall collisions leading to severe plastic deformation and disintegration of hardened powders into fine fragments. Severe plastic deformation causes grain refinement according to Hall Perch relation, resulting in strain hardening of the aluminium powders during milling process. At extreme hardening condition, powders breakage occurs with ease. Milling balls are made from hard materials like hardened steel, tungsten carbide and partially stabilized zirconia. Ball agitation is controlled by the power of the electric motor modified to the rotating speed of the vial; volume of the ball-powder mixture compared to volume of the vial, which is recommended below 50% of the vial volume for effective ball agitation and balls-powder weight (charge ratio). Besides attrition and tumbler ball mills (Figure 10.1a), other mills used for aluminium nanoparticle productions are planetary, shaker, and vibratory ball mills. The planetary ball mill has a vertical drum which upon rotations set the milling balls and powders in vibration resulting in various collisions, causing breakage of coarse grain aluminium powder into fine-grained nanoparticles. A tumbler ball mill has a horizontal drum which during rotation, the milling balls fall on the aluminium powder being miniaturised and attrition mill has a vertical drum and central shaft modified to horizontal impellers at different orientations. In this case, the drum is stationary while balls are set in motion by the rotation of the central shaft. The fixed impeller enhances collision of the balls with particles causing their fractures.

10.2.3   Synthesis of Coconut Shell Nanoparticles

Attempt to address environmental issues emanating from agricultural waste has emerged various ideas on adding values to waste to turn menaces to benefits. Reinforcing fibres extracted from waste have been used over the years across the globe for green production of materials for environmental applications. This effort has been extended towards nanoparticle synthesis from agricultural waste using ball milling techniques already discussed in the previous section of this chapter. However, differences in properties of aluminium and agricultural waste powders justify variation in parameters for obtaining their nanoparticles via the milling technique. Most of agricultural waste used as sources of reinforcing nanoparticles like pumpkin pods, *Delonix regia* pods/seeds, *Daniellia oliveri* seeds, bean pods, rice husks, maize stalk, coconut, periwinkle, palm kernel, red cherry, snail, and eggshells are hard and brittle. Therefore, milling in wet or cryogenic environment is not applicable to the nanoparticle synthesis from agricultural waste. Their breakage tendency increases with levels of dryness forming basis for drying of the agricultural waste before refinement via ball milling technique. In some cases, volatile components of the wastes are removed via carbonisation (plant waste) or calcination (animal waste like snail/eggshell). Carbonisation is the preparation of a solid, carbonaceous residue, by expulsion of volatiles and decomposition of solid masses in the absence of oxygen at high temperature. Carbonisation of plant waste like coconut shell results in production of coconut shell-based charcoal having amorphous structure. From 900°C, structural rearrangement via atomic diffusion takes place leading to transformation of charcoal to graphite. Calcination is the decomposition of carbonate to release carbon IV oxide as volatile while the CaO remains as the solid. Thermal treatment of snail shell, periwinkle, and eggshell at a

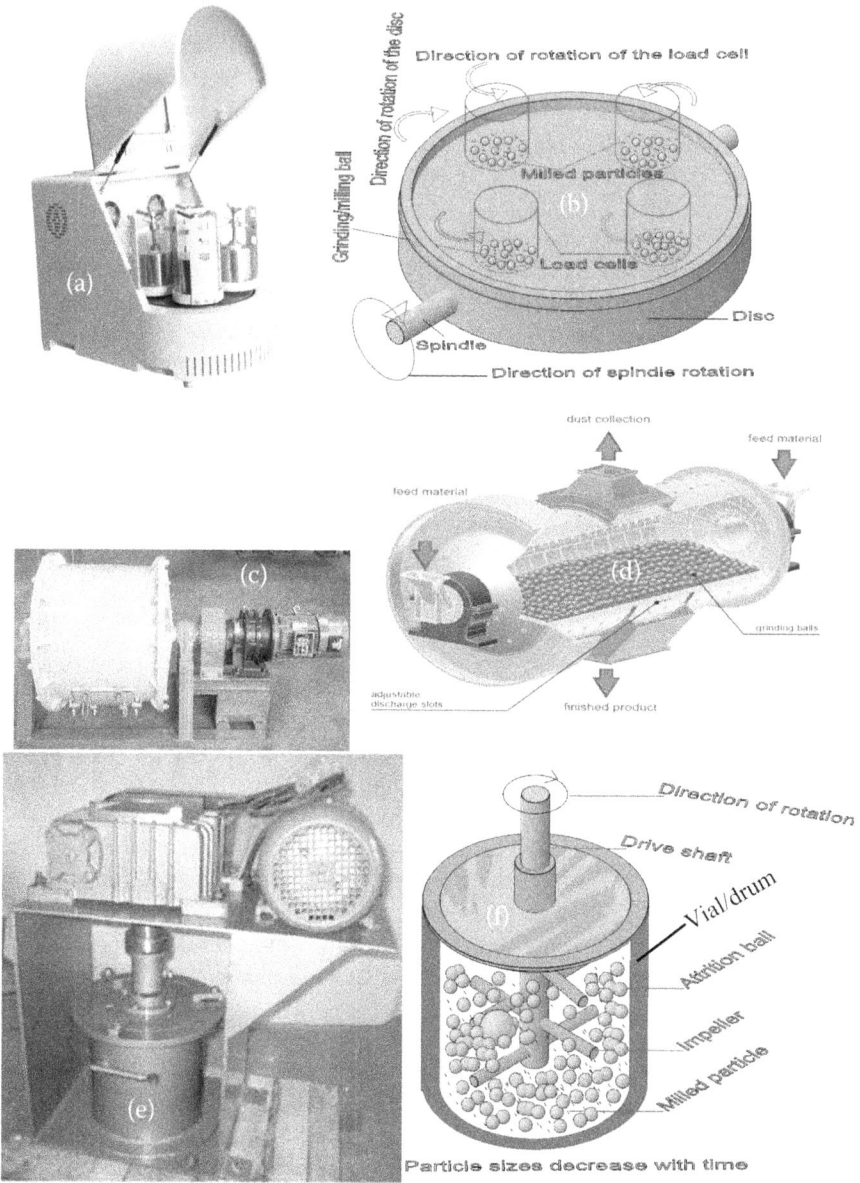

FIGURE 10.1 Grinding mills for synthesis of aluminium nanoparticles (a) planetary ball mill (Source: Landt Instruments), (b) components of planetary ball mill (c) tumbler ball mill (Source: Agico Cement), (d) components of tumbler ball mill (Source: Every single topic), (e) attrition mill (Source: Jinshibao Mining Machinery) and (f) components of attrition mill.

temperature range 750°C–800°C [30] are termed as calcination because they contain mainly carbonates which dissociate at the temperature range to produce the solid CaO as residue. Both calcination and carbonisation begin with removal of water molecules from the respective agricultural waste via vapourisation at temperature above 100°C. In calcination, drying continues until at equilibrium temperature at which the decomposition of the mineral begins at 750°C. Carbonisation involves physical-chemical changes as temperature increases. At less than 300°C, slight distillation resulting in evolution of gases like, CO, CO_2, H_2, N_2, and H_2O takes place. Between 300°C–320°C, there is a faint appearance of oil that is thin and light coloured; mainly methane (CH_4) and unsaturated hydrocarbons plus some other gases are observed. At temperature range of 320°C–360°C, there is a marked evolution of thicker oils, hydrocarbons gases like CH_4, higher alkanes (C_nH_{2n+2}), and more hydrogen; CO_2 and CO are expelled. Within the range (360°C–430°C), evolution of dark-brown viscous oil, more hydrocarbon gases, and hydrogen (H_2) take place. From 430°C to 460°C, more viscous oils distil off with tar production. Pronounced gas evolution occurs, causing the residue to swell and the carbonaceous shells starts to partially decompose, leaving a soft carbonaceous mass of maximum volume and bubble structure. The oils releasing at this temperature range are very thick and gases evolved are unsaturated hydrocarbon with CO and water. H_2 is not formed because it is not very stable at high temperature and secondly, the reaction is without air. Between 460°C and 600°C, oils and tar production reduce with solid residue (carbonised shell or charcoal) having bubble shape and a faint cellular structure. Expulsion of oils and tars completes at this temperature while evolution of hydrocarbons continues at a smaller rate. Within 600°C–900°C, the black residue (charcoal) atoms diffuse into each other causing hardening of the residue. No more hydrocarbon once the tars production has stopped. At temperature greater 900°C, there is a shrinkage of the residue by atomic diffusion and the decomposition of the carbon to form graphite lattice structure. At 1000°C, evolution of nitrogen compound stops. Above 1000°C, no further decomposition but structural arrangement leading densification of graphite continues. Therefore, agricultural wastes are hard, brittle and more susceptible to breakage after carbonisation/calcination than waste without any thermal treatment.

The presence of water in wet agricultural waste improves their energy absorption with a consequent reduction in their brittleness. This fact has been proved in the comparative synthesis of nanoparticles from both 36 μm sized carbonised and uncarbonised coconut shells via ball milling technique under the same conditions. SNH 510 … .680 tumbler ball mill, model A 50 … .43 with ceramic balls of size range (5–60 mm) operating at 192 RPM was used for refinement of both powders for 50 hours of milling. By considering structures of nanoparticles obtained from both powders after milling under the same conditions, finer particles than those of the uncarbonised coconut shells (Figure 10.2b) are observed with carbonised coconut shell (Figure 10.2a). Higher degree of fineness in respect of the carbonised coconut shell affirms their greater breakage tendency due to their brittleness. This fact is further buttressed by findings from literature. Average sizes of uncarbonised coconut shell at 16, 46, and 70 hours of

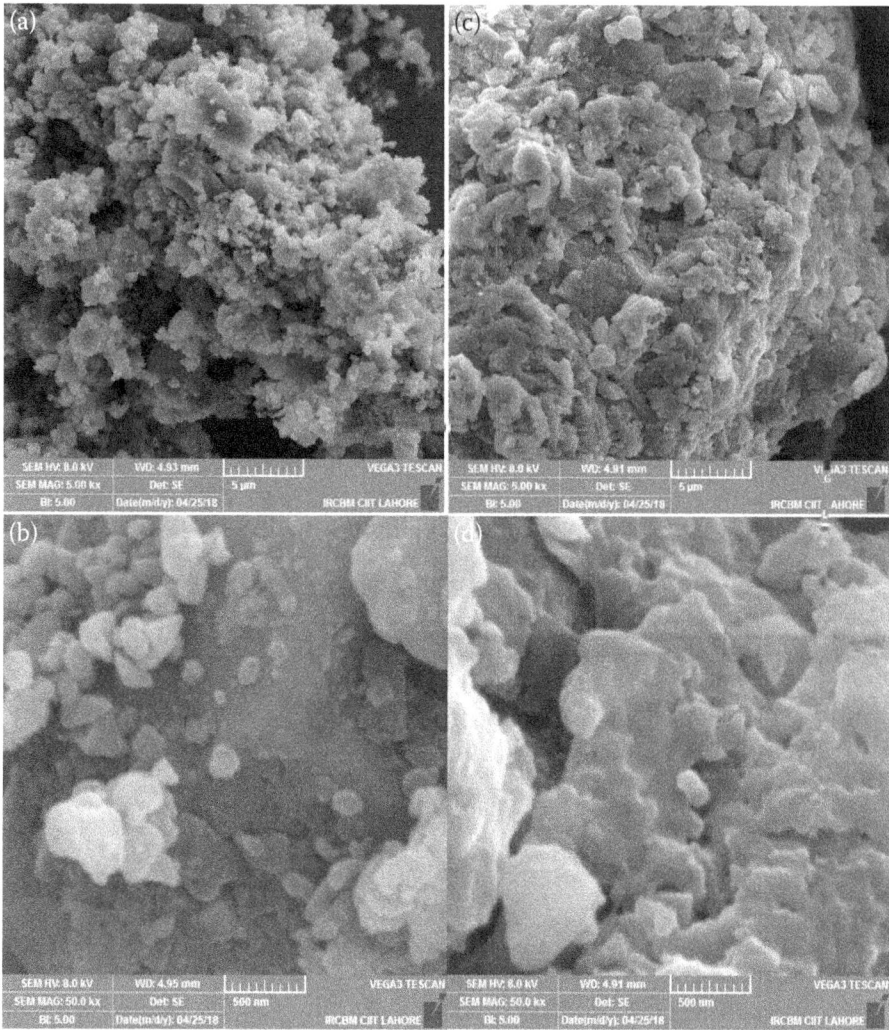

FIGURE 10.2 Microstructural properties of coconut shell nanoparticles (a–b) carbonised at low and high magnifications, (c–d) uncarbonised at low and high magnifications.

milling determined from SEM/Gwyddion software reported in [12] are 118.95, 71.78, and 49.85, respectively.

Moreover, findings disclosed in [18] on carbonised coconut shells indicates average sizes of 50.01, 31.76, and 14.29 nm. Smaller sizes of the carbonised co-conut shells at all durations of milling also buttress the greater breakage tendency of the vacuum thermally treated agricultural waste than their uncarbonised coun-terparts. This implies that the nature (dryness, wetness, brittleness, and toughness) of the agricultural waste is an important parameter to be considered in addition to other milling variables.

10.2.4 DEVELOPMENT OF COCONUT SHELL ALUMINIUM REINFORCED EPOXY NANOCOMPOSITES

Particulate nanocomposites were reported in this section containing uncarbonised coconut shell nanoparticles (UCSp) of average sizes 49.85 nm at 70 hours of milling and 10 charge ratios and 107.1 nm sized aluminium nanoparticles at 70 hours and 8.5 charge ratios. Sizes of all nanoparticles used were determined from SEM image analysed with Gwyddion software. Chemical composition both of carbonised and uncarbonised coconut shell nanoparticles are similar except the presence of K and greater properties of C, O, and Si in the carbonised coconut shells. Aluminium nanoparticles (Alp) have Al as major elements in addition to O and Ar in trace amount. Hand lay-up, filament winding bag, moulding process, pultrusion, sheet moulding bulk moulding, resin transfer moulding, injection moulding, and intercalation are various techniques employed in developing thermosetting composites. Explanations on each technique are found extensively in literature [31–34]. However, this section presents liquid intercalation approach to nanocomposite developments since the mixture of aluminium and coconut shell acting as hybrid reinforcements are nanoparticles. In the production of epoxy hybrid nanocomposites, filler additions up to 12 wt% were used by keeping one filler constant while varying the second filler, conforming to a composite design formulation in Table 10.1. Manual stirring combined with magnetic and mechanical mechanisms was used to ensure intercalation of the particles in the epoxy. In the typical production, a weighed amount of hybrid mixture equivalent to 4 wt% (2%Alp2%UCp) of epoxy resin was added to 40 cm^3 of absolute ethanol in a container and stirred mechanically for 5 minutes at an approximate 314 rpm. The mix was added to 80 cm^3 of epoxy resin and stirred for additional 10 minutes.

Then, the mixture was heated at 100°C for 30 minutes using Stuart UC 152 hot plate and stirrer to evaporate the ethanol. After heating, it was mechanically stirred for 10 minutes at 250–300 rpm. Then, air trapped in the mixture was evacuated using Shel lab vacuum oven (model SVA S2E) after which 40 cm^3 of hardener equivalent to 2:1 (volume ratio) resin/hardener ratio was added to the mixture and then stirred manually until the mixture was in a viscous state. At this state, hybrid particle-reinforced epoxy blend in the viscous or gelled state was poured into a steel die open mould. It was left to set and harden at room temperature for 48 hours. After setting, the hybrid nanocomposites were subjected to post curing at 120°C for 4 hours. The process was repeated with increases in wt% of hybrid particle additions giving rise to different grades of hybrid nanocomposites.

10.3 CHARACTERISATION OF HYBRID PARTICLE-REINFORCED EPOXY

Various grades of hybrid particle-reinforced epoxy produced were subjected to different examinations to determine their structural, tensile, and physical properties.

TABLE 10.1

Design Formulation Showing Proportions of Aluminium and Uncarbonised Coconut Shell Nanoparticles for Developing Hybrid Nanocomposites

S/N	Weight Fractions (%)	Alp	UCSp
1.	0	0	0
2.	4	2	2
3.	6	2	4
4.	6	4	2
5.	8	2	6
6.	8	4	4
7.	8	6	2
8.	10	2	8
9.	10	6	4
10.	10	8	2
11.	12	2	10
12.	12	4	8
13.	12	6	6
14.	12	8	4
15.	12	10	2

Key: Alp – Aluminium nanoparticles; UCSp – uncarbonised coconut shell nanoparticles.

10.3.1 X-Ray Diffraction (XRD) Examination

Epoxy hybrid composites were subjected to X-ray diffraction analysis to probe interaction of epoxy with additive hybrid reinforcements through identification of phases present in the epoxy matrix. Panalytical Empyrean X-ray diffractometer with Pixel detector was used for the X-ray diffraction analysis. The slits were fixed with Fe filtered Co Kα radiation at 40 kV, 30 mA, and 1°/1° scattering slit and 400 μm receiving slit. Intensity/count scores were recorded between 0° and 80° diffracting angles (2Θ) with a scintillation counter at 0.05°/minute. The phases were identified using X' Pert High score plus software; PAB-ICSD and ICDD (2014) databases. Samples were prepared using a zero-background sample holder.

10.3.2 Optical Microscopic Analysis

Microstructure of the hybrid particle-reinforced epoxy (HPRE) composites were examined using Baskar optical microscope (resolution of 640 × 840 pixels and 10–300 × magnifications). Prior to optical examination, 10 × 10 × 8 mm^3 representative samples were cut from unreinforced epoxy and the hybrid particle-reinforced epoxy nanocomposites. Surface of each sample was cleaned with soft cloth soaked in the absolute ethanol. This was done to prepare the surface for the microscopic examination.

(a)

(c)

(b)

(d)

Element	Weight%	Atom%
C	53.60	60.74
O	45.84	39.00
Si	0.55	0.27
Total	100.00	

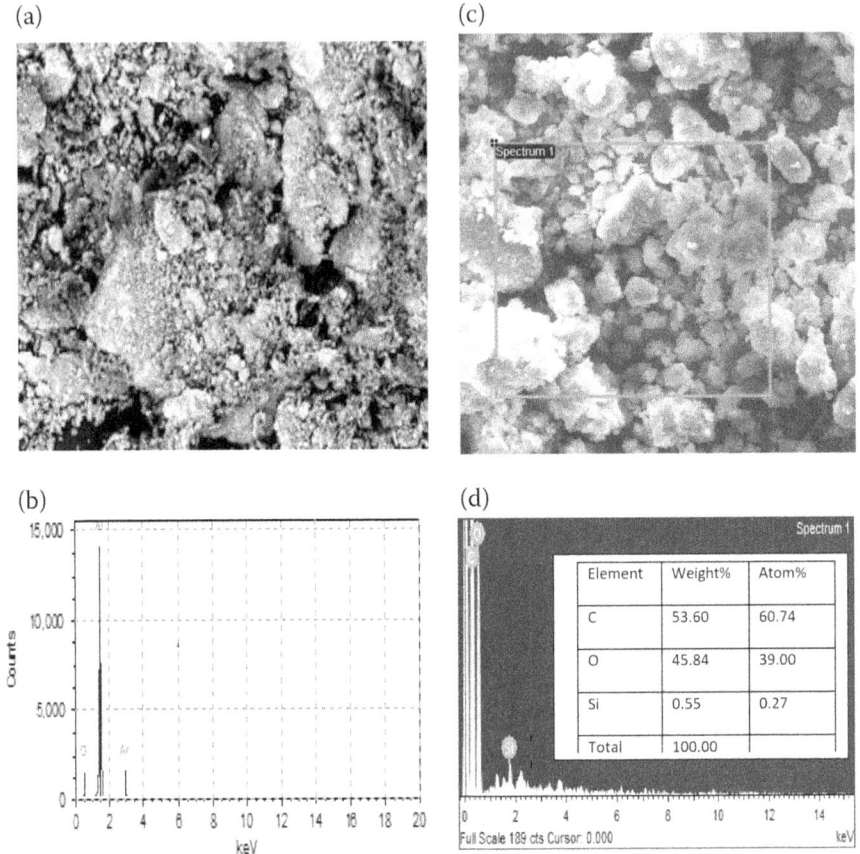

FIGURE 10.3   SEM/EDX of (a–b) aluminium nanoparticles and (c–d) uncarbonised coconut shell nanoparticles.

10.3.3   SCANNING ELECTRON MICROSCOPE (SEM) EXAMINATION

Morphology and elemental composition of the unreinforced epoxy and its hybrid nanocomposites were studied using SEM. In SEM analysis, samples were placed on SEM mount with the aids of double-sided adhesive tape. Samples held on the mount were placed inside the SEM stage and scanned at 15.1 kV accelerating voltage; liquid nitrogen was used to cool the detector. VEGAS TESCAM SEM with the attached EDX was used for the investigation.

10.3.4   TENSILE PROPERTIES DETERMINATION

Behaviour of hybrid particle-reinforced epoxy nanocomposite and unreinforced epoxy samples under load was studied to investigate their resistance to deformation. A tensile test samples (Figure 10.4) of the unreinforced epoxy and the HPRE with gauge dimensions of 80 mm in length, 10 mm in thickness, and 6 mm in width were used. An Instron Universal Testing Machine (UTM) model 3369; system ID: 3369S3457 was

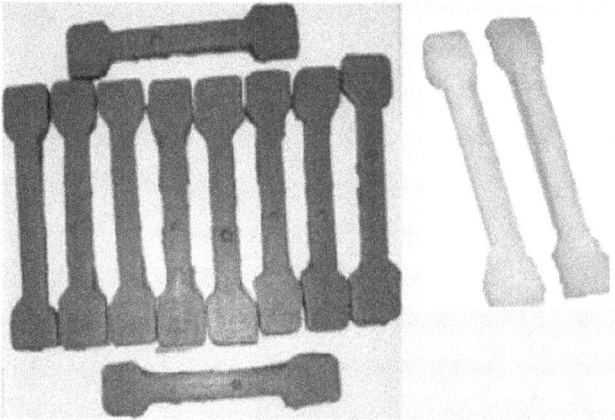

FIGURE 10.4 Tensile specimens of the developed composites.

used to gradually stretch the samples using a uniaxial tensile load at a cross head speed of 0.008 (8×10^{-3}) mms^{-1} equivalent to a strain rate of 0.0001 (1×10^{-4}) s^{-1} until fracture occurred after some plastic deformations. This analysis was carried out in accordance with [35].

10.3.5 DENSITY MEASUREMENT

Density of the unreinforced epoxy and each sample of HPRE were determined using Archimedes' principle. Mass of $50 \times 12 \times 10$ mm^3 samples of the un-reinforced epoxy and epoxy composites were determined using Pioneer weighing scale displayed in. Figure 10.5. In the volume determination, a measuring cylinder was filled with water to a certain level (h_o). Then a composite sample was gently dropped on the water surface, and this caused a rise in water level. The new water level was noted and recorded as h_1, as shown in Figure 10.6. Volume and density of the sample were determined using Equations 10.1–10.2 (Figure 10.6).

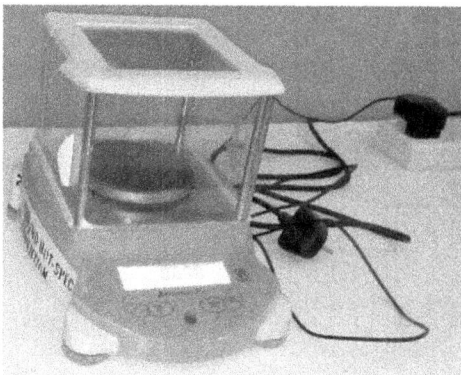

FIGURE 10.5 Pioneer weighing scale.

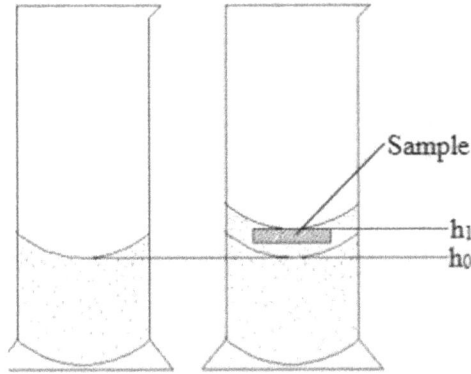

FIGURE 10.6 Volume determination using direct measurement technique.

$$Volume = h_1 - h_0 \qquad (10.1)$$

$$Density = \frac{Mass\ in\ air}{Volume\ of\ water\ displaced} \qquad (10.2)$$

10.3.6 THERMAL PROPERTY DETERMINATION

Epoxy polymer and HRPE nanocomposites were subjected to thermal analysis using DSC (model: TA 2920) in accordance with ASTM D3418-82. Small sample representatives whose mass varies from 4.0 to 6.5 mg were cut from the HPRE nanocomposites, washed with distilled water, and dried. Each sample was placed in a smooth bottom-rolled down-side aluminium pan which was hermetically sealed. The sample pan was placed in the raised platform of the cell with the aids of tweezers and the reference pan was placed in the rear. Then, the cell was covered with the lid. After connection of nitrogen and warming of DSC for 30 seconds, samples were scanned from 25°C to 500°C at a heating ramp of 10°C/minute and nitrogen flow rate of 25 mL/minute. During the scanning process, the heat flow through the sample to keep it at a zero-temperature difference with reference pan was plotted as a function of temperature and time using an endothermic down convention, which meant the endothermic process was indicated by absorption of negative energy and a downward peak.

10.4 PROPERTIES OF HYBRID PARTICLE-REINFORCED EPOXY

10.4.1 X-RAY DIFFRACTOMETRIC PROPERTIES

An X-ray diffraction profile of epoxy polymer is presented in Figure 10.7 while Table 10.2 gives structural properties of the different phases present in the epoxy polymer. The prominent molecules (Table 10.2) of the epoxy polymer are linked

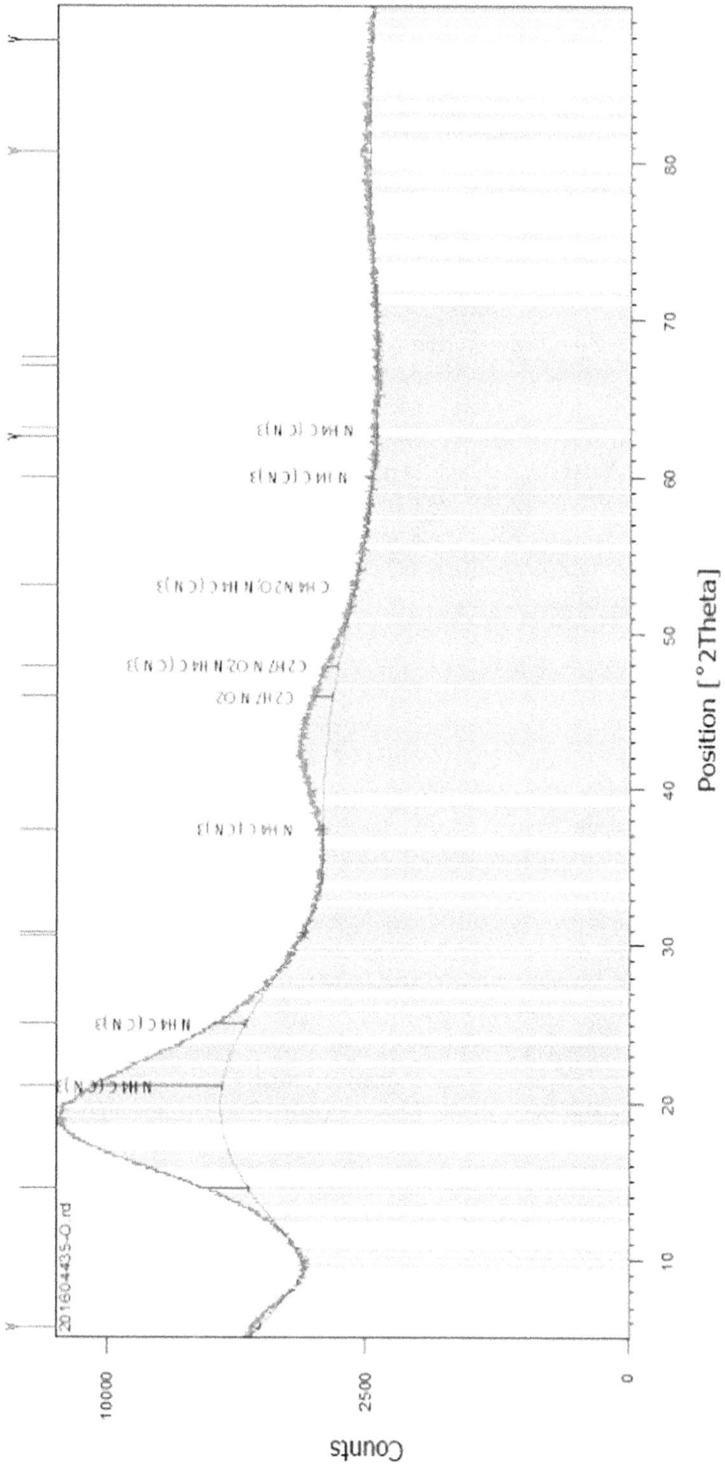

FIGURE 10.7 X-ray diffraction profiles of epoxy polymer.

TABLE 10.2
Phase Identities of Epoxy Polymer

Position [°2Th.]	Compound Name	d-Spacing [Å]	Chemical Formula
53.0162	Urea	1.72587	CH_4N_2O
47.9497	Ammonium Acetate	1.97308	$C_2H_7NO_2$
21.0846	Ammonium Carbon Cyanide	4.21366	$NH_4C(CN)_3$

to an exothermic reaction that occurred between epoxy resin (diglycidyl ether of bi-sphenol A) and the amine hardener during the cross-linking (curing) reaction. The process is exothermic due to heat released to the surroundings during the reaction at room temperature, though the samples were post cured at 130°C after a 48-hour room temperature curing. The effects of hybrid particle additions to epoxy are clear from the formation of new molecules as observed in Figures 10.8–10.12 and Tables 10.3–10.7. New molecules detected by XRD in each of the HPRE nanocomposites are facts to support the chemical reaction between the epoxy matrix and the AlpUCSp reinforcement. For HPRE containing 2%Alp2%UCSp (Figure 10.8), its XRD peaks are fitted with various compounds/second phase particles produced from chemical interaction between epoxy matrix and 2%Alp2%UCSp reinforcements and residual elements like Al and C (Table 10.3) from unreacted 2%Alp and 2%UCSp, respectively. When a proportion of hybrid AlpUCSp reinforcement sums up to 10 wt% with four different design formulations, various XRD profiles of the HPRE nanocomposite produced are displayed in Figures 10.9–10.12. Different proportions of Alp and UCSp that make up 10 wt% is noticeable from different phases detected from the XRD examinations conducted on each of the nanocomposite grades. Structural properties of the detected compounds are presented in Tables 10.4–10.7. Generally, the broadness of the XRD peaks as observed in Figures 10.7–10.12 indicates amorphous structure of the epoxy as a polymer. Moreover, sharp peaks at 38.33 and 44.56° in Figure 10.8; 38.28 and 44.53° in Figure 10.9; 38.44 and 43.31 in Figure 10.10; 38.41, 43.17°, and 65.02° in Figure 10.11; 38.26 and 43.08° in Figure 10.12 establish degrees of crystallinity due to AlpUCSp addition to the epoxy matrix that acts toward higher mechanical properties of the HPRE than those of the polymer. Moreover, variations in interplanar (d) spacings of epoxy molecules and those of the phases of the HPRE nanocomposites confirm straining of the nanocomposite structure due to hybrid nanoparticle additions.

10.4.2 MICROSTRUCTURAL PROPERTIES OF HYBRID PARTICLE-REINFORCED EPOXY

There are networks of imperfect hexagonal structure evenly distributed within the matrix (Figure 10.13a). This makes microstructure homgeneous without segregation of any second-phase compounds. The observed amorphous structure is in line with the XRD profile of epoxy polymer in Figure 10.7 that shows broad peaks between 10 and 30° and 40 and 50° and angle of refraction 2θ [36]. When the hardener was added to epoxy resin, addition and cross-linking reactions generated heat, released to the surroundings. Heat released to the surroundings reduced energy of the system

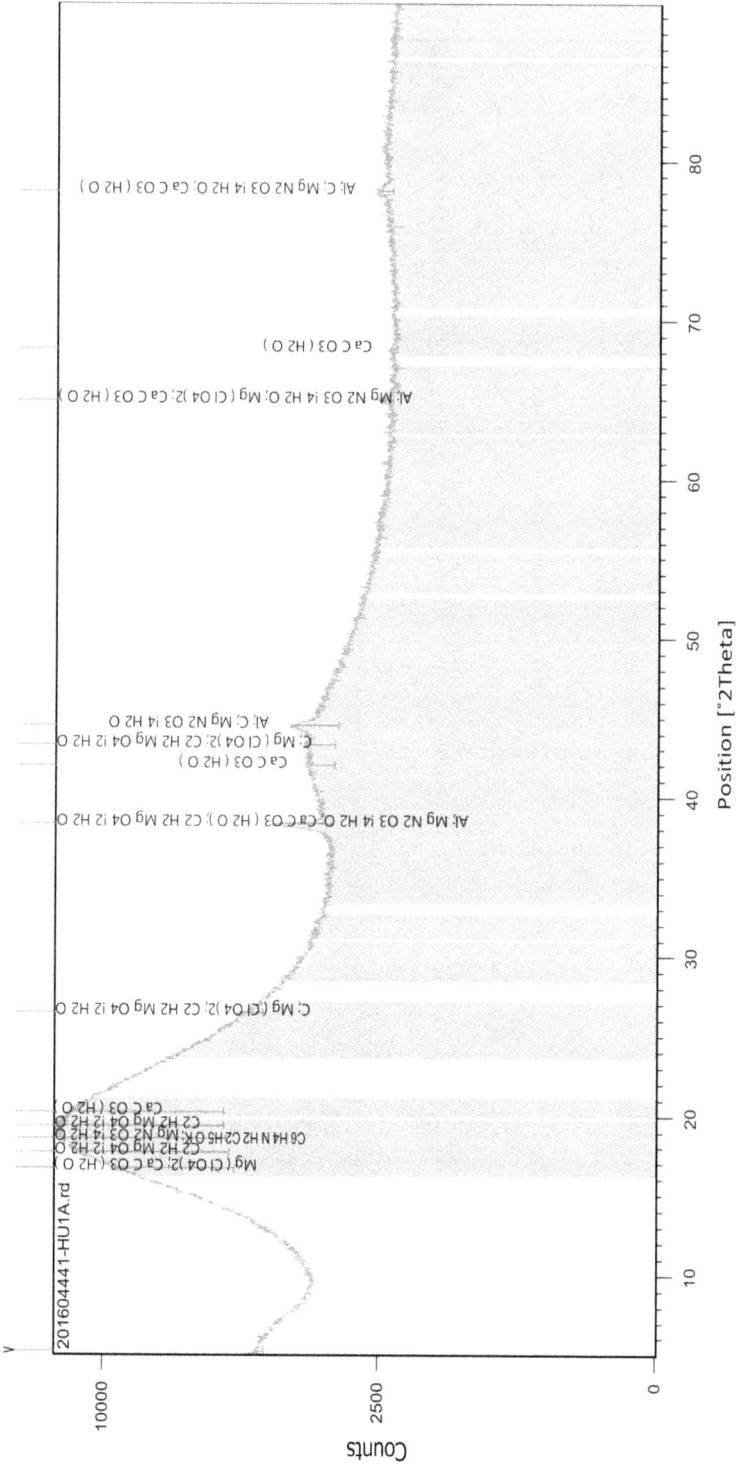

FIGURE 10.8 X-ray diffraction profiles of 2%Alp2%UCSp hybrid reinforced epoxy.

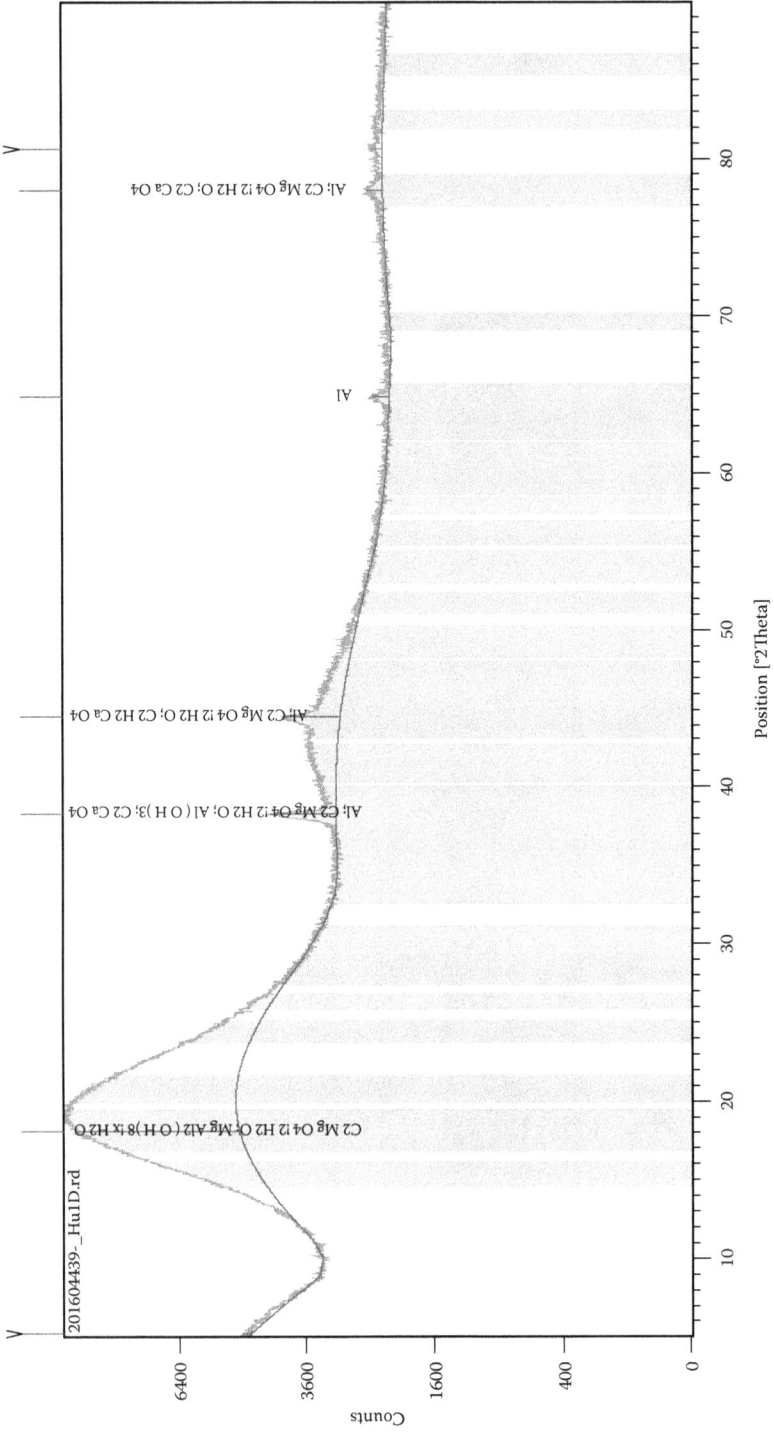

FIGURE 10.9 XRD profile of epoxy/2Alp8UCSp nanocomposite.

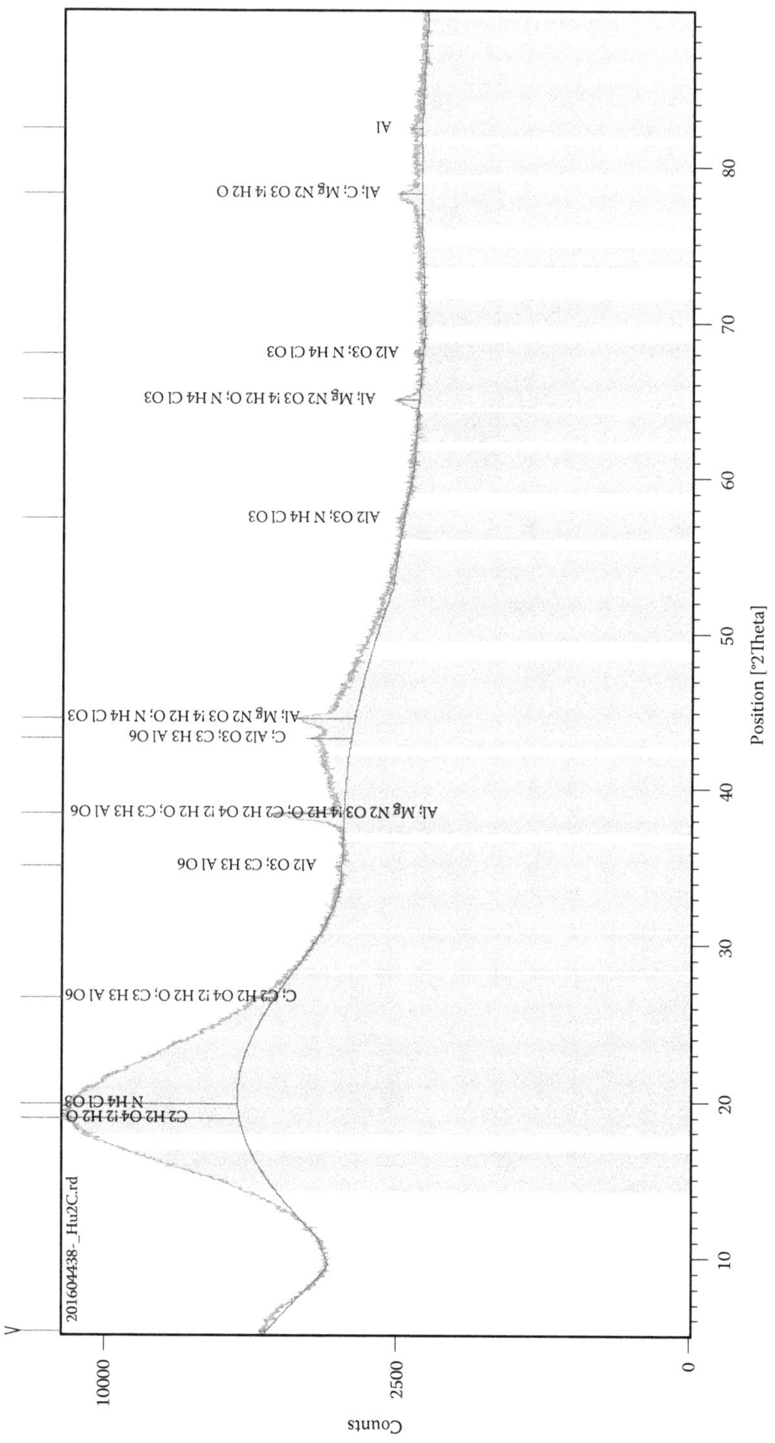

FIGURE 10.10 XRD profiles of epoxy/4Alp6UCSp nanocomposite.

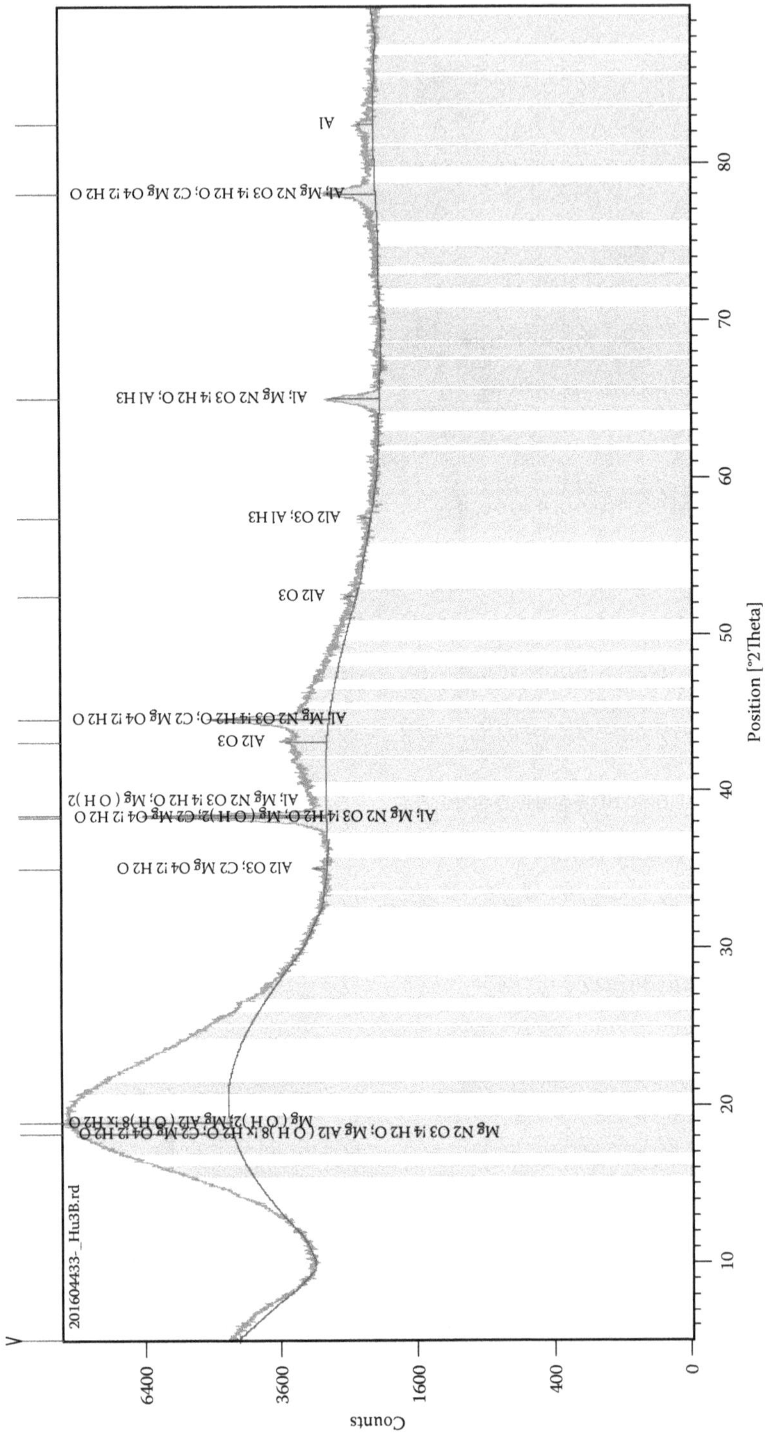

FIGURE 10.11 X-ray diffractometric profile of epoxy/6Alp4UCSp nanocomposite.

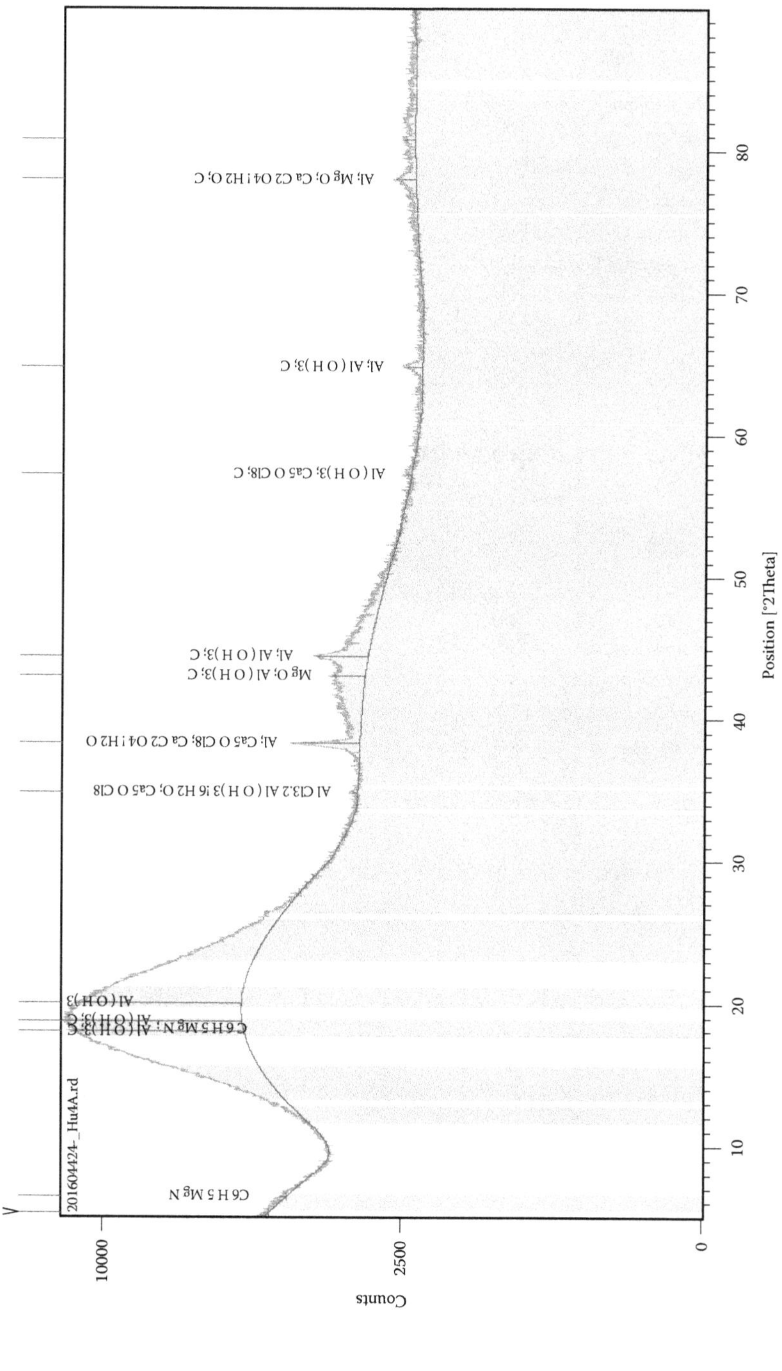

FIGURE 10.12 X-ray diffractometric profile of epoxy/8Alp2UCSp nanocomposite.

TABLE 10.3
Identities of Phases Characterising 2%Alp2%UCSp Hybrid Reinforced Epoxy

Pos. [°2Th.]	Compound Name	d-Spacing [Å]	Chemical Formula
38.33	Aluminium	2.3	Al
44.56	Carbon	2.0	C
18.60	Magnesium Nitrate Hydrate	4.8	$MgN_2O_3.4\ H_2O$
16.79	Magnesium Chlorate	5.3	$Mg(ClO_4)_2$
20.27	Calcium Carbonate Hydrate	4.4	$CaCO_3(H_2O)$
19.38	Magnesium Formate Hydrate	4.6	$C_2H_2MgO_4.2H_2O$
19.38	Benzene Amine Potassium Ethoxide	4.6	$C_6H_4NO_2C_2H_5OK$

TABLE 10.4
Identities of Epoxy/2Alp8UCSp Nanocomposite's Second-Phase Particles

Pos. [°2Th.]	Compound Name	d-Spacing [Å]	Chemical Formula
38.28	Aluminium	2.35	Al
18.13	Magnesium Oxalate Hydrate	4.9	$C_2MgO_4.2H_2O$
38.28	Bayerite	2.35	$Al(OH)_3$
38.28	Calcium Oxalate	2.35	C_2CaO_4
44.53	Calcium Formate	2.03	$C_2H_2CaO_4$
18.13	Magnesium Aluminium Hydroxide Hydrate	4.9	$MgAl_2(OH)_8.xH_2O$

TABLE 10.5
Identification of Phases Found in Epoxy/4Alp6UCSp Nanocomposite

Pos. [°2Th.]	Compound Name	d-Spacing [Å]	Chemical Formula
38.44	Aluminium	2.3	Al
43.31	Aluminium Oxide	2.1	Al_2O_3
38.44	Magnesium Nitrate Hydrate	2.34	$MgN_2O_3.4\ H_2O$
18.92	Hydrogen Oxalate Hydrate	4.7	$C_2H_2O_4.2\ H_2O$
19.82	Ammonium Chlorate	4.5	NH_4ClO_3
38.45	Aluminium Formate	2.34	$C_3H_3AlO_6$

(reacting species), leading to inreased orderliness and viscosity/gellation of the system. This implied an increase in the intermolecular forces of attraction. When the forces of attraction were strong enough, epoxy molecules were joined together in an irregular manner, as depicted in Figure 10.13. A similar explanation has been offered in literature (Josmin et al., 2012; Kandelbauer, 2014; Kandelbauer et al., 2014). However, a comparison of micrographs in Figures 10.13–10.15 shows that

TABLE 10.6
Properties of Phases Characterising Epoxy/6Alp4UCSp Nanocomposite

Pos. [°2Th.]	Compound Name	d-Spacing [Å]	Chemical Formula
38.41	Aluminium	2.3	Al
43.17	Aluminium Oxide	2.1	Al_2O_3
18.15	Magnesium Nitrate Hydrate	4.9	$MgN_2O_3.4H_2O$
18.90	Brucite	4.7	$Mg(OH)_2$
65.02	Aluminium Hydride	1.4	AlH_3
18.90	Magnesium Aluminium Hydroxide Hydrate	4.7	$MgAl_2(OH)_8.xH_2O$
18.15	Magnesium Oxalate Hydrate	4.9	$C_2MgO_4.2H_2O$

TABLE 10.7
Identified Phases Characterising Epoxy/8Alp2UCSp Nanocomposite

Pos. [°2Th.]	Compound Name	d-Spacing [Å]	Chemical Formula
38.26	Aluminium	2.4	Al
43.08	Magnesium Oxide	2.1	MgO
18.75	Gibbsite	4.7	$Al(OH)_3$
34.82	Aluminium Chloride Hydroxide Hydrate	2.6	$AlCl_3.2Al(OH)_3.6H_2O$
38.26	Calcium Oxide Chloride	2.4	Ca_5OCl_8
38.26	Calcium Carbon Oxalate Hydrate	2.4	$CaC_2O4.H_2O$
18.75	Benzene Magnesium Nitride	4.7	C_6H_5MgN

the spaces within the hexagonal structure had been filled up with AlpUCSp and/or compounds due to a chemical interaction of AlpUCSp with an epoxy system. This agrees with XRD in Figures 10.8–10.12. In addition, differences observed in molecular structures of HPRE nanocomposites at different formulations (Figure 10.15a–d) may result in a varied behaviour of the nanocomposite under loadings.

The SEM presents the epoxy polymer as a large, continuous, undulating body with numerous tiny grains on top (Figure 10.16a). At high magnification (Figure 10.16b), the tiny grains appear as rigid, rounded, ellipsoidal, and irregular bodies fusing with one another and with the large undulating body forming an infusible epoxy structure. Moreover, AlpUCSp additions to epoxy give rise to a fine-grained homogenous structure (Figure 10.16c). Refinement of epoxy grain due to AlpUCSp additions is substantial since the observed grains are still very fine even at 10,000 magnifications. This structure has a high potential to prevent formation of craze with ease when the nanocomposite is subjected to uniaxial tensile loading. Upon this, it is expected to show much higher mechanical performances than the epoxy polymer. Chemically, the epoxy contains C, O, and Au Figure 10.17). Both C and O justify various compounds detected by XRD as presented in Figure 10.7. Au

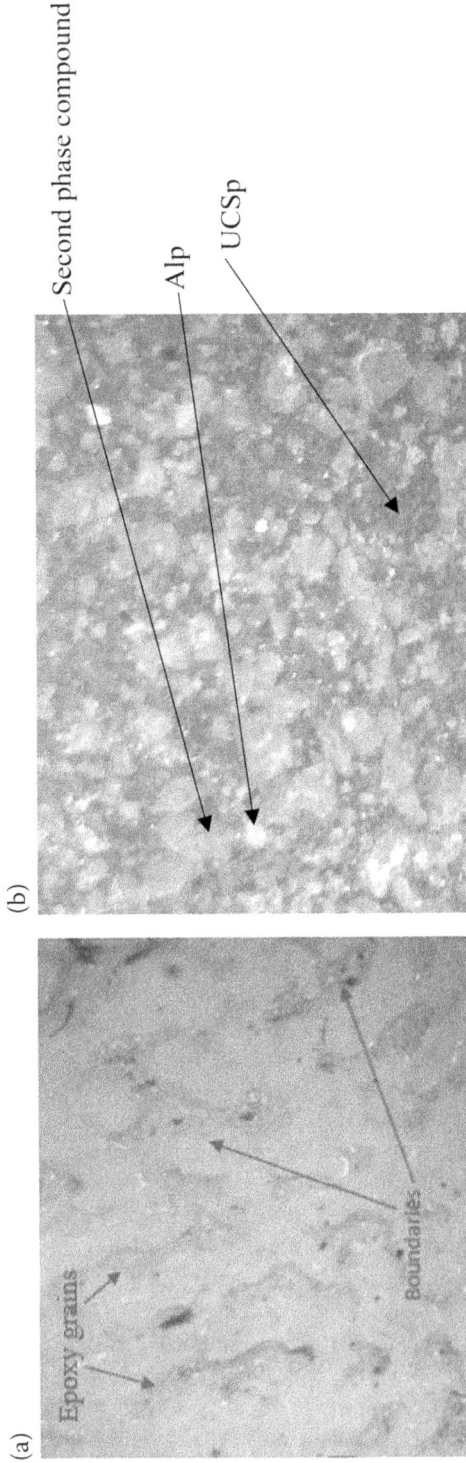

FIGURE 10.13 Optical micrograph showing (a) Epoxy grains, (b) 2%Alp (dull white); 2%UCSp (deep brown) in epoxy matrix (× 100).

(a)   (b)

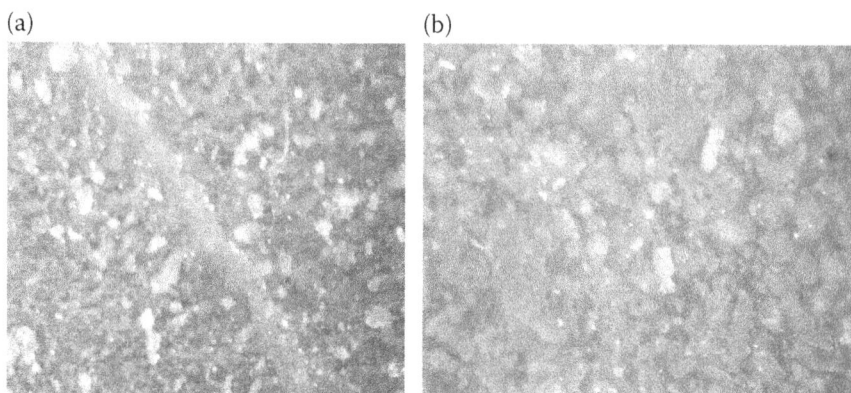

FIGURE 10.14 Optical micrographs displaying Alp (white/dull) UCSp (ash) in epoxy matrix of (a) epoxy/2Alp4UCSp, (b) epoxy/4Alp2UCSp nanocomposites (×100).

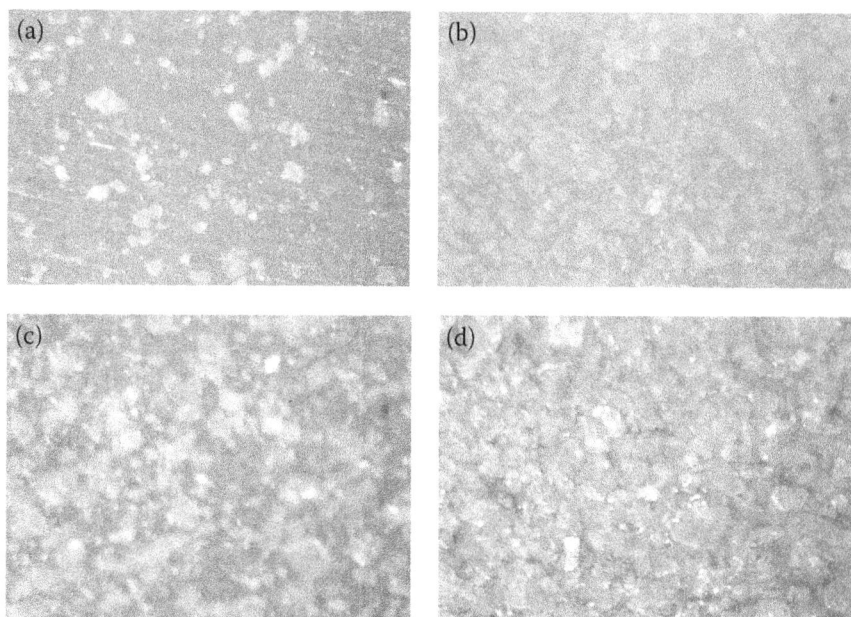

FIGURE 10.15 Optical micrographs presenting aluminium nanoparticle (white/dull); uncarbonised coconut shell nanoparticles (ash/dark) in epoxy matrix of (a) epoxy/2Alp8UCSp, (b) epoxy/4Alp6UCSp, (c) epoxy/6Alp4UCSp (d) epoxy/8Alp2UCSp composites (×100).

may form part of the flame retardant added to the resin during manufacturing. Elements (N, Al Si, Ca, and Br) were detected in the HPRE nanocomposite in addition to C and O. Aluminium (Al) is linked to Alp addition to epoxy, Ca and Si are attributed to mineral content of the UCSp, while C belongs to epoxy and UCSp.

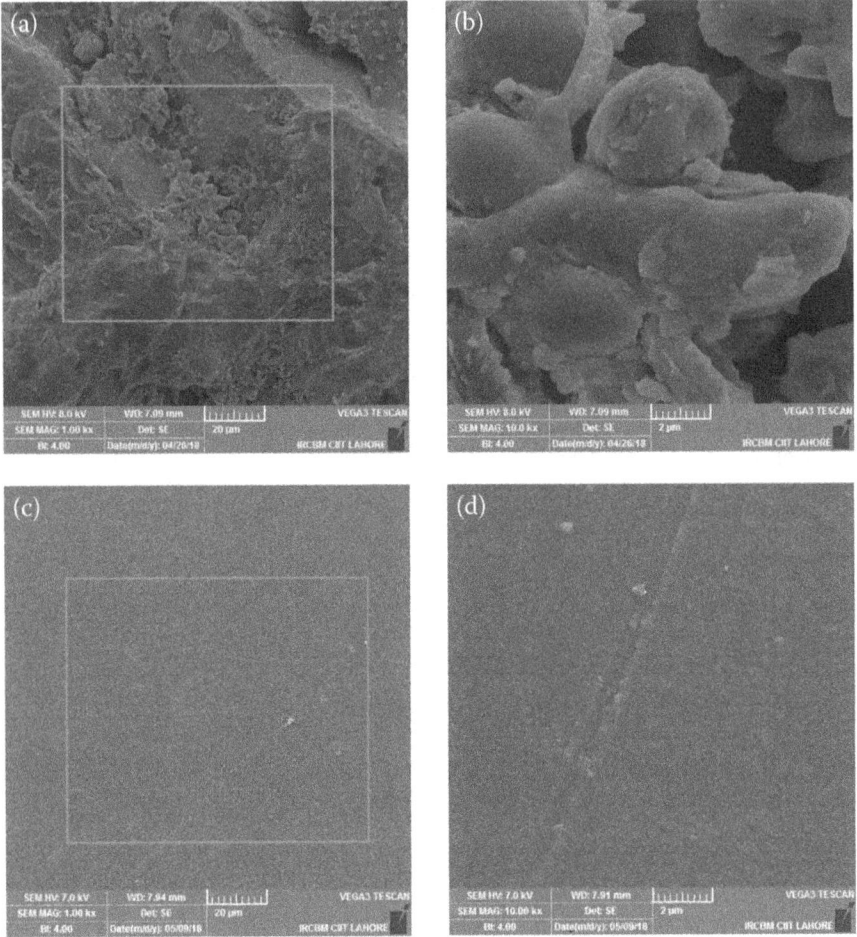

FIGURE 10.16 Micrographs of epoxy polymer at (a) 1000 magnifications, (b) 10,000 magnifications; hybrid particle-reinforced epoxy at (c) 1000 magnifications, and (d) 10,000 magnifications.

10.4.3 TENSILE PROPERTIES OF HYBRID PARTICLE-REINFORCED EPOXY NANOCOMPOSITES

Differences in tensile strengths of HPRE at 6, 8 10, and 12 wt% of AlpUCSp hybrid additions are observed as shown in Figure 10.18. Variations are linked to different design formulations used, leading to various proportions of AlpUCSp interacting with the matrix. For instance, in Figure 10.18a, epoxy/2Alp4UCSp exhibits higher tensile strength (33.69 Nmm^{-2}). In Figure 10.18b, epoxy/4Alp4UCSp shows maximum tensile strength (35.43 Nmm^{-2}) while epoxy/4Alp6UCSp and epoxy/8Alp4UCSp possess peak tensile strengths (43.2 Nmm^{-2}, 4.11 Nmm^{-2}), as observed in Figure 10.18c–d, respectively. Each value of tensile strength is the

(a)

Element	Weight%	Atomic %
C	71.50	79.83
O	23.67	19.84
Au	4.83	0.33
Totals	100.00	

(b)

Element	Weight%	Atom%
C	53.86	58.01
N	45.21	41.75
Al	0.07	0.04
Si	0.13	0.06
Ca	0.21	0.07
Br	0.52	0.08
Total	100.00	

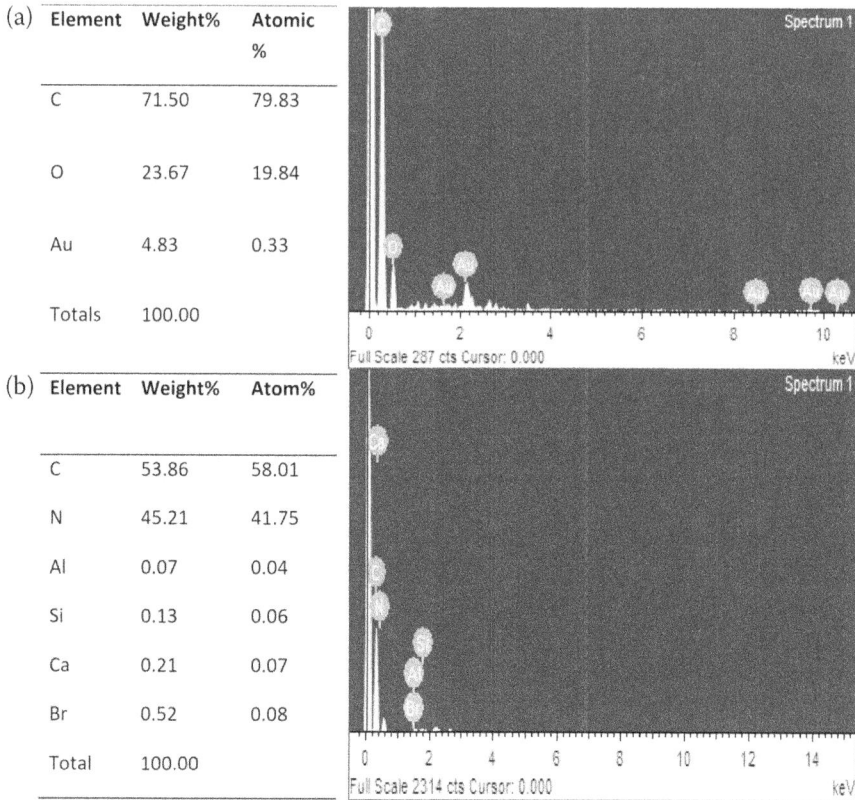

FIGURE 10.17   Energy dispersive X-ray spectroscopy revealing elemental composition of (a) epoxy polymer (the scanned region is presented in Figure 10.16a) and (b) hybrid particle-reinforced epoxy (the scanned region is presented in Figure 10.16c).

average tensile strengths of four different samples of HPRE nanocomposites produced at each formulation, as shown in Table 10.1.

A measure of variances among the sample tensile strengths are inserted in the form of error bars in Figure 10.19, explaining the variation of the optimum tensile strengths and the corresponding % elongation at each level of AlpUCSp reinforcement additions. It is observed from Figure 10.19 that tensile strength increases with an increment in wt% of AlpUCSp additions and proportionate decrease in % elongation of the HPRE nanocomposites is also found. The increase in the tensile strength may be traced to strong binding force resulting from chemical interactions between AlpUCSp reinforcement and epoxy matrix. However, reduction in % elongation may be a compensation for the increase in tensile strength due to high rigidity of the second-phase particles synthesised from the interaction of AlpUCSp reinforcement with epoxy. Similar interactions of aluminium nanoparticle and epoxy were reported in literature [37].

In addition, tensile strength at cumulative 12 wt% is lower than that at 10 wt% of AlpUCSp additions, meaning that maximum increase in the tensile strength (43.2 Nmm^{-2}) is discovered at 10 wt% of AlpUCSp additions. This is equivalent to

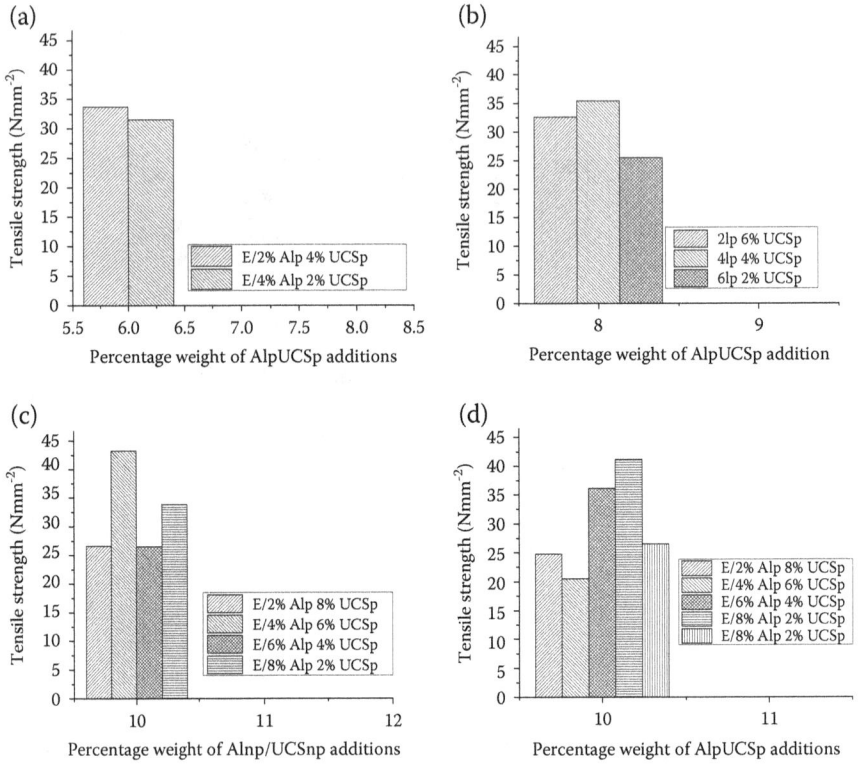

FIGURE 10.18 Tensile properties of hybrid nanocomposite with levels of reinforcements at different design formulations.

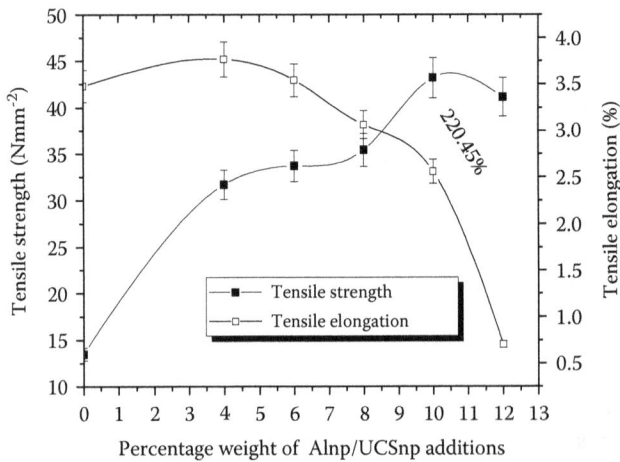

FIGURE 10.19 Optimum tensile properties of hybrid particle-reinforced epoxy nanocomposites.

about a 220% increase. Comparison of this finding (220% increase in the tensile strength) with when epoxy matrix was reinforced with either Alp [38], a 38% increase was found at 10 wt% corresponding to epoxy/10Alp nanocomposites, respectively. Moreover, a 44% increase in tensile strength was reported in [39] due to 12 wt% of 36.03 μm sized uncarbonised coconut shell microparticle additions to epoxy. Therefore, a 220% increase in the tensile strength attributed to AlpUCSp additions is much greater and is linked to particle hybridisation and refinement. Also, about a 17%, 26%, and 34% decrease in % elongation is observed with epoxy/10Alp, epoxy/4Alp6UCSp, and epoxy/10UCSp composites, respectively. Deduction from this study is that despite the peak increase in tensile strength of epoxy/4Alp6UCSp, reduction in its % elongation still falls between those of other composites. Maximum improvement in the tensile properties of epoxy/4Alp6UCSp when compared with epoxy/10Alp is attributed to synergetic effects of ductile and rigid phases produced from Alp and UCSp when combined with the epoxy matrix. The strong bonding of the phases with the matrix increased their stability and enhanced rapid load transfer/distribution from the epoxy matrix to the rigid phases. Therefore, the phases stiffened the matrix and enhanced the load bearing capacity of the HPRE nanocomposites.

Moreover, the reduction in the tensile strength above 10 wt% of AlpUCSp addition can be attached to the matrix saturation. At this condition, epoxy matrix is incapacitated to firmly bond all the phases or particles together. That is, there are colonies of freely existing AlpUCSp particles which interfere with the composite mobility during pulling by uniaxial tensile load. Consequences of this are formation of vacancies within the composites as the weakly bounded particles moved from a point to another. Vacancies acted as a stress raiser and caused the necking of the examined composites under lower pressures/stresses than the expectation. A similar explanation has been offered in literature [40–42].

10.4.4 Modelling of Tensile Strength of Hybrid Particle-Reinforced Epoxy Nanocomposites

Regression analysis is employed to quantify a relation between two or more variables, in which at least one of the variables depending on others is called a dependent variable, having the number of elements or genes from 1 to n to form a row matrix, y (Equation 10.3); n is an integer that stands for number of genes or elements of the matrix y, which depends on experimentation levels and number of independent variables. Other variables are called independent or predictor variables, X. The predictor matrix, X, may contain one or more element (Equation 10.4) that determines the number of factors on which the response variable (y) relies. The number of genes possessed by an individual predictor, x_i and x_j, are used for a design of the matrix X, as in Equation 10.5. Therefore, X is a matrix of more than one row and column depending on experimentation levels.

$$y = [y_1, \ y_2, \ ..., y_n] \qquad (10.3)$$

$$X = [x_i, \ x_j, \ ...] \qquad (10.4)$$

$$X = \begin{bmatrix} 1 & x_{i1} & x_{j1}, & \cdots & x_{z1} \\ 1 & x_{i2} & x_{j2}, & \cdots & x_{z2} \\ \cdots & \cdots & \cdots & \cdots & \cdots \\ 1 & x_{in} & x_{jn}, & \cdots & x_{zn} \end{bmatrix} \qquad (10.5)$$

When $X = x_i$, that is the dependent/response variable (Y) depends on only one variable/factor x_i, mono-variate regression model of 1, 2, 3, 4, or higher degree can be used to establish a relationship between the predictor and the response variables (Equation 10.6).

$$y = \begin{bmatrix} a_1 x + c\,(\text{linear}) \\ a_1 x^2 + a_2 x^1 + c\,(\text{quadratic}) \\ ; \\ ; \\ \text{or} \\ a_1 x^n + a_2 x^{n-1} + \ldots + a_{n-1} x + c\,(\text{polynomial}) \end{bmatrix} \qquad (10.6)$$

The regression coefficients $a_1, a_2, \ldots a_n, c$, as seen in Equation 10.6, can be determined by using a MATLAB command expression in Equation 10.7. The small letter "k" is the degree of polynomial. The choice for an appropriate mono-variate regression model for analysing the data depends on the probability (P) value, which is expected to be less than 5% (0.05) to be significantly appropriate for establishing the relationship between the dependent and independent variables. In a case in which linear, quadratic, and polynomial have P values less than 0.05, the one among others has a highest regression coefficient of determination (R^2) is the most relevant and appropriate.

$$\text{coeff} = \text{polyfit}(x, y, k) \qquad (10.7)$$

When X contains more than one variable, say x_i, x_j, and x_k, which means that y depends on three factors, multiple regression analysis is employed. The predictor matrix X in Equation 10.5 holds well with the analysis. Moreover, the multiple regression model can be linear (Equation 10.8), two functional interaction (2FI) (Equation 10.9), quadratic (Equation 10.10), and polynomial (Equation 10.11). Coefficient (β) in Equation 10.8 for the linear multiple regression model can be obtained using a MATLAB command left division of X by y in Equation 10.12.

$$y = \beta_1 + \beta_2 x_1 + \beta_3 x_2 + \ldots \qquad (10.8)$$

$$y = \beta_1 + \beta_2 x_1 + \beta_3 x_1 x_2 + \ldots \qquad (10.9)$$

$$y = \beta_1 + \beta_2 x_1 + \beta_3 x_1^2 + \beta_4 x_2^2 \ldots \qquad (10.10)$$

$$y = \beta_1 + \beta_2 x_1 + \beta_3 x_1^2 + \beta_4 x_2^2 + \ldots + \beta_5 x_1^3 + \beta_6 x_2^3 \qquad (10.11)$$

$$\beta = X\backslash y \qquad (10.12)$$

As indicated in the case of the mono-variate regression model, the choice for the appropriate multiple regression model depends on P and R^2 values. Adjusted R^2 is preferable for analysis involving few data. With few data, the R^2 value is an optimistic overestimated value. Fishers' value is also used for evaluating the model. A high Fishers' value affirms that there is a low chance for noise to occur in the model, thereby establishing significant relevance of the model to the data analysis. Moreover, to develop a multiple regression model, the following tests are to be conducted on variable data to ensure assumptions in the null hypothesis: collinearity, multicollinearity, normality, homoscedasticity, and outliers. For every independent and dependent variable, there must be a certain degree of correlation that is expected to not be less than r = 0.3, where r is the correlation ratio. Multicollinearity takes places when there is a strong correlation ($r \geq 0.7$, up to 0.9 is acceptable but with caution) between two independent variables. This violates an assumption in the null hypothesis. It implies that all independent variables are acceptable to be analysed using multiple regression when correlation ratio of every two of such variables are less than 0.7. Both tolerance and variance inflation factor (VIF) are also used to check multicollinearity. Tolerance is a parameter for determining how much a particular independent variable is not explained by another independent variable. It is estimated using the formula in Equation 10.13. The value of the tolerance (T) must conform to the range, $0.1 \leq T \leq 1$ to ascertain that multiple correlations of one independent variable with another independent variable is not high. VIF is a reciprocal of T, and its value must be between 1 and 10 for every independent variable to be included in the multiple regression.

$$Tolerance = 1 - R^2 \qquad (10.13)$$

Normality is checked by the normal probability plot and scatterplot of the standardised residuals. The normal probability plot is expected that data points should lie in a straight-line diagonal; meanwhile, points in the scatterplot of the standardised residuals should be distributed to assume a rough rectangle with most data focused on the centre along the zero point. Outliers are data that are much greater or smaller than others. They are data having standardised residuals greater than 3.3 and less than −3.3 in the scatterplot. Also, they are data having Cook's distance less than zero and leverage greater than 0.33 [38,43]. They are also identified by a critical chi-square value calculated from Mahalabonis' distance using a number of independent variables as a degree of freedom [44]. Various software like MATLAB, Design Expert, SPSS, and OriginPro are employed for running regression analysis. In exemplifying the regression analysis as a tool for establishing relationship among variables, tensile strength (Ts) of HPRE was analysed as a function of wt% of Alp (A_n) and UCSp (U_n). It means there are two factors or variables, which Ts depends on. Therefore, bivariate regression analysis is used at three different levels of A_n

and U_n, say (0, 2, 4) and (0, 2, 6) equivalent to a total weight addition of 0%, 4%, and 10%. The experimental data were analysed using a bivariate linear function expressed in Equation 10.8. The regression coefficients (β) were obtained using a MATLAB command left division of X by y, as shown in Equation 10.12.

Reporting the result of the regression analysis on Ts, Table 10.8 shows the Pearson correlation coefficients (r) between tensile strength, T_s, (responses) and A_n and U_n, independent variables. Values of r in respect of A_n and T_s are 0.776 and 0.929 between T_s and U_n. This indicates that both variable pairs exhibit a good linear relationship with T_s. The value of r between U_n and A_n is 0.614 < 0.7, which is acceptable in the multicollinearity assumption in the null hypothesis [44]. Non-violation of multicollinearity assumption is affirmed by predictor tolerance and VIF of 0.623 > 0.1 and 1.606 < 10. A function developed for this model is expressed in Equation 10.14. Values of β (coefficients) of every term of the function and their standardised errors are displayed in the third and fourth columns of Table 10.9.

The standardised β in the fifth column of Table 10.9 shows a unique contribution of A_n and U_n in the generation of response genes at each level of predictor variables. Standardised β of U_n is 0.727 > 0.33 of A_n. This implies that U_n shows a stronger unique contribution to theoretical determination of epoxy/AlpUCSp hybrid nanocomposite tensile strength (response). This contribution is expressed graphically by response surface plot in Figure 10.20. Since the model was developed at $\alpha = 0.05$, meaning a 95% confidence level; probability 0.018 < 0.05 in Table 10.10 establishes significant influences of every term of the model on the response. This is confirmed by significances of predictors U_n, A_n, and model constant in the sixth column of Table 10.9. The model constant has the smallest P value (0.012), followed by U_n (0.032). This implies that U_n exhibits a higher significant influence than A_n while the model constant is the most influencing factor. The model explains 93.2% of the expected probabilities. The remnants account for residuals (Figure 10.21). Evaluation of the model using standardised residuals and Mahalabonis' distance confirms the absence of the outlier in the response matrix and the model maximum critical value, 3.763, which is much less than 13.82, is excellent in line with [44]. The zero value of standardised residual mean (Table 10.11) is an indication of the absence of systematic error in the prediction of response. Based on these observations, it can be inferred that this model exhibits a good predictability of the relationship between response

TABLE 10.8
Pearson Correlation Coefficients Predicting a Linear Relationship between Dependent and Independent Variables

		T_s	A_n	U_n
Pearson Correlation	T_s	1.000	0.776	0.929
	A_n	0.776	1.000	0.614
	U_n	0.929	0.614	1.000

TABLE 10.9

Coefficients, β of the Model Function for Computing Epoxy/AlpUCSp Tensile Strength

Standard Order	Predictors	β	Standardised Error	Standardised β	Significance	95 Confidence Intervals for β		Tolerance	VIF
						Lower Bound	Upper Bound		
1	(Constant)	16.446	2.992		0.012	6.926	25.967		
2	A_n	1.277	0.739	0.330	0.182	−1.075	3.630	0.623	1.606
3	U_n	3.719	0.978	0.727	0.032	.607	6.832	0.623	1.606

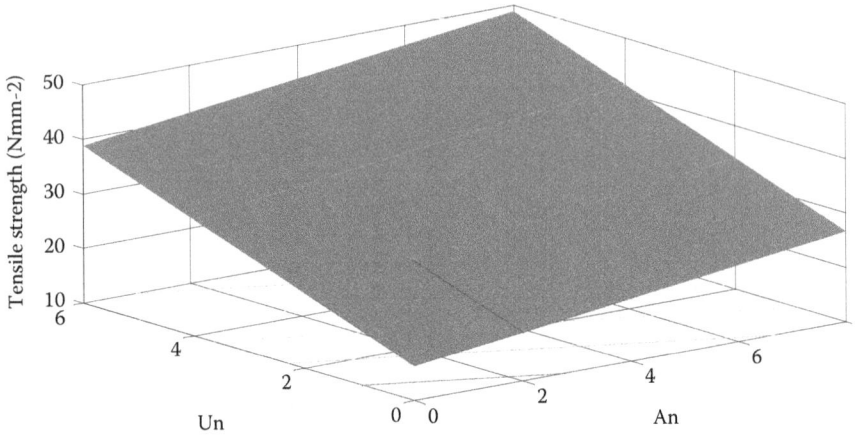

FIGURE 10.20 Response surface plot showing unique stronger contribution of Un to prediction of epoxy/AlpUCSp hybrid nanocomposite tensile strength.

TABLE 10.10
ANOVA of Epoxy/AlpUCSp Tensile Strength

Standard Order	Model	Sum of Squares	Df	Mean Square	F	P Value (sig)	R Square
1	Regression	520.725	2	260.363	20.491	0.018	
2	Residual	38.119	3	12.706			
3	Total	558.844	5				
4	Summary						0.932

Key: Df – degree of freedom; F – Fisher's value; P – Probability; R^2 – Regression coefficient of determination.

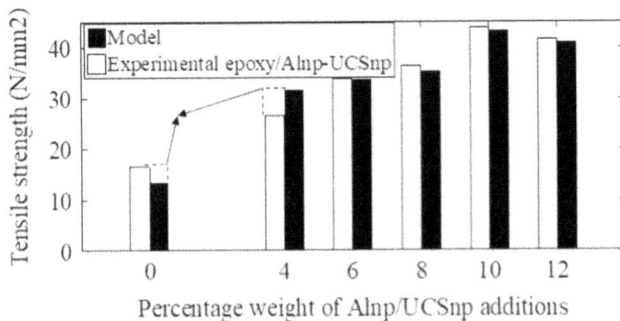

FIGURE 10.21 Matching of experimental epoxy/AlpUCSp hybrid tensile strength with theoretical data.

TABLE 10.11

Model Diagnostic Parameters for Epoxy/AlpUCSp Hybrid Nanocomposite Tensile Strength

	Minimum	Maximum	Mean	Standard Deviation
Mahalabonis distance	0.108	3.763	1.667	1.434
Standard residual	−0.832	1.475	0.000	0.775

(epoxy/AlpUCSp tensile strength) and predictors (A_n and U_n). Hence, it is statistically appropriate for the theoretical prediction of epoxy/AlpUCSp tensile strength.

This is a plot of tensile strength of the hybrid nanocomposite as a function of cumulative weight % addition of aluminium (An) and uncarbonised coconut shell nanoparticles. It validates the model via comparison of experimental values of the tensile strength with those obtained from the regression model.

$$Ts_{Epoxy/AlpUCSp} = \beta_1 + \beta_2 A_n + \beta_3 U_n \qquad (10.14)$$

10.4.5 DENSITY AND DYNAMIC THERMAL PROPERTIES OF HYBRID PARTICLE-REINFORCED EPOXY

The density of epoxy polymer determined from Archimedes' principle described in Section 10.2.5 of this chapter is 1.06 gcm^{-3}. Densities of HPRE nanocomposites containing 2%Alp8UCSp is 1.05, 4%Alp6%UCSp is 1.06, 6%Alp4%UCSp is 1.07, and 8%Alp2%UCSp is 1.08 gcm^{-3}. The density variations at cumulative sum of 10 wt% is attributed to distinct proportions of Alp and UCSp in the HPRE nanocomposites since both the reinforcements have different densities. Density of Alp is approximate 2.7 gcm^{-3}, which is much greater than that of the UCSp. Results of thermal test conducted on HPRE nanocomposites using dynamic differential scanning calorimetry are presented in Figures 10.22–10.27. In these figures, amounts of heat flux to keep the temperature of the sample under investigation with that of the reference pan was plotted against the temperature and time; hence, the term dynamics. Figures 10.22–10.26 present the thermal properties of epoxy/AlpUCSp at 4 and 10 wt% of AlpUCSp reinforcement additions. Spectrographic geometries appear different from one another indicating different thermal behaviour of the nanocomposites during the thermal test. Figure 10.22 shows little or no endothermic hook-up at the onset of thermal scanning until after 10 minutes when the upward sloping ramp shows the occurrences of some exothermic events. Between 190°C and 210°C, there is a solid-solid state transition which could be described as T_g of epoxy/2Alp2UCSp nanocomposite at exactly 210.58°C corresponding to point of inflexion between 190°C and 210°C.

Between 350°C and 390°C, there is a change from exothermic to endothermic events such as thermal degradation which begins from 381.99°C and ends at 414.26°C, equivalent to a peak of 397.65°C. However, a very short thermal

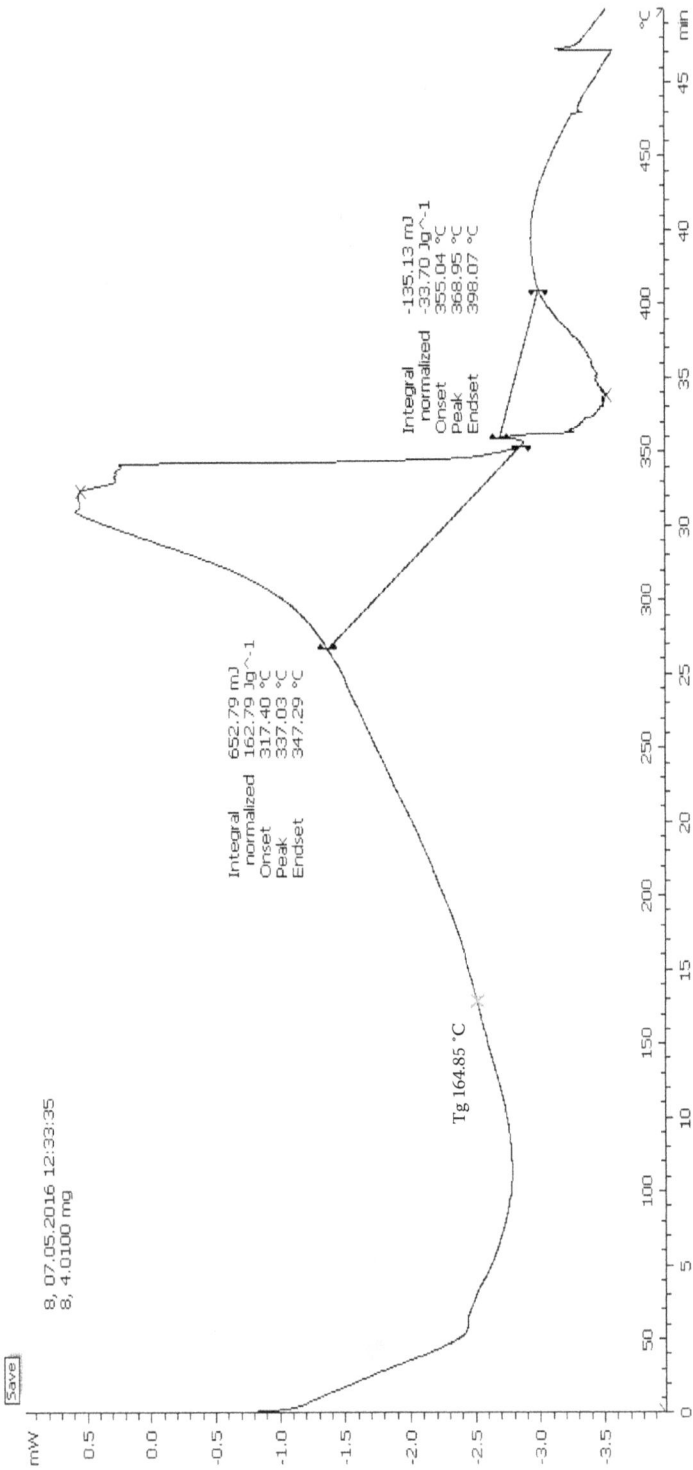

FIGURE 10.22   Dynamic differential scanning calorimetry of epoxy polymer.

2, 09.05.2016 14:10:54
2, 3.0900 mg

Integral   -11.65 mJ
normalized   -3.77 Jg^-1
Onset   477.24 °C
Peak   477.78 °C
Endset   482.26 °C

Integral   -77.93 mJ
normalized   -25.22 Jg^-1
Onset   381.99 °C
Peak   397.65 °C
Endset   414.00 °C

Tg 205.58 °C

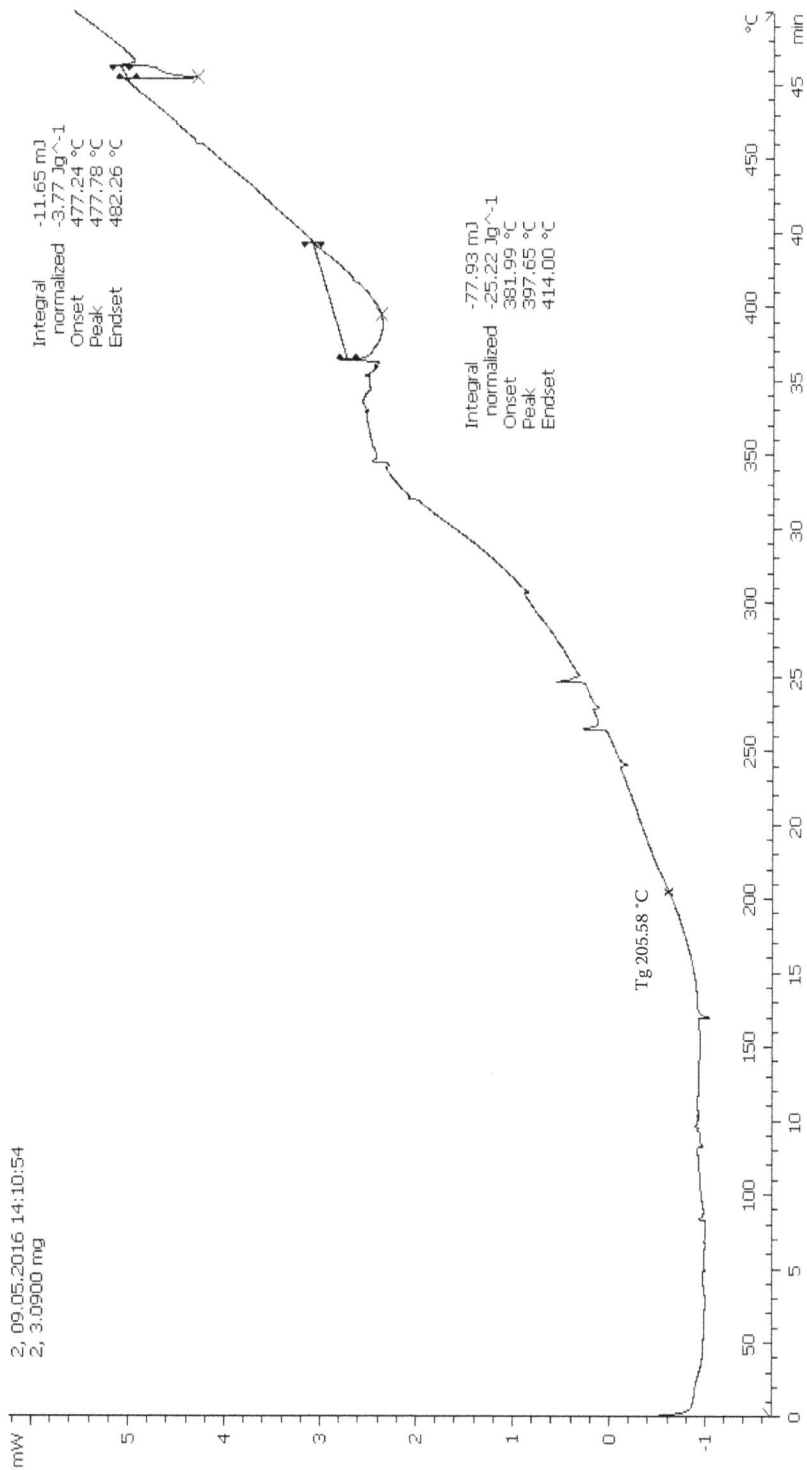

FIGURE 10.23   DSC profile of epoxy/2Alp2UCSp nanocomposite.

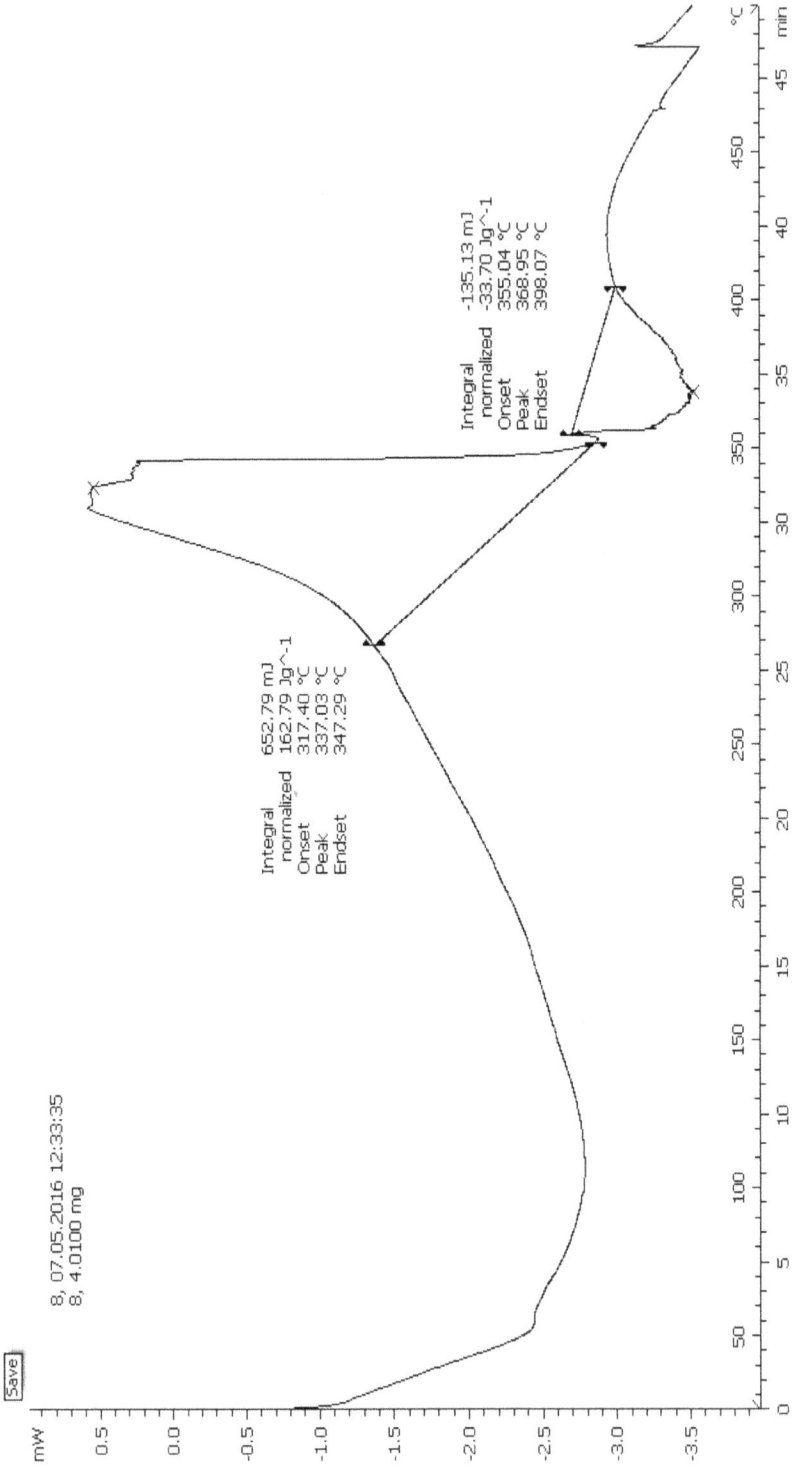

FIGURE 10.24 DSC profile of epoxy/2Alp8UCSp nanocomposite.

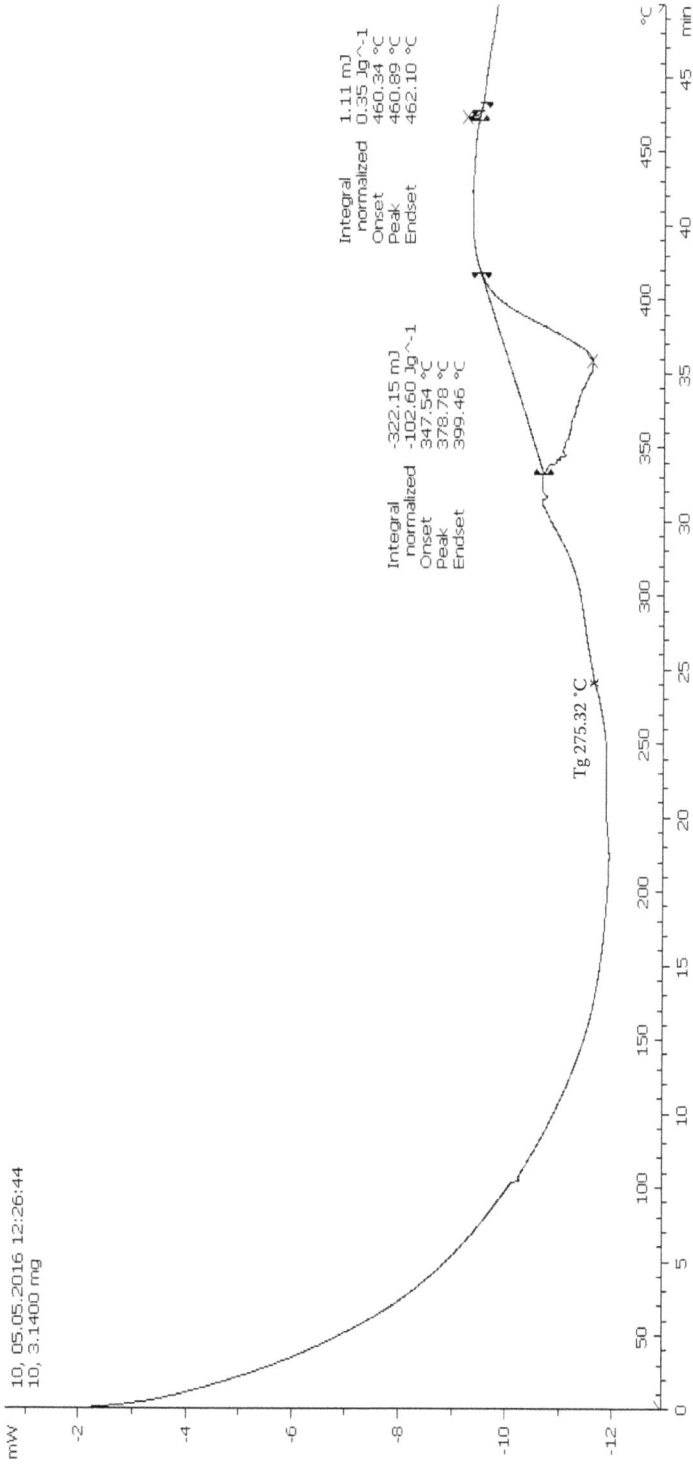

FIGURE 10.25 DSC profile of epoxy/4Alp6UCSp nanocomposite.

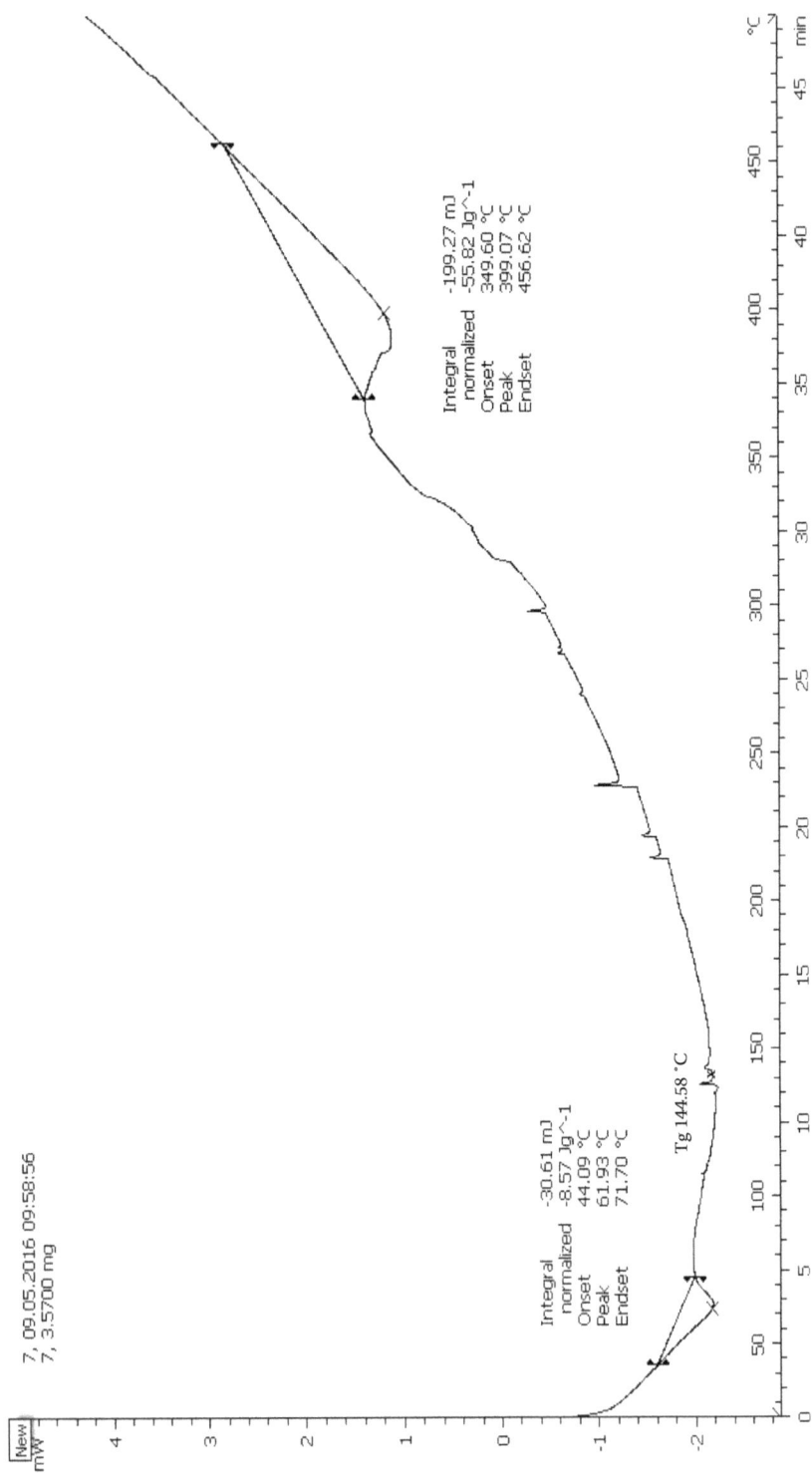

FIGURE 10.26 DSC profile of epoxy/6Alp4UCSp nanocomposite.

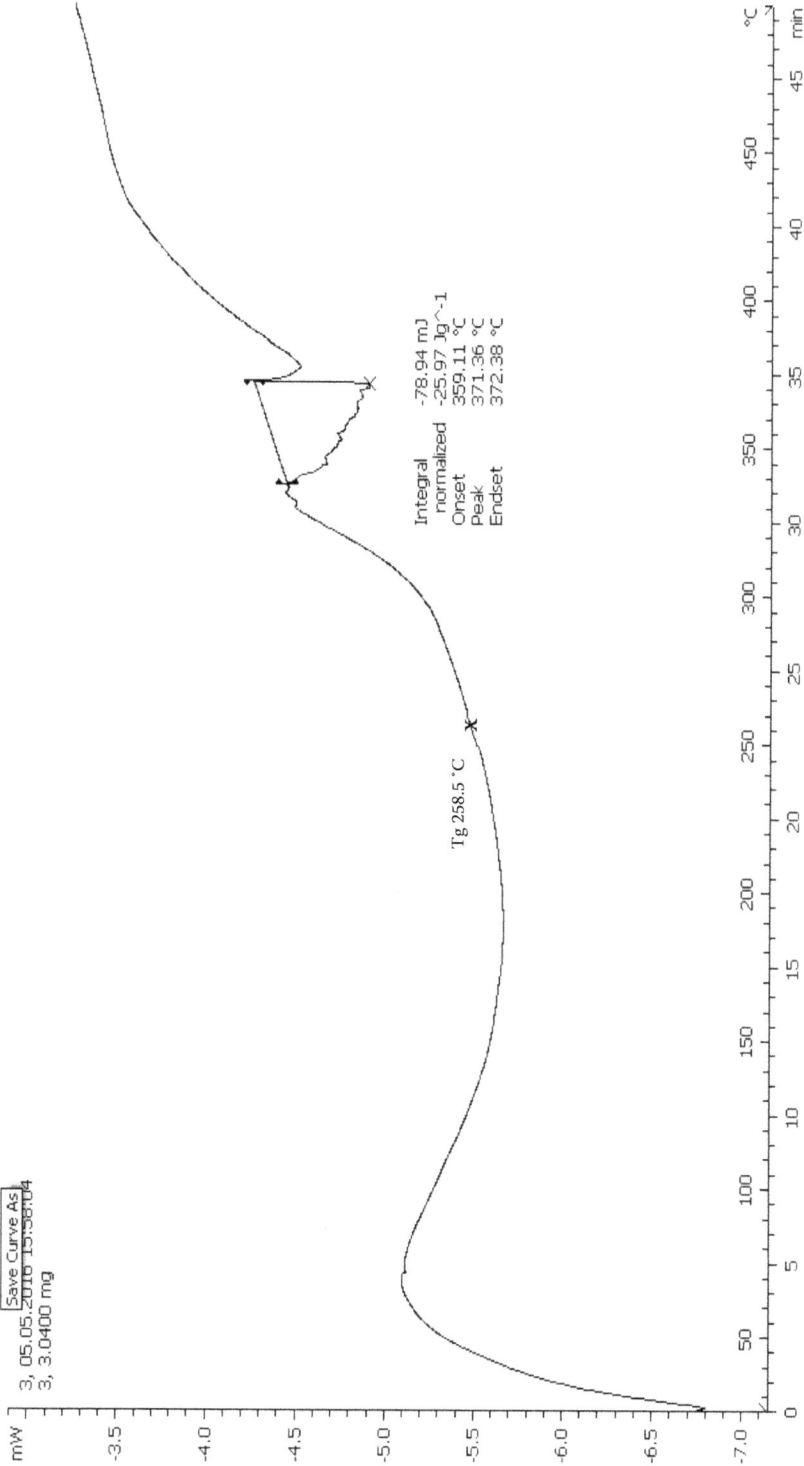

Integral	-78.94 mJ
normalized	-25.97 Jg^-1
Onset	359.11 °C
Peak	371.36 °C
Endset	372.38 °C

Tg 258.5 °C

Save Curve As
3, 05.05.2016 15:58:04
3, 3.0400 mg

FIGURE 10.27 DSC profile of epoxy/8Alp2UCSp nanocomposite.

degradation span between 477°C and 482.26°C with a peak at 477.28°C may define degradation of some stable phases at 397.65°C. The different thermal behaviour observed in Figures 10.23–10.26 can be linked with different proportions of Alp and UCSp making up the hybrid nanoparticles at 10 wt% additions. Maximum increase in thermal stability is found with epoxy/4Alp6UCSp hybrid nanocomposites.

Glass transition temperature of HPRE at each level of reinforcement are 164.85°C; 205.58°C and 275.32°C at 0%, 4% (2%Alp2%UCSp) and 10% (4%Alp6UCSp). Percentage increases, 26.52% and 67.01% are established at 4 and 10 wt% AlpUCSp additions, respectively. During heat supply, epoxy polymer is energised, causing vibration of its molecules. This causes weakness of the C-O-C bond of the polymer, leading to transformation from a brittle hard phase to rubbery ductile phase (Glass transition). However, Al, C, and other elements belonging to AlpUCSp in the epoxy matrix attach themselves to epoxy molecules via formation of new bonds in addition to covalent bond in the epoxy C-O-C ring. Formation of bidentate dative bond has been established in [43], connecting Al and C-O-C. Such additional bonds improve stability of the HPRE, thereby justifying greater glass transition temperatures of the HPRE nanocomposites. Better thermal stability of HPRE nanocomposites is reflected from AlpUCSp additions based on 67.01% enhancement at 10 wt% (4%Alp6UCSp), which is much greater than 31.93% due to 10 wt% of aluminium microparticle additions. It is noteworthy that both hybridisation and refinement of the particle contribute to the enhanced thermal stability of the HPRE nanocomposites. Similarly, improvement in thermal and mechanical properties of epoxy-based composite due to 200–300 μm peanut shells [45] and carbon fibre-nanosilica [21] particle additions. Although this study establishes hybrid AlpUCSp particles as reinforcement in epoxy, combined sisal-kenaf [46] and other hybrid particles in epoxy [47] are reported in literature with better performances than mono-particle reinforced epoxy.

10.5 FUTURE DIRECTION OF 4% ALUMINIUM 6% UNCARBONISED COCONUT SHELL PARTICLE REINFORCED EPOXY NANOCOMPOSITES

Summarily, a maximum tensile strength of 43.2 Nmm^{-2} was obtained at 4% aluminium nanoparticle and 6% uncarbonised coconut shell nanoparticle addition to epoxy. Tensile strain of the nanocomposite is 2.6%. As a complement to the tensile properties to enable specification of the epoxy nanocomposite to an application, other mechanical properties of 4% aluminium 6% uncarbonised coconut shell particle reinforced epoxy nanocomposites were determined. Its flexural strength is 193.43 ± 15.32 Nmm^{-2}; impact energy is 14.52 ± 1.3 J, and hardness value is 12.03 ± 1.5 HV. To check suitability of the 4% aluminium 6% uncarbonised coconut shell particle reinforced epoxy nanocomposite for automobile dashboard cover application, Toyota Sienna 2010 model dashboard cover was bought to use as a case study to benchmark properties of the HPRE. The procured dashboard was dissembled into parts containing inner plastic layer, middle foam layer and upper leather layer. Standard samples for various mechanical tests cut from the inner layer of Toyota Sienna dashboard cover (Figure 10.28) were tested under the same conditions with those of 4% aluminium 6% uncarbonised coconut shell particle

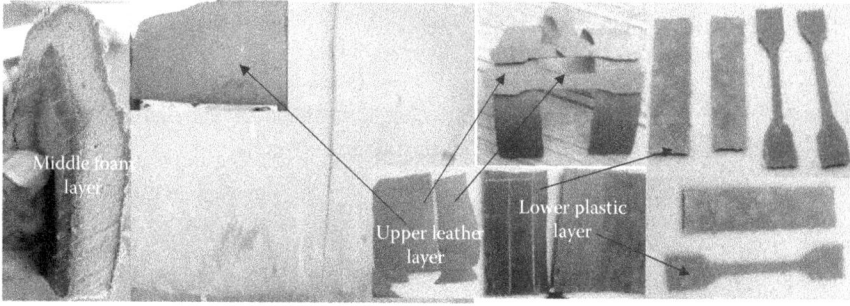

FIGURE 10.28 Procured vehicle dashboard in dissembled form with samples of standard dimensions for analysis cut from the dashboard cover.

reinforced epoxy nanocomposite. Mechanical properties of the inner layer of the Toyota Sienna 2010 model dashboard cover were compared with those of the 4% aluminium 6% uncarbonised coconut shell particle reinforced epoxy nanocomposite (Figure 10.29). It is observed from Figure 10.29 that 4% aluminium 6% un-carbonised coconut shell particle reinforced epoxy nanocomposite has tensile and flexural strengths greater than those of the material for the inner layer of the Toyota Sienna 2010 model dashboard cover, but its tensile elongation and the impact energy are slightly lower than those of the dashboard cover material. Since the tensile strength is a prerequisite of the material property for dashboard cover application and flexural strength for curvature forming without breakage, higher tensile and

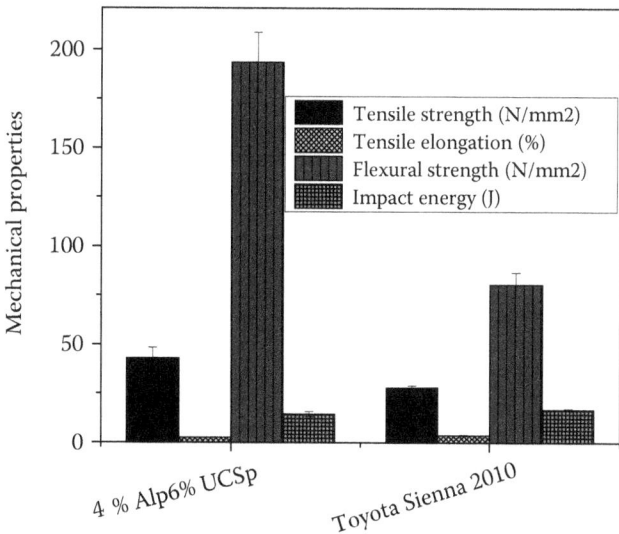

FIGURE 10.29 Benchmarking mechanical properties of 4% aluminium 6% uncarbonised coconut shell particle reinforced epoxy nanocomposite with those of the material for the inner layer of the Toyota Sienna 2010 model dashboard cover.

flexural strengths of the 4% aluminium 6% uncarbonised coconut shell particle reinforced epoxy nanocomposite than those of the existing materials affirms its suitability for the inner layer of the dashboard cover applications, since its tensile elongation and impact energy are nearly equal to those of the material for inner layer of the Toyota Sienna 2010 model dashboard cover.

10.6 CONCLUSIONS AND SUMMARY

This chapter presents epoxy as a potential matrix for emerging polymeric hybrid composites. It focuses on challenges of mono-particle reinforced epoxy, limiting its products as a structural member in engineering applications and enhancement of its properties via structural modifications using reinforcing hybrid nanoparticles sourced from agricultural and metallic wastes. Findings from experimental investigation affirm improved tensile and thermal properties of hybrid particle-reinforced epoxy when compared with epoxy polymer and aluminium particle reinforced epoxy nanocomposite. A linear regression model for predicting the tensile strength of the HPRE nanocomposites was developed. Higher tensile and flexural strengths of the 4% aluminium 6% uncarbonised coconut shell particle reinforced epoxy nanocomposite than those of the existing material for the inner layer of the Toyota Sienna dashboard cover affirms its suitability for the inner layer of the dashboard cover applications since its tensile elongation and impact energy are nearly equal to those of the material for inner layer of the Toyota Sienna 2010 model dashboard cover.

REFERENCES

[1] Jan, W.V.D.W., Martin, V.B., and Stijn, H. (2012). Future of Automotive Design & Materials Trends and Developments in Design and Materials. *Automatic Technology Centre*, acemr.eu.
[2] Bose, S. and Bandyopadhyay, A. (2013). Chapter 1 – Introduction to Biomaterials, in *Characterization of Biomaterials*, A. Bandyopadhyay & S. Bose, Editors, Oxford Academic Press, Elsevier. pp. 1–9. 10.1016/b978-0-12-415800-9.00001-2
[3] Chen, Y., Dou, J., Yu, H., and Chen, C. (2019). Degradable Magnesium-Based Alloys for Biomedical Applications: The Role of Critical Alloying Elements. *J Biomater Appl*, **33**(10), 1348–1372. 10.1177/0885328219834656
[4] Okitsu, Y. and Tsuji, N. (2012). Nanostructured Steel for Automotive Body Structures, in *Advanced Materials in Automotive Engineering*, J. Rowe, Editor, Woodhead Publishing. pp. 57–84. 10.1533/9780857095466.57
[5] Bloeck, M. (2012). Aluminium Sheet for Automotive Applications, in *Advanced Materials in Automotive Engineering*, J. Rowe, Editor, Woodhead Publishing. pp. 85–108. 10.1533/9780857095466.85
[6] Berhanuddin, N.I.C., Zaman, I., Rozlan, S.A.M., Karim, M.A.A., Manshoor, B., Khalid, A., Chan, S.W., and Meng, Q. (2017). Enhancement of Mechanical Properties of Epoxy/Graphene Nanocomposite. *Journal of Physics: Conference Series*, **914**, 012036. 10.1088/1742-6596/914/1/012036
[7] Ong, Y.T. and Tan, S.H. (2019). Carbon Nanotube-Based Biodegradable Polymeric Nanocomposites: 3rs (Reduce, Reuse, and Recycle) in the Design, in *Handbook of Ecomaterials*, L. Martinez, O. Kharissova, B. Kharisov, Editors, Springer. pp. 1–17. 10.1007/978-3-319-48281-1_69-1

[8] Agunsoye, J.O., Bello, S.A., Talabi, S.I., Yekinni, A.A., Raheem, I.A., Oderinde, A.D., and Idegbekwu, T.E. (2015). Recycled Aluminium Cans/Eggshell Composites: Evaluation of Mechanical and Wear Resistance Properties. *Tribology in Industry*, **37**(1), 107–116.

[9] Agunsoye, J.O., Bello, S.A., Yekinni, A.A., Raheem, I.A., and Awe, O.I. (2016). Effects of Reinforcement Particle Sizes on Mechanical Properties of Aluminium/ Egg Shell Composites. *Unilag Journal of Medicine, Science and Technology*, **4**(2), 133–143.

[10] Agunsoye, J.O., Talabi, S.I., Bello, S.A., and Awe, I.O. (2014). The Effects of Cocos Nucifera (Coconut Shell) on the Mechanical and Tribological Properties of Recycled Waste Aluminium Can Composites. *Tribology in Industry*, **36**(2), 155–162.

[11] Bello, S.A., Raheem, I.A., and Raji, N.K. (2017). Study of Tensile Properties, Fractography and Morphology of Aluminium (1xxx)/Coconut Shell Micro Particle Composites. *Journal of King Saud University - Engineering Sciences*, **29**, 269–277. 10.1016/j.jksues.2015.10.001

[12] Bello, S.A., Agunsoye, J.O., and Hassan, S.B. (2015). Synthesis of Coconut Shell Nanoparticles Via a Top Down Approach: Assessment of Milling Duration on the Particle Sizes and Morphologies of Coconut Shell Nanoparticles. *Materials Letters*, **159**, 514–519. 10.1016/j.matlet.2015.07.063

[13] Agunsoye, J.O., Anyanwu, J.E., Bello, S.A., and Hassan, S.B. (2018). Study of Breakage Tendencies of Palm Kernel, Coconut and Periwinkle Shells Using Ball-Milling Process. *Nigerian Journal of Technological Development*, **15**(3), 102–107. 10.4314/njtd.v15i3.5

[14] Agunsoye, J.O., Bamigbaiye, A.A., Bello, S.A., and Hassan, S.B. (2020). Mechanical Properties of Maize Stalk Nano-Particle Reinforced Epoxy Composites. *Arabian Journal for Science and Engineering*, **45**, 5087–5097. 10.1007/s13369-02 0-04345-5

[15] Bello, S.A. (2019). Agglomerates within Ball-Milled Lignocellulosic Particles Using Minimum Crystal Sizes. *Unilag Journal of Medicine, Science and Technology*, **7**(1), 15–40.

[16] Bello, S.A., Agunsoye, J.O., Adebisi, J.A., Kolawole, F.O., Raji, N.K., and Hassan, S.B. (2018). Quasi Crystal Al (1xxx)/Carbonised Coconut Shell Nanoparticles: Synthesis and Characterisation. *MRS Advances*, **3**(42-43), 2559–2571. 10.1557/ adv.2018.369

[17] Kandemir, A., Pozegic, T.R., Hamerton, I., Eichhorn, S.J., and Longana, M.L. (2020). Characterisation of Natural Fibres for Sustainable Discontinuous Fibre Composite Materials. *Materials (Basel)*, **13**(9). 10.3390/ma13092129

[18] Bello, S.A., Agunsoye, J.O., Adebisi, J.A., Kolawole, F.O., and Suleiman, B.H. (2106). Physical Properties of Coconut Shell Nanoparticles. *Kathmandu University Journal of Science, Engineering and Technology*, **12**(1), 63–79. 10.3126/kuset. v12i1.21566

[19] Agunsoye, J.O., Bello, S.A., Bello, L., and Idehenre, M.M. (2016). Assessment of Mechanical and Wear Properties of Epoxy Based Hybrid Composites. *Advances in Production Engineering & Management*, **11**(1), 5–14. 10.14743/apem2016.1.205

[20] Chikhradze, N.M., Marquis, F.D.S., and Abashidze, G.S. (2015). Hybrid Fiber and Nanopowder Reinforced Composites for Wind Turbine Blades. *Journal of Materials Research and Technology*, **4**(1), 60–67. 10.1016/j.jmrt.2015.01.002

[21] Liu, F., Deng, S., and Zhang, J. (2017). Mechanical Properties of Epoxy and Its Carbon Fiber Composites Modified by Nanoparticles. *Journal of Nanomaterials*, **2017**, 1–9. 10.1155/2017/8146248

[22] Zhang, H., Li, X., Qian, W., Zhu, J., Chen, B., Yang, J., and Xia, Y. (2020). Characterization of Mechanical Properties of Epoxy/Nanohybrid Composites by Nanoindentation. *Nanotechnology Reviews*, **9**(1), 28–40. 10.1515/ntrev-2020-0003

[23] Bledzki, A.K., Mamun, A.A., and Volk, J. (2010). Barley Husk and Coconut Shell Reinforced Polypropylene Composites: The Effect of Fibre Physical, Chemical and Surface Properties. *Composites Science and Technology*, **70**(5), 840–846. 10.1016/j.compscitech.2010.01.022

[24] Patel, R.H., Sevkani, V.R., Patel, B.R., and Patel, V.B. (2016). Properties of Glass/Carbon Fiber Reinforced Epoxy Hybrid Polymer Composites. **1728**, 020238. 10.1063/1.4946289

[25] Boopalan, M., Niranjanaa, M., and Umapathy, M.J. (2013). Study on the Mechanical Properties and Thermal Properties of Jute and Banana Fiber Reinforced Epoxy Hybrid Composites. *Composites Part B: Engineering*, **51**, 54–57. 10.1016/j.compositesb.2013.02.033

[26] Davoodi, M.M., Sapuan, S.M., Ahmad, D., Ali, A., Khalina, A., and Jonoobi, M. (2010). Mechanical Properties of Hybrid Kenaf/Glass Reinforced Epoxy Composite for Passenger Car Bumper Beam. *Materials & Design*, **31**(10), 4927–4932. 10.1016/j.matdes.2010.05.021

[27] Ulus, H., Ustun, T., Sahin, S.O., and Avci, A. (2015). Synergistic Effect of Bnnp-Cnt Hybridisation on Fracture Toughness of Carbon Fiber Reinforced Epoxy Laminates, in 8th Ankara International Aerospace Conference. Ankara TURKEY. pp. 1–6.

[28] Altuwirqi, R.M., Baatiyah, B., Nugali, E., Hashim, Z., and Al-Jawhari, H. (2020). Synthesis and Characterization of Aluminum Nanoparticles Prepared in Vinegar Using a Pulsed Laser Ablation Technique. *Journal of Nanomaterials*, **2020**, 1–5. 10.1155/2020/1327868

[29] Hanna, W., Maung, K., El-Danaf, E.A., Almajid, A.A., Soliman, M.S., and Mohamed, F.A. (2014). Nanocrystalline 6061 Al Powder Fabricated by Cryogenic Milling and Consolidated Via High Frequency Induction Heat Sintering. *Advances in Materials Science and Engineering*, **2014**, 1–9. 10.1155/2014/921017

[30] Karunadasa, K.S.P., Manoratne, C.H., Pitawala, H.M.T.G.A., and Rajapakse, R.M.G. (2019). Thermal Decomposition of Calcium Carbonate (Calcite Polymorph) as Examined by in-Situ High-Temperature X-Ray Powder Diffraction. *Journal of Physics and Chemistry of Solids*, **134**, 21–28. 10.1016/j.jpcs.2019.05.023

[31] Bulut, M., Bozkurt, Ö.Y., Erkliğ, A., Yaykaşlı, H., and Özbek, Ö. (2019). Mechanical and Dynamic Properties of Basalt Fiber-Reinforced Composites with Nanoclay Particles. *Arabian Journal for Science and Engineering*, **45**(2), 1017–1033. 10.1007/s13369-019-04226-6

[32] Chen, Y. (2019). Chapter 2 – Manufacture of Carbon Fiber Composite Core Aluminum Conductor, in *Engineering Energy Aluminum Conductor Composite Core (ACCC) and Its Application*, Y. Chen, Editor, Academic Press, Elsevier. pp. 23–46. 10.1016/b978-0-12-815611-7.00002-8

[33] Chung, D.D.L. (2017). Introduction to Carbon Composites, Elsevier. pp. 88–160. 10.1016/b978-0-12-804459-9.00002-6

[34] Kollia, E., Vavouliotis, A., and Kostopoulos, V. (2018). Thermal Ageing of Carbon Fiber-Reinforced Cyanate Ester Composites under Inert and Oxidative Environment. *Polymer Composites*, **40**(S2), E1388–E1396. 10.1002/pc.25016

[35] ASTM D638-14 (2014). *Standard Test Method for Tensile Properties of Plastics*. ASTM International, West Conshohocken, PA. https://www.astm.org/Standards/D638.

[36] Kinloch, A.J. and Taylor, A.C. (2006). The Mechanical Properties and Fracture Behaviour of Epoxy-Inorganic Micro- and Nano-Composites. *Journal of Materials Science*, **41**(11), 3271–3297. 10.1007/s10853-005-5472-0

[37] Bello, S.A., Agunsoye, J.O., Raji, N.K., Adebisi, J.A., Raheem, I.A., and Hassan, S.B. (2021). Interfacial Adhesion and Microstructure of Epoxy/Aluminium Particulate Nanocomposites. *Nano Hybrids and Composites*, **32**, 1–14.

[38] Bello, S.A., Agunsoye, J.O., Adebisi, J.A., Adeyemo, R.G., and Hassan, S.B. (2020). Optimization of Tensile Properties of Epoxy Aluminum Particulate Composites Using Regression Models. *Journal of King Saud University –Science*, **32**(1), 402–411. 10.1016/j.jksus.2018.06.002

[39] Hassan, S.B., Agunsoye, J.O., Bello, S.A., Adebisi, J.A., and Agboola, J.B. (2018). Microstructure and Mechanical Properties of Coconut Shell Reinforced Epoxy Composites in Materials Science and Technology 2018 (MS&T18). Greater Columbus Convention Center, Columbus, Ohio, USA, Materials Science and Technology.

[40] Agunsoye, J.O., Bello, S.A., and Adetola, L.O. (2019). Experimental Investigation and Theoretical Prediction of Tensile Properties of Delonix Regia Seed Particle Reinforced Polymeric Composites. *Journal of King Saud University - Engineering Sciences*, **31**(1), 70–77. 10.1016/j.jksues.2017.01.005

[41] Bello, S.A., Raheem, I.A., and Raji, N.K. (2015). Study of Tensile Properties, Fractography and Morphology of Aluminium (1xxx)/Coconut Shell Micro Particle Composites. *Journal of King Saud University – Engineering Sciences*, **29**, 269–277. 10.1016/j.jksues.2015.10.001

[42] Atuanya, C.U., Edokpia, R.O., and Aigbodion, V.S. (2014). The Physio-Mechanical Properties of Recycled Low Density Polyethylene (Rldpe)/Bean Pod Ash Particulate Composites. *Results in Physics*, **4**, 88–95. 10.1016/j.rinp.2014.05.003

[43] Bello, S.A. (2020). Wear and Thermal Resistance Properties of Aluminium Particulate Microcomposites. *Bulletin of Materials Science*, **43**(261), 1–15. 10.1007/s12034-020-02206-3

[44] Pallant, J.,(2005). *Spss Survival Manual*. 2nd ed., Allen & Unwin. i-318. 174114 478 7

[45] Prabhakar, M.N., Shah, A.U.R., Rao, K.C., and Song, J.-I. (2015). Mechanical and Thermal Properties of Epoxy Composites Reinforced with Waste Peanut Shell Powder as a Bio-Filler. *Fibers and Polymers*, **16**(5), 1119–1124. 10.1007/s12221-015-1119-1

[46] Yorseng, K., Mavinkere Rangappa, S., Parameswaranpillai, J., and Siengchin, S. (2020). Influence of Accelerated Weathering on the Mechanical, Fracture Morphology, Thermal Stability, Contact Angle, and Water Absorption Properties of Natural Fiber Fabric-Based Epoxy Hybrid Composites. *Polymers (Basel)*, **12**(10), 1–14. 10.3390/polym12102254

[47] Matykiewicz, D. (2020). Hybrid Epoxy Composites with Both Powder and Fiber Filler: A Review of Mechanical and Thermomechanical Properties. *Materials (Basel)*, **13**(8), 1–22. 10.3390/ma13081802

11 *Parquetina nigrescens*–Reinforced Polylactic Acid (PLA) Composites for Engineering Applications

Sefiu Adekunle Bello
Department of Materials Science and Engineering, Kwara State University, Malete, Kwara State, Nigeria

Stephen Idowu Durowaye
Department of Materials and Metallurgical Engineering, University of Lagos, Lagos State, Nigeria

Winfred Emoshiogwe Aigbona
Department of Materials Science and Engineering, Kwara State University, Malete, Kwara State, Nigeria

Babatunde Olumbe Bolasodun
Department of Materials and Metallurgical Engineering, University of Lagos, Lagos State, Nigeria

Kemi Audu
Department of Materials Science and Engineering, Kwara State University, Malete, Kwara State, Nigeria

Soliu Oladejo Abdul Ganiyu
Department of Civil and Environmental Engineering, University of Alberta, Edmonton AB, Canada

CONTENTS

DOI: 10.1201/9781003170549-15

11.1   INTRODUCTION

Matters such as global climate change and the depletion of fossil fuels that are inevitable outcomes of increased usage of energy have become issues of discussion among scientists and politicians worldwide [1]. Due to the depletion of petroleum resources and growing environmental concerns, the industries are seeking environmentally friendly materials that are independent of fossil fuels [2]. Nevertheless, petroleum-based products have found applications in many areas due to their desirable properties. However, the use of petroleum-based products that are non-biodegradable materials, especially plastics for various packaging applications, has raised environmental pollution concerns [3].

The petrochemical industry is responsible for most plastics produced. Their production and sales have proven to be very lucrative because of the immense global demand, especially in developing countries and this trend is expected to continue and even rise [1]. Despite the suitability of plastics for a wide variety of applications, organisations are confronted with the growing problem of finding alternative methods for disposing large volumes of plastic waste. Disposal of plastic waste in the environment is considered a big problem due to their non-biodegradability and presence in large quantities [4]. In the USA, about 30 million tonnes of plastic waste were produced in 2009 and only about 7% was recycled. Plastic waste ends up in landfills, beaches, rivers, and oceans, thereby causing environmental problems [5]. According to the European Commission, about 5 billion pounds of plastic wastes are generated every year in the United Kingdom [6]. In Japan, plastic wastes are found to be the third major municipal and industrial waste [7] but second in Nigeria, which is a developing country [8].

The extensive use of petroleum-based polymers, especially plastics, has buttressed the fact that many plastics products, after expiration, form major parts of solid wastes because they are non-biodegradable. Despite the advantage of low cost of production and huge versatility of plastics, the sustainability of this synthetic material is still a challenge. Thus, because of the global environmental issues and the growing complications of solid waste treatment, bio-based or renewable polymeric materials could be the appropriate alternatives to petroleum-based products. Furthermore, the planet's fossil fuel capacity is expected to run out in the near future, and humans are expected to lose their dependency on the aforementioned unsustainable resources [1]. Development of technologies to produce renewable polymers instead of petroleum-based materials that will reduce CO_2 emissions resulting from fossil fuel consumption is the fundamental reason behind global warming and climate change [9] is a welcome development.

The problems associated with products of petroleum-based resources have necessitated the advocacy for finding good alternatives, effective management of the wastes, and most importantly, converting the non-hazardous ones to useful engineering materials for biomedical and industrial applications. In the light of these, attention is on the development and utilisation of bio-based polymer composites, which are renewable, using natural fillers and biopolymers (bioplastics) as replacement for the conventional petroleum-based composites like polyethylene for packaging. Use of PLA-based composites for packaging applications is proposed to replace nonbiodegradable materials known to have environmental issues. Moreover, the use of PLA-based composites as prothesis materials for replacing metallic implants is emerging. After installation of metallic artificial teeth, their interaction with human fluid releases corrosion products that are harmful to body. Polymers are stable to human fluid and release no toxic substance, but they are less strong to meet up with the strength requirement for artificial implant applications. Modification of their structures with reinforcement is imperative to produce biomedical composites. Polylactic acid is safe to the body, but being biodegradable requires reinforcement with a high potential for antibacterial activity. Therefore, the reinforcement serves to protect the PLA composite from decomposition by bacteria and improves strength. Therefore, special reinforcement is needed to achieve this, creating a novel path into materials developments.

Biopolymer composites, which are regarded as renewable or green composites are biodegradable. In biopolymer composites, natural fibre (reinforcement agent) in discontinuous phase is added to the biopolymer matrix in continuous phase to improve the stiffness and tensile strength of the prepared composite. The purpose of preparing this kind of composite is to obtain a product with good mechanical behaviour and durability performance imparted by natural fibre and biopolymer, respectively. Usually, maximum stiffness and tensile strength of bio-composites ranges from 1 to 4 GPa and 20 to 200 MPa, respectively [10]. The natural fibre reinforcements also affect the thermal properties, electrical conductivity, morphological characteristics, crystallinity, degradability, and production cost of biopolymer composites. The primary applications of natural fibre–reinforced polymer composites are in the fields of automobiles, aerospace, and structural constructions [11]. Some noticeable benefits of biopolymer composite applications are sustainability, cost effectiveness, lightweight characteristics, appreciable specific strength, biodegradability, environmental friendliness of renewable materials, and health and safety of manufacturer and consumers [12].

11.2 BIOPOLYMERS/BIOPLASTICS: PRODUCTION, MARKETS, BENEFITS, AND APPLICATIONS

Living organisms produce biopolymers. They are also derived from biomass and after serving their purpose, and they degrade within a reasonable time without causing environmental waste problems [13,14]. They are made up of long-chain compounds that contain molecular subunits, and they make up much of living bodies and most of the biosphere. These biopolymers can either be thermosetting or thermoplastic polymers [15]. Bio-based or renewable polymers are materials derived from renewable sources such as corn, potato, molasses, tapioca, cane sugar, and rice [16]. The terms "renewable polymer" or "biopolymer" apply not only to naturally occurring polymeric materials but also to natural substances that have been polymerised into high molecular weight materials by chemical and/or biological methods. Therefore, bio-based or renewable polymers include various synthetic polymers derived from renewable resources and CO_2. A few examples of bio-based polymers are polylactide or polylactic acid (PLA), polynucleotides, polyoxoesters, polythioesters, polyamides, polysaccharides, polyanhydrides, polyisoprenoides, and polyphenols, their derivatives, and their blends and composites [9].

Food packaging is generally viewed as an unnecessary cost [17]. It is considered as an additional economic and environmental cost instead of rightfully being seen as an added value for the reduction of food loss by improving shelf life. Sustainability of green packaging development using edible or biodegradable materials and their composites, plant extracts, and nanomaterials can help to bring down the environmental impacts of food packaging [18]. Bio-based materials can be particularly valuable in the three main areas of food-related applications, namely food packaging, food coating, and edible films for food and encapsulation [19]. The biopolymers that are used to replace petrochemical plastics are also referred to as bioplastics. Adoption of biopolymers as food packaging materials should no longer be viewed as an option but as a necessity [20]. The widespread use of biodegradable materials for packaging can result in reduced plastic waste, lower greenhouse gas emissions, and guarantee the sustainable exploitation of environmental resources [21–24].

The bioplastics industry is reported to be a young and innovative sector possessing a great economic and ecological potential for a low-carbon, circular bioeconomy that utilises resources more efficiently [25,26]. The market for biopolymers is expected to show some growth and new demands are envisioned from applications that have a clear benefit to the consumer and the environment [18]. Biopolymer global production has grown from just under 300,000 tonnes in 2009 to 2.11 million tonnes in 2019 and is predicted to grow to 2.42 million tonnes by the year 2024 [27,28]. Bioplastics produced in 2015 globally had the shares of bio-based non-degradable and biodegradable plastics at about 64% and 36%, respectively [21,29]. There is envisioned growth for the share of biodegradable polymers to meet environmental demands of reduced pollution, and in 2019, the share for biodegradable plastics grew to about 56%. Figure 11.1 illustrates the drop in nonbiodegradable share and rise of biodegradable share up to 2019. Beyond 2019, there are increases in the production capacities of both biodegradable and non-biodegradable bioplastics, confirming growth in the biodegradable bioplastic productions, which, in future, their production could be large enough to replace the nonbiodegradable bioplastics.

Figure 11.2 shows the global production capacities of bioplastics in 2019. Asia was the major producer, accounting for 45% of global production. Land used to grow renewable stocks for bioplastic production stands at 0.79 million hectares, which is 0.02% of the global agricultural area and this will remain the same despite the predicted growth of bioplastic production. As such, this will have no impact on the 97% of global agricultural area, which is being used for pasture, feed, and food [30,31]. Africa is yet to make its mark as it is still lagging behind in terms of production of bioplastics, despite the availability of raw material resources.

In developed countries, biopolymer materials have managed to replace conventional counterparts as food packaging material for mostly organic, natural, and

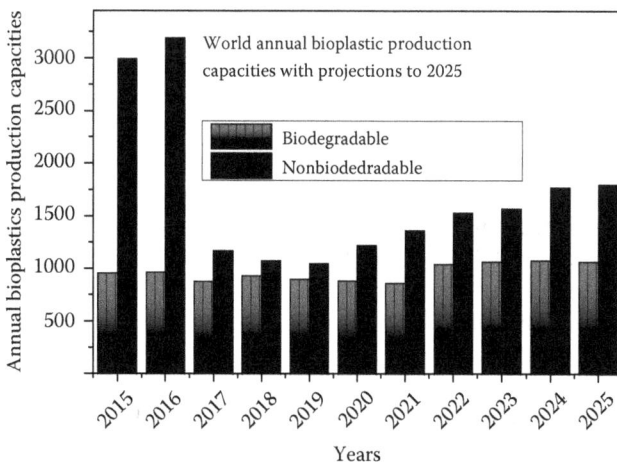

FIGURE 11.1 Share of biodegradable and nonbiodegradable bioplastics production capacities from 2015 to 2021 with projections to 2025.

FIGURE 11.2   Global production capacities of bioplastics in 2019 [31].

functional foods [32]. Biopolymers have greatly influenced the packaging sector because of increasing importance of environmental responsibility to both consumers and industry. As a result, packaging is the biggest consumer of biodegradable bioplastics, which in 2019 accounted for 1.14 million tonnes, which is more that 53% of the total bioplastics market [28]. Biopolymers limit the emission of carbon dioxide during production and degrade to organic matter after disposal. Biodegradable polymers can, without any threat to the environment, retard deterioration, extend shelf life, and maintain quality and safety of packaged food [33]. Despite their high cost of production, bio-based products are undoubtedly the fundamental concept of a circular economy [34]. As a plus, the use of biopolymers to remedy the environmental pollution from fossil-based nonbiodegradable polymers can result in an efficient waste management system envisioned globally.

Environmental benefits that come with the use of biopolymers alone have been reported not to be enough to create a big market [35]. Biodegradable polymers have replaced only about 1% of global plastics [36,37]. Therefore, there is still great need for basic and applied research to improve their performance, reduce cost and improve their ease of production [21]. The adoption of biodegradable polymers for packaging can provide a solution to waste-disposal problems associated with the traditional petroleum-derived packing materials as well as address the fact that crude oil and natural gas resources are limited [38,39]. It is expected that, by 2022, the use of biopolymers in food packaging should have grown by 25% [40].

Several biopolymers that can be used for packaging purposes include starch, cellulose, chitosan, polylactic acid (PLA), polycaprolactone (PCL), and polyhydroxybutyrate (PHB) among many others. Products like bottles, jars, cans, buckets, food containers, disposable cups, packaging films, and refuse bags can be manufactured using biodegradable polymers [33]. In packaging applications, PLA has the advantage of being compostable together with organic waste, even though it would degrade at a slower rate [41]. This means that food-contaminated packaging can be sent for composting with no need for cleaning. The challenge comes when the packaging waste enters water bodies into the oceans where there are low temperatures and high pressures presenting conditions that are not conducive for PLA degrading microorganisms [42]. Packaging waste is one of the biggest polluters of water bodies and, as such, there is a need to curb the mismanagement of bioplastic products marketed as biodegradable as they may have similar effects as fossil-based counterparts on marine organisms and biological accumulation [43]. Among bioplastics, PLA polymer has been shown to possess the properties that enable it to compete with petroleum-based plastics in various application areas [44,45]. Thus, this review is concentrated on PLAs and their bio-composites.

11.3 POLYLACTIC ACIDS: PRODUCTION, PROPERTIES, ADVANTAGES, AND CHALLENGES

A polylactide or polylactic acid (PLA) is an aliphatic polyester biopolymer that can be derived from renewable sources, such as corn, potato, molasses, tapioca, cane sugar, and rice [16]. From these renewable sources, lactic acid is basically produced by a fermentation process and used as a monomer to synthesised PLA through different polymerisation routes, such as ring opening polymerisation (ROP), polycondensation, and other direct methods (e.g., azeotropic dehydration and enzymatic polymerisation) [46]. PLA is a thermoplastic biopolymer, which can be semi-crystalline or totally amorphous in nature. PLA is produced from lactic acid through fermentation of agricultural products like corn. A PLA can be prepared by both direct condensation of lactic acid and ring opening polymerisation of the cyclic lactide, as shown in Figure 11.3.

In PLA synthesis, corn (or rice, potatoes, sugar beets, agricultural wastes, etc.) is converted into dextrose. Lactic acid is obtained through fermentation of dextrose, which is converted into lactide in the presence of catalyst. After purification by vacuum distillation, lactide is converted into a PLA polymer through polymerisation in the presence of suitable catalyst. A PLA is a fully sustainable polymer as it is derived from annually renewable raw materials, and it is fully biodegradable. After composting, PLA-based materials are converted into water and carbon dioxide, which are used in growing more agricultural products for further conversion to PLA [47]. Steps of the PLA synthesis and the life cycle of its materials are shown in Figure 11.4.

The starting material to produce PLA is derived from plant materials, which are renewable. It means that principally, this aliphatic polyester can be produced in a

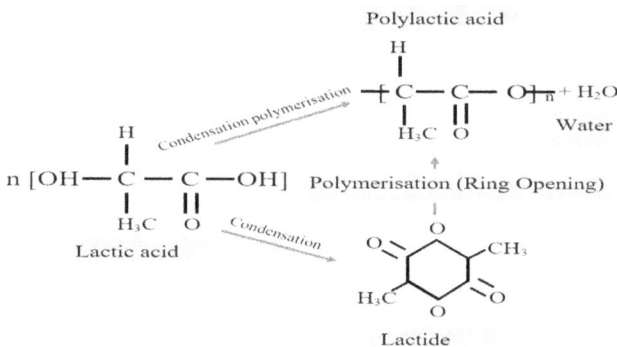

FIGURE 11.3 Polylactic acid (PLA) polymerisation.

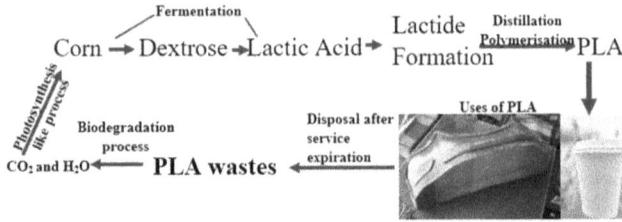

FIGURE 11.4 Polylactic acid (PLA) synthesis.

sustainable way. Since this polymer is biodegradable, it can be converted back to CO_2, which is then fixed by plants via photosynthesis. Therefore, the production of PLA is regarded as "carbon neutral" or "zero-emission" processes, which the net amount of carbon in the environment remains constant over the extended time and on a global scale. Since there is a decrease in the consumption of fossil resources, the application of bio-based and biodegradable plastics such as PLA is mostly referred to as being eco-friendly and sustainable [1].

Although PLA has inferior mechanical properties to polyethylene terephthalate and PP which are the most common materials used in automobiles, its mechanical properties can be improved with filler additions. The temperature at which PLA melts for processing is safe for natural fibres because natural fibres do not degrade at the processing temperature [47]. PLA is a hydrophobic polymer because of the incorporation of the CH_3 side chain [47]. Because of all these favourable properties, a PLA has strong candidacy among the biopolymers for the matrix material to be used in green composites.

Increasing the D-lactide concentration produces high-cost PLA polymers with a more crystalline structure and films with better thermal stability, mechanical strength, and barrier properties. Poly(D, L-lactide) with 88%–90% L-lactide has been widely used for packaging purposes as it is easy to process by thermoforming and displays properties similar to polystyrene [48,49]. PLA can be processed into films, sheets, or moulded products by employing methods such as extrusion, thermoforming, injection- or blow-moulding, and film blowing or stretching. Biodegradable and non-biodegradable plasticizers have been used with PLA to improve processability, although only nontoxic ones are suitable for food contact applications [50]. Packaging products from PLA include shopping bags, cups, trays, wrapping, films, containers, foams, and bottles [51]. Over the years, the production of PLA has been steadily on the rise, as can be seen from Figure 11.5, and estimates indicate that it would rise to just over 559,650 tonnes by 2025.

A polylactic acid (PLA) is the most promising material with a lot of research potential. It presents packaging applications for a wider array of products including films, forms, food containers and coatings among many others [3,24]. PLA is an environmentally friendly, biodegradable, thermoplastic polyester. When disposed into the environment, pure PLA can slowly degrade over a period of months to two years, whilst petroleum-based plastics take more that 500–1000 years [52,53]. Composting with other biomasses such as compost soil, PLA can completely degrade within 3–4 weeks [23]. PLA possesses a wide range of desirable properties

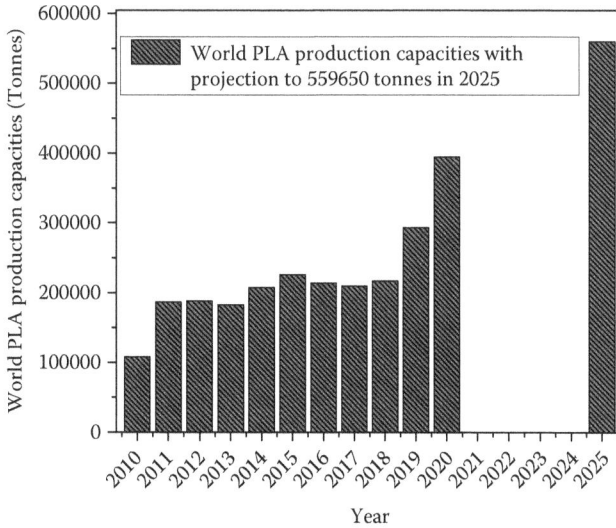

FIGURE 11.5 World production capacity for PLA from 2010 to 2020 with projection to 2025.

including biocompatibility, favourable mechanical properties and it can be moulded into various shapes making its performance comparable to petroleum-based plastics [23,54]. Although, its modulus is lower than that of the petroleum-based polymer, incorporation of additives into it has tendency to enhance its properties, meanwhile PLA consumes less energy in processing than that of the petroleum-based polymer. For example, production of the PLA bottle consumes 36% less energy compared to the manufacture of the same bottle using PET and produces 44% less carbon dioxide [48].

Thus, the use of PLA brings in environmental benefits with similar performance as traditional PET. A comparison of mechanical properties (modulus and elongation at break) of some plastics from bio-based polymers with those of fossil-based plastic polymers is shown in Figure 11.6. In the figure, fossil-based polymers are shown in bold.

PLA is one of the most investigated biopolymers because of its desirable mechanical characteristics, which are like the ones exhibited by petroleum-based thermoplastics like polypropylene, polyethylene, and polystyrene. For example, because of its transparency, recyclability, biocompatibility, exhibition of high tensile strength of 70 MPa, and Young's modulus of 3 GPa, it is used for composites production and it is also used in various fields such as agriculture, civil engineering, packaging, etc [55,56]. The importance of bio-composites has made PLA receive the attention of the industrial sector as a very good substitute for petroleum-based polymers. Despite the above positive features, its application is faced with some challenges. For instance, its commercial application is limited because it is brittle and exhibits low impact energy. In addition, there has not been much production of PLA in large amounts because it is expensive. Hence, it is imperative to boost its impact energy and make PLA much more available and

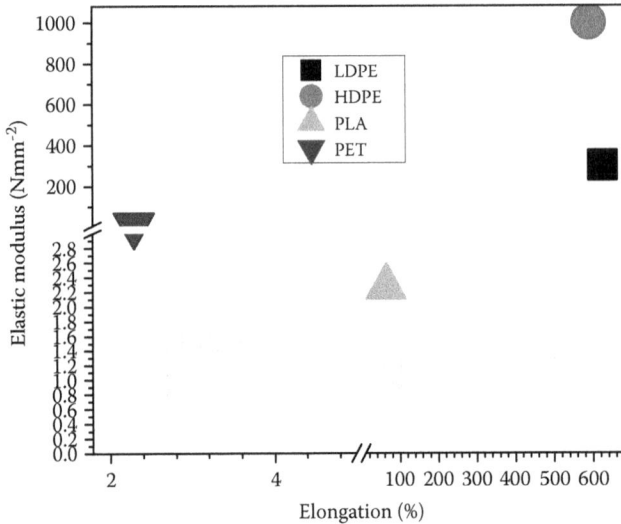

FIGURE 11.6   Mechanical properties of biopolymers and fossil-based polymers.

affordable, as this will promote its utilisation as a biopolymer with natural fillers to make green composites for application in various fields.

11.4   LIFE CYCLE OF PLA

Bioplastic PLA generally follows the same route as all other biodegradable bio-plastics, and it is converted into consumer products. In food packaging, these products are mostly single-use primary packaging plastics for ready meals. They are characterised with short service lifetime and these products include cups, bottles, clamshells, and wrappers. Thus, their degradability is an important factor as they easily find their way into the environment because of the human throwaway culture. Replacing nonbiodegradable plastics with biodegradable PLA will go a long way in environmental protection. Figure 11.7 depicts the life cycle of the polymer. Generally, waste management calls for products that have served their purpose to be re-used, recycled, incinerated, and disposed into landfills. Recycling of PLA is environmentally beneficial, as shown by investigations in almost all environmental impact categories [57]. Thus, from an environmental point of view, composting is the most favourable end-of-life option for PLA as it is considered as an organic or biological recycling option for biodegradable plastics [58]. Composting is the controlled aerobic or biological degradation of organic materials to generate a carbon- and nutrient-rich compost that acts as a natural fertiliser than can be used to grow crops and thus reducing the demand for chemical fertilisers [59,60]. PLA degradation happens mainly through the scission of the backbone ester bonds and in humans and animals PLA undergoes hydrolysis to form soluble oligomers that can be metabolised by cells [14,61]. The use of PLA results in an intrinsic zero material carbon footprint value [41].

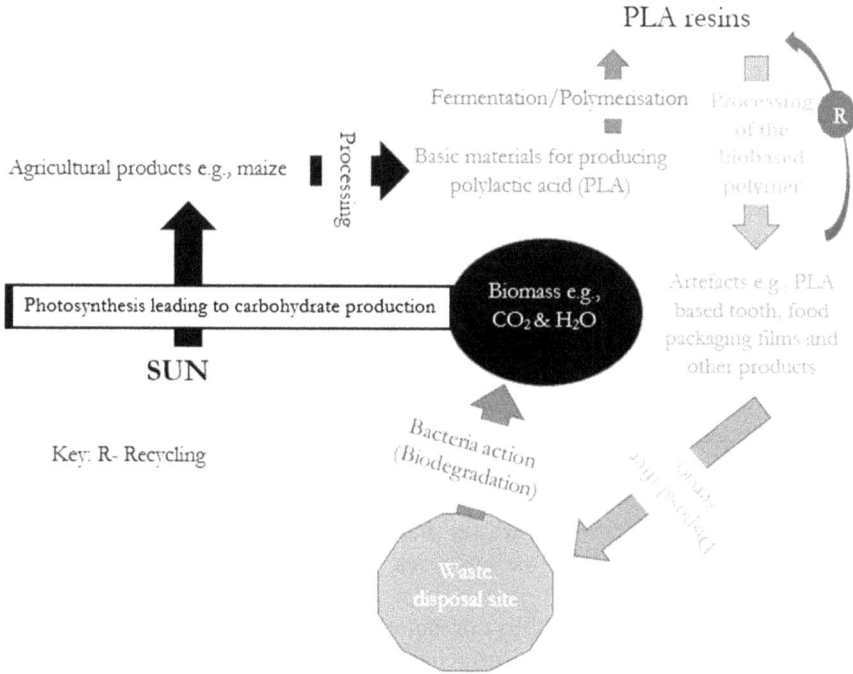

FIGURE 11.7 Life cycle of PLA bioplastic.

11.5 AGRICULTURAL WASTE MANAGEMENT AND UTILISATION FOR GREEN COMPOSITES DEVELOPMENT

In many countries, agricultural wastes are generated in large quantities and are the leftovers from growing and processing of raw agricultural products such as fruits, plants, dairy products, meat, and crops. In many instances, animal bones and crop wastes like corn stalks, sugarcane bagasse, drops and culls from fruits, etc. have been used as fillers to develop composites. The rate at which these wastes is being generated has greatly increased because of the advent of modern technology. The fact is that technology has positively influenced agricultural development in terms of output, which has also increased the amount of waste generated. In many places, much an increase in wastes (animals, plants, and agricultural-industrial by-products) has made the effective management and utilisation of the waste imperative [62].

Composites represent a combination of two of more chemically and structurally different constituent materials whose combination produces a synergistic effect and aggregate characteristics that are better than those of its constituents. Generally, for the development of polymer matrix composites, fillers like silica, graphite, kaolin, carbonate, glass fibres, talc, mica, and synthetic fillers have been blended with a polymer matrix. However, the reuse of petroleum-based polymer composites is very difficult because they are not degradable, leading to their dumping on-site or being burnt, causing environmental pollution [63].

Adverse impact of petroleum-based materials on the environments has necessitated the focus of attention on the development of biopolymer composites,
which are renewable, recyclable, and biodegradable. In developing green composites, some natural agricultural fillers like hemp, flax, jute, and kenaf, which are not
hazardous, have been used as substitutes for conventional synthetic fillers. For
instance, Kenaf-PLA composites have been used to develop automotive parts including, which is claimed to be the first 100% natural automotive product in the
world [64]. Wastes as natural fillers and biopolymers to develop green composites
have both economic and environmental benefits.

11.6   MODIFICATIONS OF POLYLACTIC ACIDS (PLA) USING
AGRICULTURAL WASTE FIBRES/NANOPARTICLES

To find solution to the disadvantages/limitations of PLA and develop materials that
will exhibit unique characteristics to serve as substitute for conventional petroleum-
based composites, blending PLA with agricultural wastes as natural fillers was
explored. In Africa, agricultural wastes such as hemp seeds, flax seeds, jute seeds,
kenaf seeds, hazelnut seeds, rice husks, etc., shown in Figure 11.8, are readily
available in large quantities.

Blending PLA with micro-fillers or nanofillers to form PLA-based composites
has attracted the attention of researchers because it is easy and economical
[65]. In addition, reinforcing PLA with biodegradable fillers like starch and nanocellulose to develop green composites will yield improved thermal and mechanical characteristics without constituting any environmental hazard because
PLA and nanocellulose are biodegradable [66]. Many of these agricultural
wastes have potential properties, which can be leveraged on to produce PLA
composites that are biodegradable and promotes the sustainability of a society
that is productive and generates waste that can be utilised without adverse environmental effects.

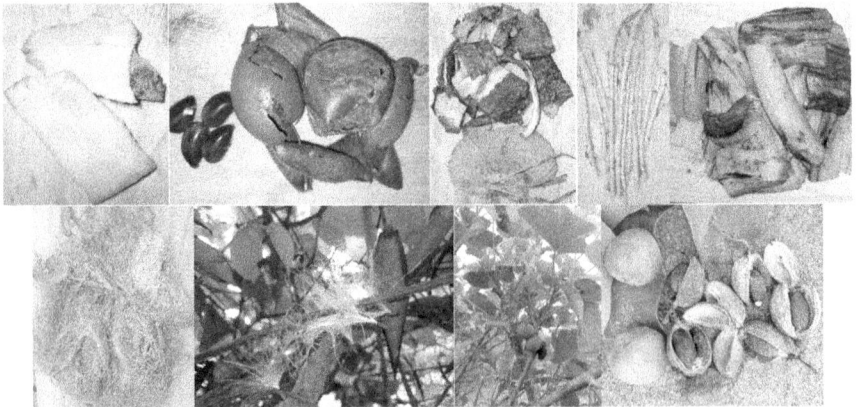

FIGURE 11.8   Agricultural waste as sources of particles for PLA blending.

11.6.1 PRODUCTION OF PLA-BASED COMPOSITES

Conventional plastics are resistant to biodegradation, as the surfaces in contact with the soil in which they are disposed are characteristically smooth [47]. Currently, biodegradable polymers are attracting a great attention from researchers and industries as these polymers are designed to degrade upon disposal by the action of living organisms. Biopolymers derived from renewable resources such as corn, cellulosic, soy protein, and starch are attracting the attention of scientists to replace traditional petro-based plastics in designing green composites [47]. The adverse effects of synthetic-based composites on the environment have necessitated the shift in focus to the development of biopolymer composites of which PLA-based composites have attracted much attention using materials from natural sources. Industries that are producing composites are focusing on using natural reinforcements as substitutes for synthetic fillers. This is because of their desirable attributes like low cost, biodegradability, availability in large quantities, and valuable mechanical and physical characteristics. Lignocellulosic fibres and natural fillers from agricultural and industrial crops, such as corn, wheat, and bagasse, have been used in the production of composites for application in areas such as packaging, construction, and car manufacturing [67]. In addition, cellulose obtained from various sources, modified or unmodified wood flour, Kraft lignin, and waste obtained from olive pit, wine, and coffee production [68] have been used in the production of PLA-based composites. Similarly, other natural fillers have been blended with PLA to produce biopolymer composites.

Composites are made by combining different materials to improve their physical, chemical, mechanical, and processability properties [44,69]. Cost reduction can also be achieved using this blending of materials. The common method of composite manufacture using PLA is melt blending with the desired fillers and the final products are produced using extrusion, blow film moulding, and compression- and injection-moulding [40]. For modification of functionalities to improve their properties, they can be incorporated with other materials including starch, other bioplastics, nano-clays, carbon nanotubes, and cellulose [48,60].

A growing area of polymeric composites is in the development of ecologically viable materials with less environmental impact using raw materials from natural sources [44,70,71]. The composite manufacturing industry is now more inclined to the adoption of natural fibre reinforcements to replace synthetic fibres, and fibres such as flax, hemp, jute, sisal, kenaf, pineapple, abaca, coir, and banana have been used. A wide variety of lignocellulosic fibres and natural fillers from agricultural and industrial crops, such as corn, wheat, and bagasse, have been employed in the production of composites for areas including packaging, automotive and building industry. This is because of their low cost, biodegradability, availability, and desirable mechanical and physical attributes [44,67,72]. Plastics used for composite production account for only 4%–5% of the total plastics produced [26]. As such, opportunities exist to complement the drawbacks of biodegradable polymers as well as reduce the cost of the composite materials for the packaging industry.

Natural fibres are generally non-toxic and can be safely utilised without posing any health hazards or environmental damage. Plant-based fibrous materials used with

Materials/Compositions of Polymeric Bio composites

```
Continuous Random

    Fibrous        Fillers/Reinforcements                    Biodegradable polymers
                          Forms
 Particulate            Origins                        Natural polymers e.g., rubber,
                                                       gums, gelatin, starch, shellac,
                                                         lignin, chitin, cellulose,
        Minerals        Plants        Animals                   proteins
      e.g., limestone,  (Different parts)  e.g., fish
        barite, aluminate              stone, feather    Synthetic/manufactured e.g.,
                                                        polylactic acid (PLA), polymethyl
  Seed/Shells  Leaves    Grasses  Stem/peel  Pods        methacrylate (PMMA),
   e.g., nuts,  All plants have  Sugarcane,  Banana,  Cocoa, black  synthesised starch, manufactured
   moringa,    leaves which    elephant   orange, red   seed,      protein, produced cellulose
   melon, red  are sources of  grass, rice,  cherry     pumpkin,
   cherry,     reinforcement  wheat, maize,            calabash
   pumpkin     when dry.       wheat,
```

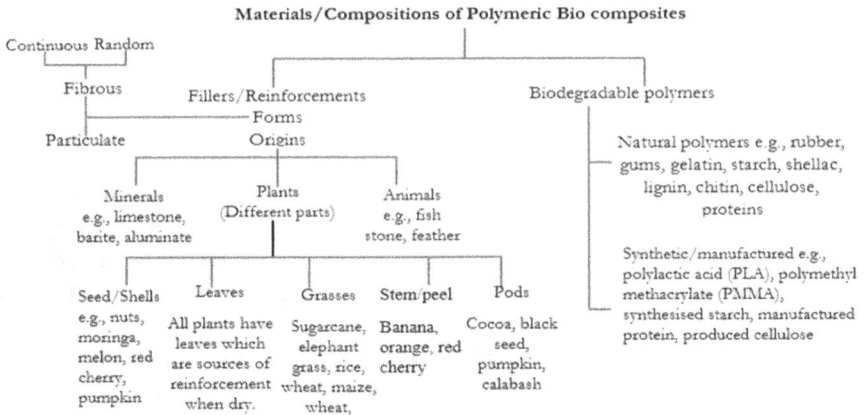

FIGURE 11.9   Constituents of biopolymer composites.

biopolymers/biomass-derived polymers can yield environmentally friendly and bio-degradable green composites with sufficient flexibility and mechanical strength comparable to commercially available petroleum-based polymers composites [73,74]. Some of the general constituents of biopolymer composites are shown in Figure 11.9.

PLA has good mechanical properties that are comparable to polyethylene ter-ephthalate (PET) and PP which are the most common materials used in automobiles. The temperature at which PLA can be melt processed with available standard pro-cessing equipment is safe for natural fibres because natural fibres do not degrade at the processing temperature [47]. PLA is a hydrophobic polymer because of the in-corporation of the CH_3 side chain [47]. Because of all these favorable properties, PLA has strong candidacy among the biopolymers as the matrix material to be used in green composites. Due to the drawbacks or disadvantages of PLA, it can be used with numerous organic fibres and fillers to improve some of its properties such as me-chanical characteristics and reduced gas or moisture permeability. These fillers can include cellulose in different forms and of different origins, as well as untreated or modified wood flour, Kraft lignin, or wastes derived from olive pits, or waste from wine production as well as waste derived from coffee production among many others [68,72]. The complete biodegradation of PLA in soil occurs at a relatively slow rate and the addition of natural biodegradable fillers can catalyse a speedy biodegradation of the resultant PLA-based composites [75]. For example, blending with Kraft lignin accelerates biodegradation of PLA in garden soil [53].

11.6.2 TYPICAL PRODUCTIONS OF *PARQUETINA NIGRESCENS* FIBRE-REINFORCED POLYLACTIC ACID COMPOSITES

Polylactic acid (PLA) was purchased from China through the Aliexpress online store. *Parquetina nigrescens* fibres were sourced locally from Amuloko, Ibadan. Vegetable oil used as coupling agent was bought at a main market in Malete, Kwara State, Nigeria. *Parquetina nigrescens* discontinuous fibre-reinforced PLA composites were produced to document a typical fabrication of the PLA-based composites in this chapter. Dried *Parquetina nigrescens* pods were carefully opened (Figure 11.10) to release the fibres that were held together by tiny seeds. The fibres held at one end by hand were beaten at the other end to separate the seeds from fibres. Then, the fibres were cut into smaller pieces using a pair of scissors. The piece fibres used for reinforcing PLA are presented in Figure 11.10. Different grades of PLA-based composites were produced with proportions of PLA and *Parquetina nigrescens* fibres in Table 11.1.

The procedure for production of the composites is as follows:

    I. PLA was measured (100 g)
   II. Fibre was measured (0.1 g) (Figure 11.11a)
  III. Fibre was poured into a bowl filled with the PLA
  IV. Mixture was stirred with a spatula
   V. Two drops of vegetable oil were added into the mixture
  VI. Mixture was stirred continuously till they blend into each other
 VII. Blended mixture was poured into the mould (Figure 11.11b)
VIII. Mould is placed in the oven and temperature was at 200°C (Figure 11.11c)

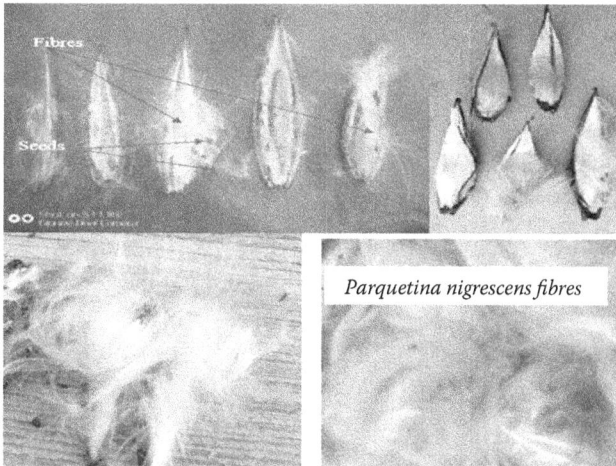

FIGURE 11.10   Pods opening to release *Parquetina nigrescens* fibres held together by seeds.

TABLE 11.1
Design Formulations for PLA Composite Fabrications

S/N	Mass of PLA (g)	Mass of Fibres (g)	Proportion (%)
	100g	0.1	0.1
	100g	0.3	0.3
	100g	0.5	0.5
	100g	0.7	0.7

(a)   (b)   (c)   (d)

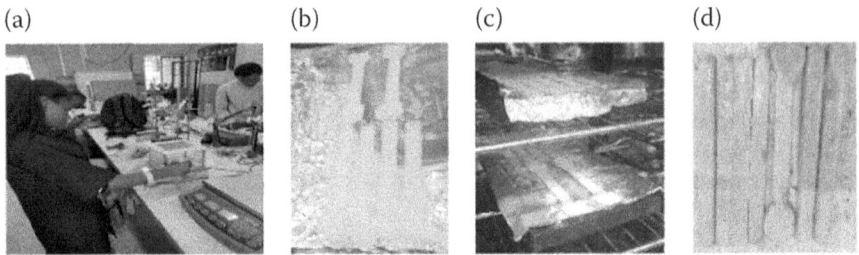

FIGURE 11.11 Procedures for PLA composite developments (a) weighing of fibres, (b) composite blend already poured into the die mould ready for processing, (c) placing of mould inside the oven, (d) processed PLA composites.

IX. At 200°C, temperature was held for 30 minutes for complete fusion of the composite blend

X. While the mould and its contents were left in the oven, it was switched off and the processed composites cooled to room temperature were removed from the die mould after 16 hours (Figure 11.11d)

XI. Repetition of processes to produce other composites containing greater fractions of the fibres (0.3, 0.5, and 0.7 g)

XII. Control samples without fibre addition were also produced

11.6.3 Characterisations of *Parquetina nigrescens* Fibres and PLA Composites

11.6.3.1 Transmission Electron Microscope (TEM) Analysis

Transmission electron microscope (TEM) analysis was carried out on *Parquetina nigrescens* fibres to study their morphology to know their shape and sizes, which are needed to predict their distribution within the PLA matrix. In analysing the fibres using TEM, electrons were transmitted through the fibres, and images formed were captured by the charged coupled device camera. TEM-ARM200F-G Verios 460 L was used.

11.6.3.2 X-Ray Diffraction Analysis

An X-ray diffraction test was performed on the pristine PLA and composites obtained by directing an X-ray beam at each sample under analysis and measuring the scattered intensity as a function of the positions (2theta). The riveted refinement technique was then used to characterise the crystal structure. Before the analysis, samples were placed inside the sample holder. Scanning was carried out carried in a continuous mode from angle 3° to 80°(2θ) at a step size of 0.0026°. at 25°C with the aid of Malvern PanAlytical Empyrean Diffractometer set at 45 kV and 40 mA using Cu anode with Goniometer radius, distant focus divergence slit and irradiated length of 240, 91, and 4 mm, respectively. Peaks were fitted using X'pert High Score software. The slits were fixed with Fe-filtered Cu Kα radiation. The phases were identified using X'Pert High Scores Plus software and PAB-ICSD and ICDD (2014) databases.

11.6.3.3 Tensile Tests

Tensile test samples of the PLA composites with gauge dimensions of 80 mm in length, 10 mm in thickness, and 6 mm in width were used. An Instron Universal Testing Machine (UTM) model 3369; system ID: 3369S3457 was used to gradually stretch the samples using a uniaxial tensile load at a cross head speed of 0.008 (8×10^{-3}) mms^{-1}, equivalent to a strain rate of 0.0001 (1×10^{-4}) s^{-1} until a fracture occurred after some plastic deformations. This analysis was performed per American Society for Testing of Materials D 3039 (ASTM D 3039).

11.6.3.4 Impact Energy Determination

With $75 \times 4 \times 10$ mm^3 notched samples of *Perquitina nigrescen,* they were subjected to impact energy test using the House Balance Izod impact tester (serial no 3915) in accordance with ASTM D 3763. Prior to the analysis, the pendulum hammer was released to set the scale to zero. Each of the samples was impacted with the double-sided pendulum hammer at 3.81 ms^{-1}. Energy absorbed by the sample was read from the scale. The analysis was carried out at Department of Mechanical Engineering, University of Ilorin.

11.7 STRUCTURAL PROPERTIES OF *PARQUETINA NIGRESCENS*–REINFORCED POLYLACTIC ACID COMPOSITES

11.7.1 TRANSMISSION ELECTRON MICROSCOPIC IMAGE OF *PARQUETINA NIGRESCENS* FIBRES

Parquetina nigrescens have hollow tubes like the flexible pipes. Naturally, they occur as random fibres because of their relative shortness. They are about 5–7 cm long, which is too short to be used as continuous fibres for fabricating full-size artefacts. However, they can be used like that or cut into smaller pieces as random or discontinuous fibres for polymer reinforcement as it is used in this study. Preliminary investigation conducted on *Parquetina nigrescens* fibres reported in [76] confirms the hollow tube-like structure of the fibres, which pairs with the findings from the TEM in the study reported in this chapter (Figures 11.12).

11.7.2 X-RAY DIFFRACTOGRAMS OF THE *PARQUETINA NIGRESCENS*–REINFORCED POLYLACTIC ACID COMPOSITES

Both yeast ($C_{10}H_{14}N_5O_7P.H_2O$) and 3$GB,17$GA-Dihydroxy-5$GA-pregnan-20-one were detected in the pristine polylactic acid (Figure 11.13). Features of peaks having both compounds (Table 11.3) are expressed in Table 11.2. Peak detection from XRD indicates that a polylactic acid polymer without any reinforcement is semi-crystalline. To ensure that the PLA polymer is safe, the toxicity level of each compound detected was examined using Toxtree-3.1.0.1815 (Figure 11.14) software. According to Toxtree software, the yeast was classified as simple branched aliphatic hydrocarbon or a common carbohydrate belonging to toxicity class level known as Low Class I. The implication is that yeast causes a minor threat when in

(a)   (b)

(c)

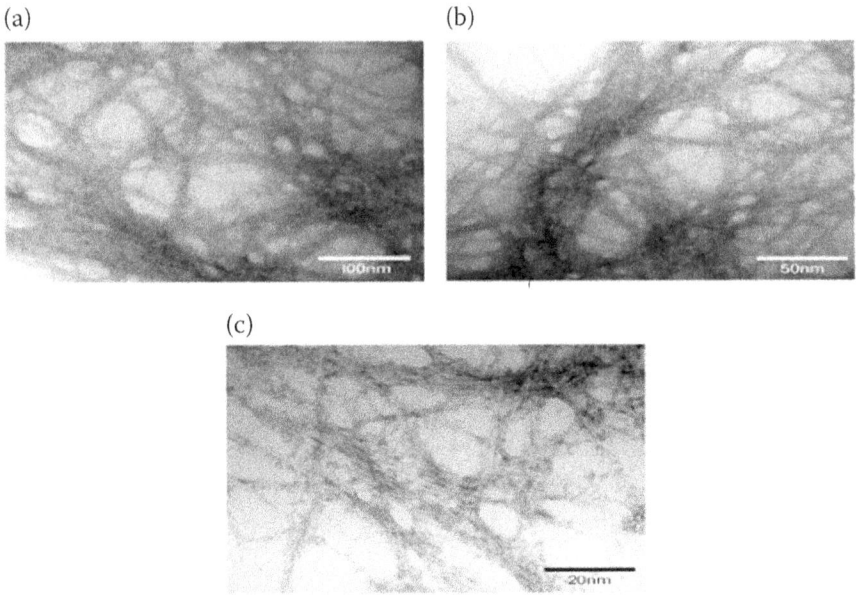

FIGURE 11.12 Transmission electron microscopic images of *Parquetina nigrescens* fibres at (a) 100 nm, (b) 50 nm, and (c) 20 nm resolutions.

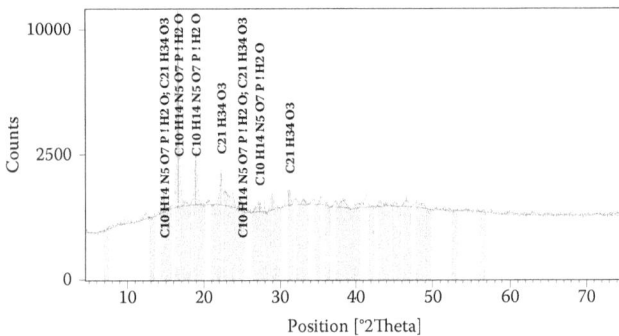

FIGURE 11.13 X-ray diffractogram of pristine polylactic acid.

contact with the body. Moreover, 3$GB,17$GA-Dihydroxy-5$GA-pregnan-20-one was put under compound with exception toxicity. It belongs to none of the Class 1, Class II, and Class III, which means that this compound does not have any iota of toxicity. Generally, PLA polymer can be considered safe for the body. Moreover, new compounds (Figure 11.15) were detected in *Parquetina nigrescens* fibre-reinforced PLA composites. The new compounds could be linked with the fibre addition to the PLA. Both compounds do not belong to any of the toxicity classes. This implies that the fibre addition does not aggravate the low toxicity level of the PLA (Table 11.4 and 11.5).

TABLE 11.2
Peak Features of the Pristine Polylactic Acid Diffractogram

Pos. [°2Th.]	Height [cts]	FWHM [°2Th.]	d-Spacing [Å]
14.6896	417.90	0.2047	6.03048
16.6093	10802.31	0.2047	5.33756
18.8782	1608.09	0.2814	4.70086
22.2558	1043.41	0.2558	3.99450
24.9949	223.40	0.5117	3.56262
27.2939	229.48	0.4093	3.26753
28.9711	477.85	0.3070	3.08208
31.1180	417.43	0.5117	2.87415

TABLE 11.3
Features of Phases Detected in the Pristine Polylactic Acid

Ref. Code	Score	Compound Name	Displacement [°2Th.]	Chemical Formula	Toxicity Level
07–0702	41	yeast	0.000	$C_{10} H_{14}N_5O_7P. H_2O$	Low Class I
15–1040	25	3$GB,17$GA-Dihydroxy-5$GA-pregnan-20-one	0.000	$C_{21}H_{34}O_3$	–

11.8 MECHANICAL PROPERTIES OF PLA-BASED GREEN COMPOSITES

The effect of different reinforcements in terms of natural fibres and fillers is studied through various characterisation techniques that may open new areas for further development to assess the effectiveness of its processing, the effect of different environments on its properties, and to find suitable areas of application. Mechanical characterisation is one of the most important techniques to predict the mechanical behavior of the materials. The mechanical properties of bio-composites depend on several parameters such as percentage of fibre content, interfacial characteristics between fibre and matrix, fibre aspect ratio, surface modification of fibres, and addition of various additives (coupling agents) to enhance the compatibility between fibre and matrix [47]. Mechanical characterisation involves determination of tensile strength (maximum stress in tension that a specimen can sustain without fracture), Young's modulus (the stiffness of the material obtained from the slope of the line tangent to the stress–strain curve), and compressive strength. It also involves flexural strength (stress at fracture from a three- or four-point bend), flexural modulus (the stiffness of the material, which is the ratio of applied stress on a test specimen in bending to the corresponding strain in outermost fibres of the specimen within elastic

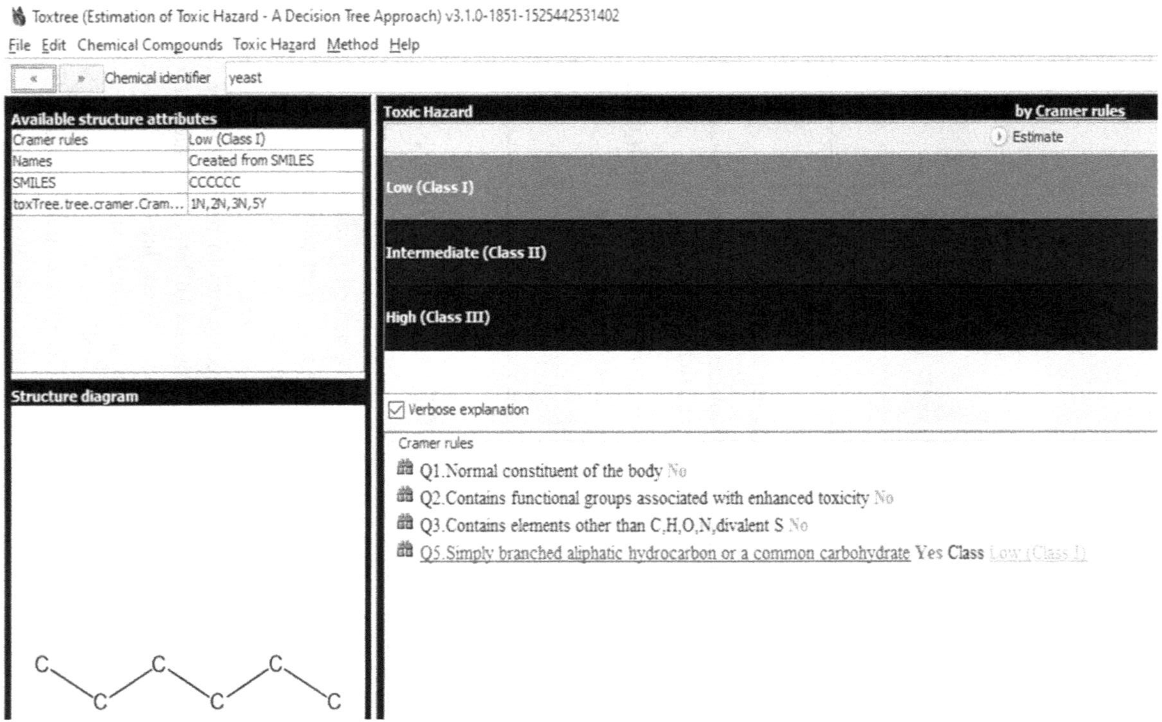

FIGURE 11.14 Toxtree software on analysing the toxicity level of the yeast.

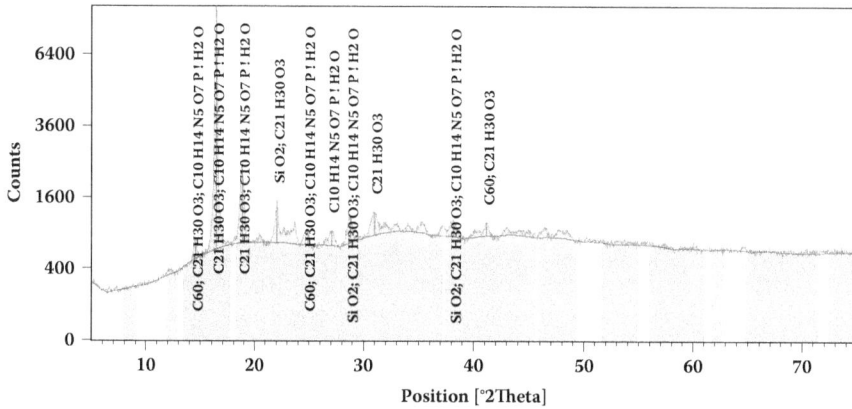

FIGURE 11.15   X-ray diffractogram of *Parquetina nigrescens* fibre-reinforced polylactic acid.

TABLE 11.4
Peak Features of the *Parquetina nigrescens* Fibre-Reinforced Polylactic Acid Diffractogram

Pos. [°2Th.]	Height [cts]	FWHM [°2Th.]	d-Spacing [Å]
14.5202	219.18	0.6140	6.10043
16.4158	7846.94	0.2558	5.40002
18.7809	1674.94	0.2558	4.72498
22.0604	772.16	0.2558	4.02944
24.8817	199.78	0.6140	3.57857
27.1001	234.36	0.4093	3.29046
28.7656	464.26	0.4093	3.10362
30.9668	418.53	0.4093	2.88785
38.1871	237.53	0.6140	2.35680
41.1845	235.31	0.4992	2.19012

TABLE 11.5
Features of Phases Detected in the *Parquetina nigrescens* Fibre-Reinforced Polylactic Acid

Ref. Code	Score	Compound Name	Displacement [°2Th.]	Chemical Formula	Toxicity Level
49–1717	38	Fullerite	0.000	C_{60}	--
82–1404	31	Cristobalite $GB, syn	0.000	SiO_2	–
24–1880	29	4,17(20)-Pregnadien-11$GB,21-diol-3-one	0.000	$C_{21}H_{30}O_3$	–
07–0702	35	Yeast	0.000	$C_{10}H_{14}N_5O_7 \ P.H_2O$	Low Class I

limit). It also involves impact strength (resistance of any material to impact loading with or without notch), and inter-laminar shear strength (shear strength of the matrix layer between the plies) [47].

11.8.1 Tensile and Compressive Strength of the PLA-Based Composites

Many studies have been done on tensile strength, tensile modulus, and elongation at break of green composites than any other mechanical property. The tensile strength of green composites can be enhanced by improving the matrix properties and reducing the stress concentration and choosing proper fibre orientation. The fibre properties like wettability of fibres into matrix and fibre loading are responsible for improving tensile stiffness of bio-composites. Literature shows that less attention has been paid to shear and compressive properties of green composites. It was reported that [76] while making composite using flax fibre (30%–40%) with PLA resin, strength was about 50% better compared to similar PP/flax fibre composite. Stiffness of the PLA was increased from 3.4 to 8.4 GPa in the fabricated green composites [77] with fibre recycling from disposable chopsticks and PLA matrix by melt-mixing method. Mechanical tests showed that the tensile strength of the composites markedly increased with the fibre content, reaching 115 MPa in the case of being reinforced with 40 phr fibres, which was about three times higher than the pristine PLA. Authors [78] developed mercerised sisal fibre–reinforced soy protein resin–based composites and concluded that mercerisation improved the fracture stress and Young's modulus of the sisal fibres while their fracture strain and toughness decreased. Developed composites showed improvement in both fracture stress and stiffness by 12.2% and 36.2%, respectively, compared to the unmercerised fibre-reinforced composites. It was reported in [79] that tensile strength of jute fibre-reinforced PP matrix composites increased with increase in the fibre size and fibre percentage; but after a certain size and percentage, the tensile strength decreased again. It was shown in [80] that the tensile strength of the PLA/denim fabric composites was improved by piling layer of denim fabrics. The three-layer denim-reinforced composite showed best results among all specimens having its tensile strength and tensile modulus 75.76 MPa and 4.65 GPa, respectively. Report in [81] showed that PLA/hemp composite with 40% volume fraction of alkali-treated fibre had the best mechanical properties. The tensile strength and elastic modulus were 54.6 MPa and 8.5 GPa, respectively, which were much higher than those of PLA alone. Authors [82] showed extremely high tensile strength (365 MPa) of manila hemp (70 vol%)/starch-based resin composite due to the novel fabrication technique proposed in the study in which the composites were fabricated with an emulsion-type biodegradable resin. It was demontrated in [83] that the tensile strength and stiffness of PLA could be approximately doubled when jute fibre reinforcement (40 wt%) was used. Results of electron microscopy showed brittle failure of jute fibres under tension and void spaces between fibre and polymer matrix, indicating that the strength of the PLA/jute interface could be improved. Authors [84] developed the biodegradable composites with maleic anhydride-grafted PLA reinforced with rice husks and kenaf fibres and concluded that compressive strength and compressive modulus of PLA/kenaf (345 + 3.11 MPa and 174 + 0.11 GPa, respectively) was higher than that of PLA/rice hull (216 + 2.67 MPa and 146 + 0.07 GPa, respectively). Researchers [85] found that neat PLA had a lower tensile strength than PLA-based ramie and jute short fibres composites. The tensile strength increased with the addition of ramie fibre or jute fibre to PLA matrix, showing that the stress was expected to transfer from the matrix to the strong fibre. When the addition of fibres was more than 30%, the tensile strength of composites decreased and was even lower than that of neat PLA. Investigators [86] fabricated PLA/sugar beet pulp composites by compression-heating and reported that the resultant composite had a lower density and tensile strength similar to that of pure PLA specimens as well as the same geometric properties. Tensile properties were dependent on the initial water content of sugar beet pulp and the process by which composites were manufactured. Authors [87] reported development of PLA/spruce wood flour (SWF) composites with different surface treatments of WF. Incorporation of 40 wt% SWF resulted in an increase in the Young's modulus (3.73 + 0.247 GPa) and a decrease in the tensile strength (37.2 + 2.0 MPa) as well as of the percentage elongation at break (1.1 + 0.2). The composites containing hydrothermally, and silane treated WF induced a tensile strength increase, along with higher elongation at break and a higher Young's modulus, respectively, which reflected the stiffening effect of the employed silane treatment, which clearly improved interfacial adhesion between the PLA matrix and WF. Researchers [88] developed composites consisting of PLA and rice straw fibre (RSF), modified by poly butyl acrylate (PBA). A morphological study of PLA/RSF (7.98 wt%) via scanning electron microscope (SEM) showed good interfacial adhesion between PLA and RSF and good dispersion of RSF in the polymer. The poor interfacial adhesion between PLA and RSF was observed when PBA content was high. These were well confirmed in the tensile test, which showed the tensile strength of PLA/RSF composites increased significantly to 6 MPa. The tensile strength of PLA/RSF rapidly decreased, while the content of PBA was more than 7.98 wt%. The addition of PBA to PLA led to the decrease the tensile strength while the elongation at break was slightly increased. [89] found a tensile strength of 82.9 MPa, Young's modulus of 10.9 GPa with 30 wt% long aligned alkali-treated industrial hemp fibre–reinforced PLA composites produced by the film stacking technique. It was revealed in [90] that addition of up to 40 wt% of wood flour particles into PLA has little influence on the tensile strength (due to poor interfacial adhesion), a significant reduction in its elongation at break and an increase of up to 95% in the tensile modulus of the micro composites. The introduction of methylene diphenyl-diisocyanate (MDI) resulted in a 10% increase in tensile strength and 135% increase in tensile modulus, showing that the addition of MDI resulted in an increase in the strength of interfacial adhesion between the PLA matrix and the surface of the wood flour particles. The incorporation of polyethylene-acrylic acid (PEAA) in PLA caused a substantial decrease in tensile strength of the matrix of up to 35%, an increase in the break elongation and peak load values due to blending of the rubbery PEAA chains into the PLA matrix. Investigators [91] evaluated tensile properties of surface-treated (alkali and silane treatments) ramie fibre–based PLA composites. Results revealed that neat PLA had a lower tensile strength than PLA-based composites due to better ramie fibre and polymer surface compatibility and good stress transfer between the fibre and the matrix. Fibre treatment by alkali and silane further improved the tensile strength and strain of the composites, and the maximum strength was 64.24 MPa (composite treated by NaOH) due to bonding at the interface between the ramie fibre and PLA matrix. Authors [92] showed the effect of isocyanate group (NCO) content on the tensile properties of the PLA/bamboo fibres (BF; 30 wt%) composites. As NCO content increased to 0.33%, tensile strength and Young's modulus increased rapidly from 29 to 42 MPa and from 2666 to 2964 MPa, respectively, and then leveled off. There was no significant effect of lysine-based diisocyanate (LDI) addition on the elongation at break, showing the value of less than 5%. Moreover, additions of the *Parquetina nigrescens* fibres to the polylactic acids reported in this chapter shows 357%, 432%, and 34.21% increases in tensile strength, Young's modulus, and tensile strain (Figures 11.16–11.18). The increase in tensile properties could be linked to strong adhesive bonding between the fibres and the PLA matrix, this requires confirmation with SEM. Although tensile strengths of the composites are all greater than that of the control, 0.1% *Parquetina nigrescens* fibre addition is found to be the best weight fraction that gives optimal tensile strength. It can be affirmed the PLA binding capacity has reached the saturation at 0.1% *Parquetina nigrescens* fibre additions.

This is regarded as the saturation level. It implies that further increase in the weight fraction of the *Parquetina nigrescens* fibres impairs the tensile strength because of inability of the PLA matrix to join them together. Similar observations were noticed with tensile strain (Figure 11.17) and Young's modulus (Figure 11.18). This implies that the *Parquetina nigrescens* fibre addition improves tensile strength, strain, and rigidity of the PLA with a maximum enhancement at 0.1% fibre additions. Enhancement of the tensile properties due to *Parquetina nigrescens* fibre additions is in line with literature [79–82,88].

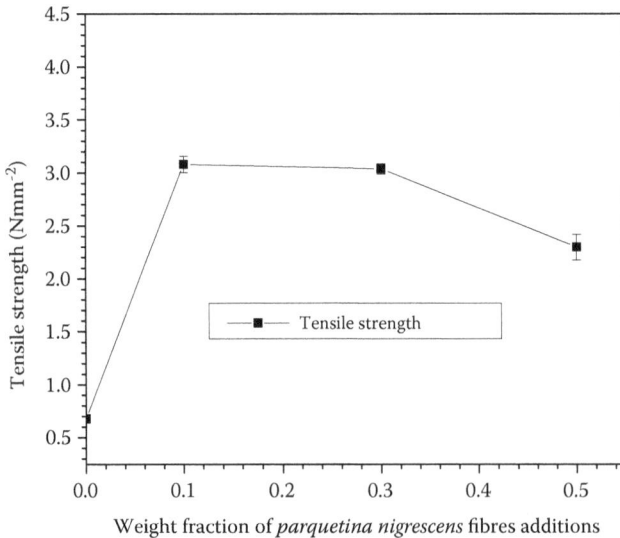

FIGURE 11.16 Tensile strength of the *Parquetina nigrescens* fibre-reinforced polylactic acid composites.

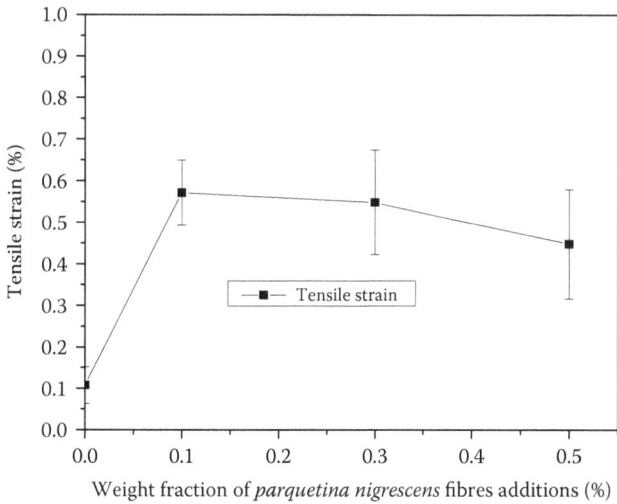

FIGURE 11.17 Tensile strain of the *Parquetina nigrescens* fibre-reinforced polylactic acid composites.

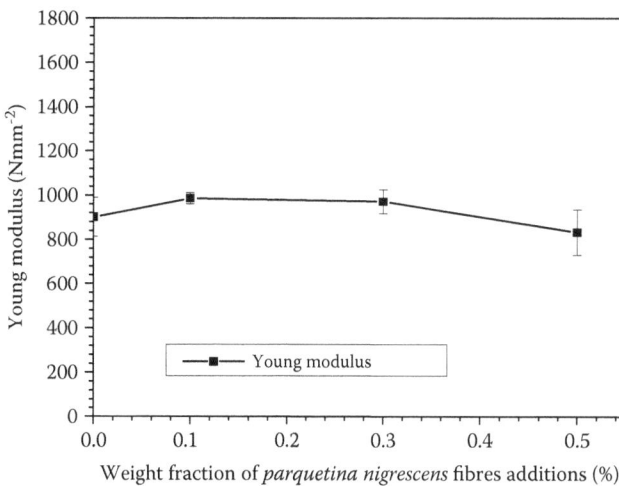

FIGURE 11.18 Young's modulus of the *Parquetina nigrescens* fibre-reinforced polylactic acid composites.

11.8.2 IMPACT, FLEXURAL, AND SHEAR STRENGTH OF THE PLA-BASED COMPOSITES

The capacity of a material to withstand fracture under a sudden load impact is the impact energy of the material [93]. Since natural fibres innately possess low impact strength, it is required to perform impact energy tests on biopolymer composites. Several factors like fibre and biopolymer type, particle size, interfacial adhesion and test, and specimen condition (unnotched or notched) affect the impact energy of composites [94]. Reinforcing PLA with bamboo fibres resulted in a decrease of the impact strength of the composite. Modification of the fibres by silane treatment caused a 33% increment in impact strength of the composite. The impact8strength of biopolymer composites can also be enhanced by hybridisation technique [94]. Contradictorily, at 0.1% *Parquetina nigrescens* fibre addition, the PLA composite has maximum impact energy, affirming its greater energy absorption than that of the pristine PLA. Other composites have impact energy lower than that of the control. Error bars in the graphs are standard deviations of three different values of impact energies determined from various samples produced at each level of reinforcement. Low height of the error bar, as seen in Figure 11.19, affirms closeness of the impact energies of the examined samples at each level. Improvement in impact energy at 0.1% could be attributed to the hollow tube-like structure of the *Parquetina nigrescens* fibre, which dampens the effects of sudden impacts on the PLA composites during the test. The damping effect is achieved through bending of the fibres during load impact and fibres straightening after the impact leading to greater impact energy absorption at 0.1%. At higher fibre reinforcement level, it is possible that a saturation level has been reached, which implies that PLA matrix has insufficient binding capacity to connect all fibres to form a single network. This is described as a discontinuity in the *Parquetina nigrescens* fibre-reinforced PLA, which has a consequent effect on poor impact energy absorption. Generally, improvement in tensile properties and impact energy absorption due to *Parquetina nigrescens* fibre additions to the PLA can be linked with vegetable oil added to fibres before mixing with PLA, which couples the fibres with the PLA. Improvement in mechanical properties due to fibre additions are linked with good

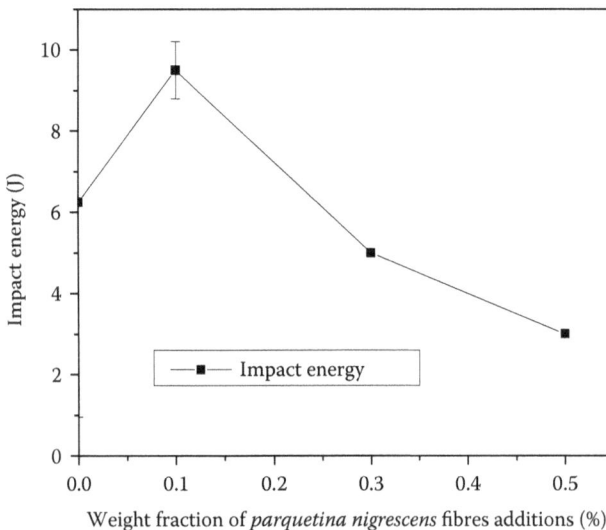

FIGURE 11.19 Impact energy of the *Parquetina nigrescens* fibre-reinforced polylactic acid composites.

interfacial adhesion exists between fibres and the PLA matrix phases. The adhesion delays debonding or pullout of the fibre when the composite stretches under load applications during the mechanical investigations. The delay in debonding and resistance of the fibres to the craze movement during deformation are proposed to be the mechanism of the strength and impact energy enhancements noted in this study. Mechanical property improvements agree with many previous studies reviewed in this chapter.

The flexural strength of green composites is slightly lower than that of the synthetic fibre composites but comparable or better if specific properties are considered. An important mechanical property of green composites is impact strength, which is generally lower when compared with glass fibre composites. For good impact strength, a most favourable bonding level is necessary. The level of adhesion, favourable bonding, fibre pull-out. and energy absorption are some of the parameters that govern the impact strength of bio-composites. It was reported that impact resistance as measured by an unnotched Izod test did not increase in PLA/jute (40 wt %) composite [83].

Authors [80] showed that denim fabric–reinforced PLA composites exhibited outstanding impact strength due to the retarded crack propagation as well as large energy dissipation. It was concluded in [81] that flexural strength of alkali-treated hemp fibre (40 vol%)–reinforced PLA composite was higher (112.7 MPa) than that of PLA alone due to the improved interfacial adhesion between fibre and matrix. Investigators [82] proposed a new fabrication technique in which the composites were fabricated with an emulsion-type biodegradable resin with Manila hemp fibres (70 vol%) and found very high flexural strength of 223 MPa. The flexural strength and flexural modulus increased linearly with increasing fibre content up to 70%. Researchers [85] reported that the impact strength of PLA/ramie composites was higher than that of PLA/jute composites due to the higher strength of ramie fibre. The flexural strength of the composites increased compared with the neat PLA matrix. When the content of fibres was over 30%, the flexural strength of composites decreased and was even lower than that of neat PLA due to the poor dispersion of fibres in the matrix. Investigators [95] evaluated the flexural and impact properties of PLA/recycled newspaper cellulose fibre (RNCF; 30 wt%)/talc (10 wt%; with and without silane treated) hybrid composites. The flexural and impact strength of these hybrid composites were reported to be significantly higher than that made from either PLA or RNCF. The hybrid composites showed improved flexural strength of 132 MPa and flexural modulus of 15.3 GPa, while the un-hybridized PLA-/RNCF-based composites exhibited flexural strength and modulus values of 77 MPa and 6.7 GPa, respectively. Authors [89] produced short (random and aligned) and long (aligned) industrial hemp fibre–reinforced PLA composites by compression molding. The best overall properties were achieved with 30 wt% long aligned alkali-treated fibre–reinforced–PLA composites produced by film-stacking technique, leading to flexural strength of 142.5 MPa, flexural modulus of 6.5 GPa, impact strength of 9 kJ/m^2. Researchers [90] found an improvement in impact strength (up to 30%) of the PLA/WF (20 wt%)-micro composites due to crack propagation from the wood flour particles. The incorporation of PEAA caused a slight improvement in impact strength (up to 15%) due to the blending of the rubbery PEAA chains into the PLA matrix. Experimenters [91] evaluated the effect of surface treatments (by alkali and silane treatments) of ramie fibres on flexural and impact strength of PLA/ramie fibre composites. When ramie fibres were treated with alkali, the flexural strength of the composites was higher than that of untreated fibre or treated with silane. The impact properties of the composites with surface-treated ramie fibres were higher than that of the composites with untreated ramie fibres. Composites with alkali treatment exhibited the highest impact strength, which showed that the alkali treatment provided effective resistance to crack propagation. It was reported in [95] that the flexural and impact properties of PLA/kenaf fibre bio-composites with alkalization and silane treatment of fibres. All surface-treated kenaf fibres showed the tendency to significantly increase the flexural modulus compared to neat PLA. The flexural strength of the PLA composites decreased with the addition of kenaf fibres probably due to poor adhesion between the kenaf fibres and PLA. With 40 wt% kenaf fibre contents, the flexural modulus increased from 5.6 GPa for untreated fibre (FIB) to 8.3 GPa for alkali-treated fibre. The composite with silane-treated fibres showed a higher increase in modulus than that of alkali-treated fibres. The flexural modulus is increased from 5.6 GPa for FIB to 9.5 GPa for silane-treated fibre. The composite with alkali followed by silane-treated fibre (FIBNASI) contents exhibited the best flexural modulus (80% improvement). The surface treatments enhanced the impact strength of the composites. The impact strength of the PLA improved nearly 45% for FIB, 50% increment for the FIBNA composites with 40 wt%. In contrast, FIBSI composites showed a decrease in impact strength; but for FIBNASI composites, the impact strength improved by 38% over neat PLA.

Authors [96] blended the mixture of PLA and unsaturated polyester resin with modified hemp fibres and the mechanical characteristics of the developed composites were investigated. The hemp fibres were treated with sodium hydroxide, acetic anhydride, maleic anhydride, and silane. The results indicated that surface modification of hemp fibres enhanced the interfacial shear strength of the bio composites. It is noteworthy here that other mechanical properties of the *Parquetina nigrescens* fibre-reinforced PLA like flexural properties, compressive properties, and hardness values are left undiscussed in this chapter. They are left as gaps to be filled by interested researchers across the globe to complement the reported properties to be able to ascertain the biomedical applications (like artificial tooth materials) of the developed *Parquetina nigrescens* fibre-reinforced PLA composites.

Table 11.6 shows some studies on PLA-based composites with natural fillers to improve the properties as well as reduce the cost of the final composite. With a high amount of agricultural and food waste, novel composites were fabricated using the waste with biopolymers, resulting in a reduced environmental burden.

Upon literature reviewed for documentation in this chapter, it is observed that incorporations of agricultural waste as reinforcement in PLA have some benefits including improvement in tensile, flexural, and compressive strength. Moreover, saturation level of reinforcement leading to impairment in the itemised properties was also disclosed, including 0.1 wt% of *Parquetina nigrescens* fibres documented in this chapter. Reports on impact energy is very few. This may be due to inherent brittleness of the PLA. However, knowledge of impact toughness of the PLA-based

TABLE 11.6
Summary of Effects of Reinforcement Additions and Methods of Fabrications on Properties of Polylactic-Based Composites

Filler	Composite Fabrication Process	Observation	Reference
Hemp fibres	Twin extruder, compression and injection moulding	• Increased tensile strength, Young's modulus and impact strength • Fibre treatments with alkali and saline increased tensile and impact strength	[96]
Fine grain filler of native cellulose	Melt-mixing	• Good dispersion of the filler giving an aesthetic appearance, creamy colour, and glossy surface • Good thermal stability	[72]
Ceramic food waste from grinding egg shells and mussel shells	Melt-mixing	• High filler amounts of 140 over 100 parts of PLA • Thermoplastic, biodegradable, and low carbon footprint composites • Composites do not release volatiles typical of fossil-based plastics that are hormone disruptors or priority air pollutants that pose public health	[34]
Silver skin (waste from roasting coffee beans)	Melt-mixing	• Up to 30% wt% of filler content can be added • Increase in Young's modulus	[97]
Waste from wine production (grape skins, seeds, and stalk fragments)	Moulding	• Up to 20% of filler can be added • High elastic modulus and impact strength but lower tensile strength • Increase in moisture absorption with increase in filler content	[68]
Cocoa bean shells	Solution casting	• Improved physical properties of the composites • Low levels of food migration and improved barrier properties	[98]

composites are gaps to be filled to complement reports on PLA-based composites for their possible structural applications, perhaps bumper material application where impact energy is a prerequisite.

Application of PLA-based composites for food packaging is sought to address environmental problems from discarded nonbiodegradable polymeric composites especially in African countries, where citizens find it difficult to observe proper management of their wastes. Properties of PLA are compared with those of other

TABLE 11.7
Comparison of PLA Properties to Other Common Polymers Used in Food Packaging

Property/Polymer	LDPE	PET	PLA	PP	PS
Strength (MPa)	6.3–12	55–79	0.75–66	14.99–27.00	24–60
Elongation at Break (%)	200–500	15–165	0.5–9.2	100–600	1.6–2.5
Oxygen barrier (permeation at 30°C [$*10^{-10}$ cm^3 (STP)-cm/cm^2.S.cm Hg])	6.89	0.04	3.3	1.5	2.6
Moisture vapour transmission rate (g-mil/10 in.^{2}/24 h)	1.0–1.5	2	18–22	0.5	10
Water absorbance (%)	0.005–0.015	0.1–0.2	3.1	0.01–0.1	0.01–0.4
Thermal properties [Glass Transition Temperature-Tg (°C)]	−110	73	55	−20	90
Transparency (Clarity)	High	Excellent	High	Poor	Excellent
Carbon dioxide barrier (permeation)	28	0.2	10.2	5.3	10.5
Chemical Attack	Resistant	Resistant	Poor	Resistant	Resistant

materials (Table 11.7), their strength is okay being within the range of other materials, but there are needs to improve ductility and water absorbance. Ductility requires improvement while the water absorbance requires reduction to reduce bacterial activity on the packaged food from early spoilage [99–103]. Ductility is needed in shaping of the PLA to final form to be used as food packaging. This is a gap that justifies incorporation of foreign materials to cause improvement on existing properties of the PLA. This forms the basis for the previous studies on the PLA-based composites. Beside the listed properties, toxicity level of the PLA-based composites is necessary to ensure that the materials are saved from contaminating the food with toxic substances. A report on toxicity of the *Parquetina nigrescens* fibre-reinforced PLA is disclosed in Section 11.6.2 of this chapter. Since only one phase of the *Parquetina nigrescens* fibre-reinforced PLA is ranked among low Class I toxicity level, it implies the level of risk is very low and the materials can be safely used as food packaging or biomedical tooth prothesis materials, depending on mechanical properties to ascertain its strength suitability.

11.9　CRYSTALLINITY AND THERMAL CHARACTERISTICS OF PLA AND PLA-BASED COMPOSITES

Neat PLA crystallises on heating (cold crystallisation temperature-Tcc) at 130°C [104]. The crystallisation rate of talc/PLA composites was much faster than pure PLA and the crystallization temperatures (Tc) reduced from 107°C–100°C as the concentration of talc increased in the PLA matrix indicating that talc increased the PLA crystallisation rate [105]. A similar thing was reported where neat PLA crystallised at 131.5°C and the crystallisation temperatures (Tc) of coffee silverskin/ PLA composites decreased from 95.4°C–96.5°C as the concentration of coffee silverskin particles increased in the PLA matrix [1]. This indicates that coffee skin particles increased the PLA crystallisation rate. Good thermal properties protect products against thermal damage during storage and/or resist deformation and degradation of packaging materials as they store products with elevated temperatures. Synthetic fibres and fillers generally have better thermal stability (resistance to thermal decomposition) and thus their composites have better heat resistance.

Flame/fire retardancy of biopolymer composites is important as it influences their areas of application. When biopolymer composite materials are chosen for thermal insulation applications or as building materials, it is mandatory to determine the flammability and flame retardancy property of composites by flame retardancy tests. Since natural fibres alone are highly flammable, they are more likely to ignite and combust exothermically and burn vigorously on combustion. Several factors influence the flammability of biopolymer composites, like type of natural fibre and biopolymer, type of reinforcement, adhesion between reinforcement and matrix, and the structure of the composite [106]. As the chemical composition and microstructure differs from one fibre to another, the flammability also varies accordingly. At elevated temperatures, the major fibre components, cellulose, hemicelluloses, and lignin markedly behave in different ways and thus the composition of a natural fibre greatly influences the composite performance [10]. The flammability of natural fibres increases with increase in cellulose content, and their decomposition emits flammable volatiles, noncombustible gasses, and tars, leading to char formation. Lignin content is also responsible for char formation, and it acts as a protection of the underlying biopolymer composite integrity by serving as an insulation layer. The natural fibres with high crystallinity and low polymerisation are less flammable [10]. Usually, the flame retardancy test for natural fibre–reinforced biopolymer composites is performed in a vertical Bunsen burner test, in line with the federal aviation regulation [106].

Natural fibres degrade at elevated temperatures, affecting their range of application, choice of matrix, and processing routes [15]. Natural fibres are also more flammable than their synthetic counterparts and as such make their composites more flammable [23]. Introduction of flame-retardants to counter this problem can lead to reduced durability and as well as to toxic products after thermal degradation of the composite. The constitution and sequence of PLA monomers can give PLA with higher tolerance to temperature [51]. Generally, natural fibres exhibit lower thermal stability than petroleum-based fibres [107]. It is expected that biopolymer composites will exhibit lower thermal stability than conventional petroleum-based composites. However, treatment of the fibres or incorporating silica or ash into natural fibres and blending with PLA is being adopted to enhance the thermal stability of the bio-composites.

To enhance the heat resistance of natural fibres, they could be blended with some heat retardants. It was reported in [108] that highly refractory/heat retardant materials like ammonium polyphosphate, nano-clay, etc were blended with natural fibre/PLA bio composite during melting. They also reported that excellent materials for obtaining heat resistance are compounds that contain chlorine, bromine, or phosphorous. However, the resistance of natural fibre composites to ultraviolet (UV) light is worrisome because UV radiation can lead to decay or modification of surface chemistry in the composites, a phenomenon known as photodegradation [89], which may have adverse effects on the properties of the composites.

11.10 BIODEGRADABILITY OF POLYLACTIC-BASED COMPOSITES

Various methods are used to study the biodegradability of natural fibre–reinforced polymer matrix composites. Some methods are natural weathering, accelerated weathering, burying in garden soils/composts, and degradation by moisture or chemicals and microorganisms [109]. Usually, after weathering, the strength of PLA-based composites reduces because the fibre and polymer are degraded. Degradation of natural fibres occurs when the lignin that is in the fibres absorbs UV radiation and forms quinoid structures, Norrish reactions and reactions of photo yellowing that take place in lignin [110]. A flowchart that shows the UV degradation process of natural fibre–reinforced polymer matrix composites is presented in Figure 11.20. Natural weathering condition, which is the environment where the composite is used,

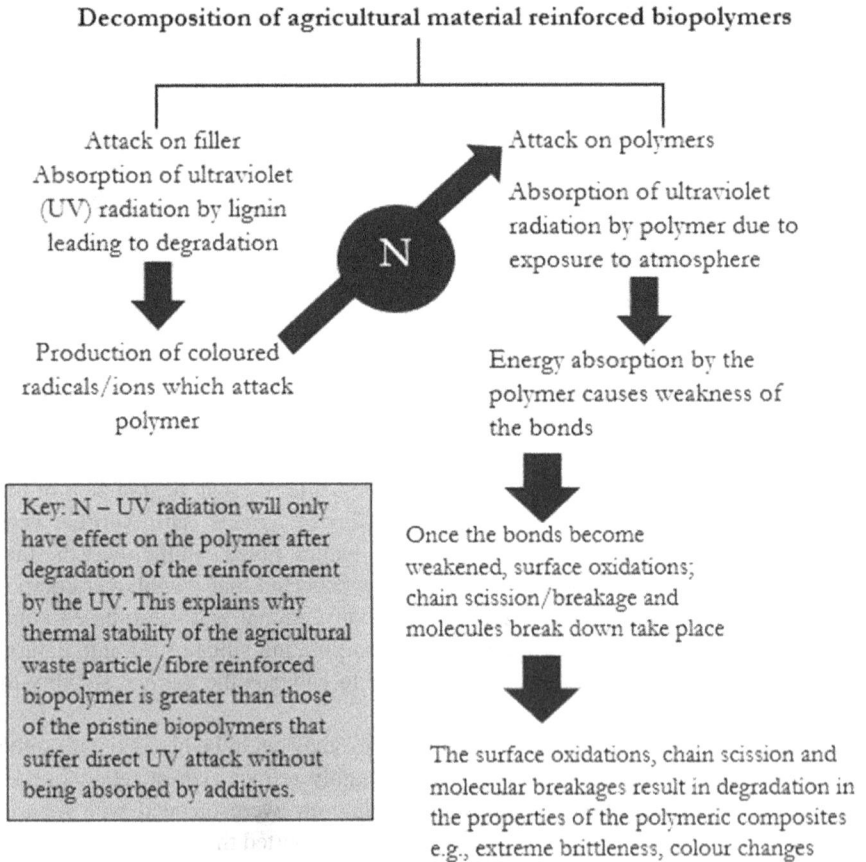

Decomposition of agricultural material reinforced biopolymers

Attack on filler
Absorption of ultraviolet (UV) radiation by lignin leading to degradation

N

Attack on polymers
Absorption of ultraviolet radiation by polymer due to exposure to atmosphere

Production of coloured radicals/ions which attack polymer

Energy absorption by the polymer causes weakness of the bonds

Key: N – UV radiation will only have effect on the polymer after degradation of the reinforcement by the UV. This explains why thermal stability of the agricultural waste particle/fibre reinforced biopolymer is greater than those of the pristine biopolymers that suffer direct UV attack without being absorbed by additives.

Once the bonds become weakened, surface oxidations; chain scission/breakage and molecules break down take place

The surface oxidations, chain scission and molecular breakages result in degradation in the properties of the polymeric composites e.g., extreme brittleness, colour changes

FIGURE 11.20 Ultraviolet degradation process of natural fibre–reinforced polymer matrix composites.

influences the natural degradation of the composite. Checking/observation for many years is used to determine the long-term performance of composites when exposed in the environment [111]. A two-year natural weathering study on jute/phenolic bio-composites by exposure to natural degradation was studied [112], which revealed polymer cracking, contraction of fibres, and over 50% tensile strength reduction. However, accelerated weathering was done in chambers that reproduced conditions of the natural environment, and the destruction caused by long-time exposure to UV radiation, moisture, and regulated temperature. The results obtained were like that of natural degradation but were faster and it has been established that this method is more convenient than natural weathering.

PLA was reinforced with treated coir fibre to develop biopolymer composites and their degradability was investigated [113]. The results indicated that the bio-composites degraded much faster than PLA with highest weight loss of 34.9% recorded in the reinforced bio-composites compared to 18% in neat PLA after being buried in the soil for 18 days. This was because the coir fibres are hydrophilic, with an affinity for water, which indicated that increasing fibre concentration in the matrix improved the degradability of the composites.

In biopolymer composites, adhesion of natural fibre to a polymer matrix can be by inter-diffusion, adsorption, chemical bonding, reaction bonding, electrostatic attraction, or mechanical bonding. The mechanisms of adhesion depend on many factors such as chemical composition and molecular conformation, morphological property of natural fibre, diffusivity of element materials, and the atomic arrangement of fibre and matrix [114]. The biopolymer composite properties are affected by the type of fibre, percentage of fibre content, moisture absorption of fibre, surface modification method of fibre, structure, and design of composite, interfacial adhesion between fibre and matrix, presence of voids, and incorporation of additives like plasticisers, compatibilisers, nanofillers, and binding agents [115,116]. Various reinforcement materials and plasticisers affect the density, water sensitivity, gas permeability, degradability, and shelf life of biopolymer composites. Thus, the hydrophilicity of natural fibre and hydrophobicity of biopolymer drives the researchers to surface modification of fibres to improve the adhesion between them in a composite. The performance of biopolymer composites is enhanced by chemical modification depending on the processing type, processing requirements, and environmental conditions [117]. Biocompatibility and durability of biopolymer composites are of major concern, as there is no proper solution adopted till now to completely control these two factors.

11.11 CONCLUSIONS

The need to intensify efforts and commitments towards the development of green or renewable composites cannot be overemphasised. This is because of the immense benefits that are obtainable from their development and application as substitutes for petroleum-based composites. Composites obtained from PLA and natural fillers (particles and fibres) from agricultural waste should be studied and for possible application in various areas. Blending PLA with natural fillers (particles and fibres) from agricultural waste will yield green composites at low cost, low greenhouse gas emissions, and improved performance for various applications. In view of the increasing demands for renewable and recyclable materials, and environmental pollution by synthetic materials, efforts should be intensified to develop biodegradable composites. PLA and natural fillers can serve as substitutes for synthetic polymers and conventional fillers, which are difficult to recycle and are not sustainable. Natural filler–reinforced PLA composites have prospect for extensive use in many areas because of their desirable attributes. These bio-composites with diverse unique characteristics may soon be competitive enough to replace the existing petroleum-based composites. However, developing biodegradable composites from natural fillers and biopolymers such as *Parquetina nigrescens* fibre-reinforced PLA that will completely replace petroleum-based composites has not recorded tremendous breakthrough as expected. Studies have been carried out on PLA-based composites to know their characteristics when subjected to various standardised tests. Such studies have mostly resulted in similar findings, which confirmed previous results that natural fillers and biopolymers complement each other in many ways. Some findings indicated reduced characteristics, but the characteristics can be boosted by adopting some modifications. Through commitment and concerted efforts, it is possible to develop fully biodegradable composites with unique multifunctional characteristics for applications in various areas in the near future. Adopting more effective modification techniques for reinforcing the agricultural waste materials could be immensely beneficial. This may promote or induce stronger interfacial bonding of the natural fillers and the PLA matrix, which can yield bio-composites with unique or superior physical, mechanical, and thermal characteristics when compared to the petroleum-based composites. Finally, leveraging on the sustainability of agricultural wastes due to their availability in large quantities worldwide and on the fact that it is economical to produce PLA-based composites could be greatly beneficial if maximally explored.

REFERENCES

[1] Ghazvini, A.K. (2018). *Evaluation of Exploiting Coffee Wastes in Production of PLA Green Composites*. M.Sc. Thesis, Faculty of Civil, Chemical, Environmental and Material Engineering, Alma Mater Studiorum – University of Bologna.

[2] Bajracharya, R.M., Bajwa, D.S., and Bajwa, S.G. (2017). Mechanical Properties of Polylactic Acid Composites Reinforced with Cotton Gin Waste and Flax Fibres. *Procedia Engineering*, **200**, 370–376.

[3] Nurul, M.F., Jayaraman, K., Bhattacharyya, D., and Mohamad, M.H. (2016). Green Composites Made of Bamboo Fabric and Poly (Lactic) Acid for Packaging Applications. *A Review. Materials*, **9**, 435.

[4] Jassim, A.K. (2017). Recycling of Polyethylene Waste to Produce Plastic Cement. *Procedia Manufacturing*, **8**, 635–642.

[5] Weber, R., Gaius, C., Tysklind, M., Johnston, P., Forter, M., and Hollerti, H. (2008). Dioxn and POP-Contaminated Sites-ContemPorary and Future Relevance and Challenges: Overview and Background, Aims and Scope of the Series. *Environmental Science and Pollution Research International*, **15**(5), 363–393.

[6] European Commission (EC) (2009). White Paper on Climate Change. *SEC*, 386–388.

[7] El-newehy, M. (2016). *Plastic Waste Management*, King Saud University, pp. 1–35.

[8] Aderogba, K.A. (2014). Polymer Wastes and Management in Cities and Towns of Africa and Sustainable Environment: Nigeria and European Experiences. *Social Sciences*, **3**(4–1), 79–88.

[9] Sudesh, K. and Iwata, T. (2008). Sustainability of Bio-Based and Biodegradable Plastics. *J. Clean*, **36**(5-6), 433–442.

[10] Christian, S.J. (2016). Natural Fibre-Reinforced Non-cementitious Composites (bio composites). *Nonconventional and Vernacular Construction Materials*, Harries & B. Sharma, Editors, Woodhead Publishing, pp. 111–126.

[11] Hosseini, S.B. (2020). Natural Fibre Polymer Nanocomposites. *Fibre Reinforced Nanocomposites: Fundamentals and Applications*, Elsevier, pp. 279–299.

[12] Mohamed, S.A.N., Zainudin, E.S., and Sapuan, S.M. (2018). Introduction to Natural Fibre Reinforced Vinyl Ester and Vinyl Polymer Composites. *Natural Fibre Reinforced Vinyl Ester and Vinyl Polymer Composites*, Woodhead Publishing, pp. 1–25.

[13] Hu, B. (2014). Biopolymer-Based Lightweight Materials for Packaging Applications. In *Lightweight Materials from Biopolymers and Biofibres*, ACS Publications, pp. 239–255.

[14] Nesic, A., Castillo, C., Castaño, P., Barjas, G.C., and Serrano, J. (2020). Bio-Based Packaging Materials. *Biobased Products and Industries*, C.M. Galanakis, Editor. Elsevier, pp. 279–309.

[15] Dicker, M.P., Duckworth, P.F., Baker, A.B., Francois, G., Hazzard, M.K., and Weaver, P.M. (2014). Green Composites: A Review of Material Attributes and Complementary Applications. *Composites Part A Applied Science and Manufacturing*, **56**, 280–289.

[16] Hu, Y., Daoud, W.A., Cheuk, K.K.L., and Lin, C.S.K. (2016). Newly Developed Techniques on Polycondensation, Ring-Opening Polymerization and Polymer Modification: Focus on Poly (Lactic Acid). *Materials*, **9**, 133.

[17] Nemat, B., Razzaghi, M. Bolton, K., & Rousta, K. (2019). The Role of Food Packaging Design in Consumer Recycling Behavior – A Literature Review. *Sustainability*, **11**(16), 1–23.

[18] Murariu, M. and Dubois, P. (2016). PLA Composites: From Production to Properties. *Advanced Drug Delivery Reviews*, **107**, 17–46.

[19] Fabra, M.J., Rubio, A.L., and Lagaron, J.M. (2014). Biopolymers for Food Packaging Applications. *Smart Polymers and Their Applications*, M.R. Aguilar and J. San Román, Editors. Woodhead Publishing, pp. 476–509.

[20] Adeyeye, O., Sadiku, E.R., Reddy, A.B., Ndamase, A.S., and Makgatho, G. (2019). Use of Biopolymers in Food Packaging. *Green Biopolymers and Their Nanocomposites*, Springer, pp. 137–158.

[21] Yadav, A., Mangaraj, S., Singh, R., Kumar, N., and Arora, S. (2018). Biopolymers as Packaging Material in Food and Allied Industry. *International Journal of Chemical Studies*, **6**, 2411–2418.

[22] Krepsztul, J.W., Rydzkowski, T., Borowski, G., Szczypinski, M., and Klepka, T. (2018). Recent Progress in Biodegradable Polymers and Nanocomposite-Based Packaging Materials for Sustainable Environment. *International Journal of Polymer Analysis and Characterization*, **23**, 383–395.

[23] Siakeng, R., Jawaid, M., Ariffin, H., Sapuan, S., and Asim, M. (2018). Natural FIbre Reinforced Polylactic Acid Composites: A Review. *Polymer Composites*, **40**, 446–463.

[24] Gonçalves de Moura, I., de Sa, A.V., Abreu, A.S.L.M., and Alves-Machado, A.V. (2017). Bioplastics from Agricultural-WastEs for Food Packaging Applications. *Food Packaging*, A.M. Grumezescu, Editor. Academic Press, pp. 223–263.

[25] Payne, J., McKeown, P., and Jones, M.D. (2019). A Circular Economy Approach to Plastic Waste. *Polymer Degradation and Stability*, **165**, 170–181.

[26] Biron, M. (2018). *Thermoplastics and Thermoplastic Composites*, William Andrew. p. 1164.

[27] Vilela, C., Sousa, A.F., Fonseca, A.C., Serra, A.C., and Coelho, J.F. (2014). The Quest for Sustainable Polyesters–Insights into the Future. *Polymer Chemistry*, **5**, 3119–3141.

[28] European Bioplastics. Bioplastics Facts and Figures. Available online: https://docs. european-bioplastics.org/publications/EUBP_Facts_and_figures.pdf (accessed on 11 August, 2021).

[29] Mangaraj, S., Yadav, A., Bal, L.M., Dash, S., and Mahanti, N.K. (2018). Application of Biodegradable Polymers in Food Packaging Industry: A Comprehensive Review. *Journal of Packaging Technology and Research*, **3**, 77–96.

[30] https://www.european-bioplastics.org/news/publications/#MarketData (accessed on 10 August, 2021).

[31] European Bioplastics (2019). *Global Production Capacities of Bioplastics 2019–2024*. Europe Bioplastics.

[32] Grujic, R., Vujadinovic, D., and Savanovic, D. (2017). Biopolymers as Food Packaging Materials. *Advances in Applications of Industrial Biomaterials*, Springer, pp. 139–160.

[33] Pawar, P. and Purwar, A.H. (2013). Biodegradable Polymers in Food Packaging. *American Journal of Engineering Research*, **2**, 151–164.

[34] Cecchi, T., Giuliani, A., Iacopini, F., Santulli, C., and Sarasini, F. (2019). Unprecedented High Percentage of Food Waste Powder Filler in Poly Lactic Acid Green Composites: Synthesis, Characterization, and Volatile Profile. *Environmental Science* and *Pollution Research*, **26**, 7263–7271.

[35] Ashok, A., Rejeesh, C., and Renjith, R. (2016). Biodegradable Polymers for Sustainable Packaging Applications: A Review. *Indian Journal of Biochemistry and Biophysics*, **1**, 11.

[36] Muncke, J. (2016). Chemical Migration from Food Packaging to Food. *Reference Module in Food Science*, Elsevier, p. 285.

[37] Kosior, E. and Mitchell, J. (2020). Chapter 6-Current Industry Position on Plastic Production and Recycling. *Plastic Waste and Recycling*, T.M. Letcher, Editor. Academic Press. p. 133–162.

[38] Pan, Y., Farahani, M.F., OHearn, P., Xiao, H., and Ocampo, H. (2016). An Overview of Bio-Based Polymers for Packaging Materials. *Journal of Bioresources and Bioproducts*, **1**, 106–113.

[39] Muller, J., Martínez, C.G., and Chiralt, A. (2017). Combination of Poly (lactic) Acid and Starch for Biodegradable Food Packaging. *Materials*, **10**, 952.

[40] Jacob, J., Lawal, U., Thomas, S., and Valapa, R.B. (2020). Chapter 4-Biobased Polymer Composite from Poly (Lactic Acid): Processing, Fabrication, and Characterization for Food Packaging. *Processing and Development of Polysaccharide-Based Biopolymers for Packaging Applications*, Y. Zhang, Editor. Elsevier, pp. 97–115.

[41] Castro-Aguirre, E., Auras, R., Selke, S., Rubino, M., and Marsh, T. (2018). Enhancing the Biodegradation Rate of Poly(lactic acid) Films and PLA Bio-Nanocomposites in Simulated Composting through Bio-Augmentation. *Polymer Degradation and Stability*, **154**, 46–54.

[42] Emadian, S.M., Onay, T.T., and Demirel, B. (2017). Biodegradation of Bioplastics in Natural Environments. *Waste Management*, **59**, 526–536.

[43] Guillermo, A. and Noa, S. (2020). Potential Effects of Biodegradable Single-Use Items in the Sea: Polylactic Acid (PLA) and Solitary Ascidians. *Environmental Pollution*, **268**(115364), 1–11.

[44] Aliotta, L., Gigante, V., Coltelli, M.B., Cinelli, P., and Lazzeri, A. (2019). Evaluation of Mechanical and Interfacial Properties of Bio-Composites Based on Poly (lactic acid) with Natural Cellulose Fibres. *International Journal of Molecular Sciences*, **20**, 960.

[45] Jabeen, N., Majid, I., and Nayik, G.A. (2015). Bioplastics and Food Packaging: A Review. Cogent Food & Agriculture, **1**(1117749), 1–6.

[46] Scaffaro, R., Botta, L., Lopresti, F., Maio, A., and Sutera, F. (2017). Polysaccharide Nanocrystals as Fillers for PLA Based Nanocomposites. *Cellulose*, **24**, 447–478.

[47] Bajpai, P.K., Singh, I., and Madaan, J. (2012). Development and Characterization of PLA-Based Green Composites: A Review. *Journal of Thermoplastic Composite Materials*, **27**(1): 52–81. doi:10.1177/0892705712439571

[48] Byun, Y. and Kim, Y.T. (2014). Chapter 142-Bioplastics for Food Packaging: Chemistry and Physics. *Innovations in Food Packaging*, 2nd ed., J.H. Han, Editor. Academic Press. pp. 353–368.

[49] Siracusa, V., Rocculi, P., Romani, S., and Dalla-Rosa, M. (2008). Biodegradable Polymers for Food Packaging: A Review. *Trends in Food Science and Technology*, **19**, 634–643.

[50] Darie-Nita, R.N., Vasile, C., Irimia, A., Lipsa, R., and Rapa, M. (2016). Evaluation of Some Eco-Friendly Plasticizers for PLA Films Processing. *Trends in Food Science and Technology*, **133**, 11.

[51] Ncube, L.K., Ude, A.U., Ogunmuyiwa, E.N., Zulkifli, R., and Beas, I.N. (2020). Environmental Impact of Food Packaging Materials: A Review of Contemporary Development from Conventional Plastics to Polylactic Acid Based Materials. *Materials*, **13**, 1–24.

[52] Kumar, N., Kaur, P., and Bhatia, S. (2017). Advances in Bio-Nanocomposite Materials for Food Packaging: A Review. *Nutrition & Food Science*, **47**, 591–606.

[53] Silva, T.F.D., Menezes, F., Montagna, L.S., Lemes, A.P., and Passador, F.R. (2019). Effect of Lignin as Accelerator of the Biodegradation Process of Poly(lactic acid)/Lignin Composites. *Materials Science and Engineering B*, **251**, 114441.

[54] Karamanlioglu, M., Preziosi, R., and Robson, G.D. (2017). Abiotic and Biotic Environmental Degradation of the Bioplastic Polymer Poly (lactic acid): A Review. *Polymer Degradation and Stability*, **137**, 122–130.

[55] Imam, S., Glenn, G., and Chiellini, E. (2012). Utilisation of Biobased Polymers in Food Packaging: Assessment of Materials, Production and Commercialization. *Emerging Food Packaging Technologies*, Yam K.L. & Lee D.S., Editors, Elsevier, pp. 435–446.

[56] Gonzalez, A. and Igarzabal, C.I.A. (2013). Food. *Hydrocoll*, **33**, 289.

[57] Hiebel, M., Maga, D., Kabasci, S., Lieske, A., and Jesse, K. (2017). *PLA in the Waste Stream*. Federal Ministry of Food and Agriculture, p. 8.

[58] Havstad, M.R. (2020). Biodegradable Plastics. *Plastic Waste and Recycling*, T.M. Letcher, Editor. Academic Press. pp. 97–129.

[59] Marsh, K. and Bugusu, B. (2007). Food Packaging-Roles, Materials, and Environmental Issues. *Journal of Food Science*, **72**, R39–R55.

[60] Song, J., Kay, M., and Coles, R. (2011). Bioplastics. *Food and Beverage Packaging Technology*, 2nd ed., Richard Coles & Mark J. Kirwan, Wiley-Blackwell. pp. 295–319.

[61] Badia, J.D., Gil-Castell, O., and Ribes-Greus, A. (2017). Long-Term Properties and End-of-Life of Polymers from Renewable Resources. *Polymer Degradation and Stability*, **137**, 35–57.

[62] Obi, F.O., Ugwuishiwu, B.O., and Nwakaire, J.N. (2016). Agricultural Waste Concept, Generation, Utilization and Management. *J. Nigerian Journal of Technology*, **35**(4), 957–964.

[63] Mantia, F.P.L. and Morreale, M. (2011). Green Composites: A Brief Review, *J. Composites, Part A*, **42**, 579–588.

[64] Nishimura, T. (2004). *Development of Car Components Using Kenaf and a New Evolution in Biomaterials*. Sus Comp Net 7, University of Bath.

[65] Raquez, J.M., Habibi, Y., Murariu, M., and Dubois, P. (2013). Polylactide (PLA)-Based Nanocomposites. *Progress in Polymer Science*, **38**, 1504–1542.

[66] Mokhena, T.C., Sefadi, J.S., Sadiku, E.R., John, M.J., Mochane, M.J., and Mtibe, A. (2018). Thermoplastic Processing of PLA/Cellulose Nanomaterials Composites. *Polymers*, **10**, 1363.

[67] Madhuri, K.S., Rao, H.R., and Reddy, B.C. (2018). Experimental Investigation on the Mechanical Properties of Hardwickia Binata Fibre Reinforced Polymer Composites. *Materials Today: Proceedings*, **5**, 19899–19907.

[68] Saccani, A., Sisti, L., Manzi, S., and Fiorini, M. (2019). PLA Composites Formulated Recycling Residuals of the Winery Industry. *Polymer Composites*, **40**, 1378–1383.

[69] Wilfred, O., Tai, H., Marriott, R., Liu, Q., and Tverezovskiy, V. (2018). Biodegradation of Polylactic Acid and Starch Composites in Compost and Soil. *International Journal* of *Nano Research*, **1**, 1–11.

[70] Hammiche, D., Boukerrou, A., Azzeddine, B., Guermazi, N., and Budtova, T. (2019). Characterization of Polylactic Acid Green Composites and Its Biodegradation in a Bacterial Environment. *International Journal of Polymer Analysis and Characterization*, **24**, 236–244.

[71] Mann, G.S., Singh, L.P., Kumar, P., and Singh, S. (2020). Green Composites: A Review of Processing Technologies and Recent Applications. *Journal of Thermoplastic Composite Materials*, **33**, 1145–1171.

[72] Cichorek, M., Piorkowska, E., and Krasnikova, N. (2017). Stiff Biodegradable Polylactide Composites with Ultrafine Cellulose Filler. *Journal of Polymers and the Environment*, **25**, 74–80.

[73] Fidan, I., Imeri, A., Gupta, A., Hasanov, S., and Nasirov, A. (2019). The Trends and Challenges of Fibre Reinforced Additive Manufacturing. *International Journal of Advanced Manufacturing Technology*, **102**, 1801–1818.

[74] Sydow, Z. and Bienczak, K. (2018). The Overview on the Use of Natural Fibres Reinforced Composites for Food Packaging. *Journal of Natural Fibers*, **16**, 1189–1200.

[75] Lv, S., Zhang, Y., Gu, J., and Tan, H. (2017). Biodegradation Behavior and Modelling of Soil Burial Effect on Degradation Rate of PLA Blended with Starch and Wood Flour. *Colloids and Surfaces B: Biointerfaces*, **159**, 800–808.

[76] Oksman, K., Skrifvars, M., and Selin, J.F. (2003). Natural Fibres as Reinforcement in Polylacticacid (PLA) Composites. *Composites Science and Technology*, **63**, 1317–1324.

[77] Shih, Y.F., Huang, C.C., and Chen, P.W. (2010). Biodegradable Green Composites Reinforced by the Fibre Recycling from Disposable Chopsticks. *Materials Science and Engineering: A*, **527**: 1516–1521.

[78] Kim, J.T. and Netravali, A.N. (2010). Mercerization of Sisal Fibres: Effect of Tension on Mechanical Properties of Sisal Fibre and Fibre-Reinforced Composites. *CompositesPart A*, **41**, 1245–1252.

[79] Rashed, H.M.M.A., Islam, M.A., and Rizvi, F.B. (2006). Effects of Process Parameters on Tensile Strength of Jute Fibre Reinforced Thermoplastic Composites. *Journal of Naval Architecture* and *Marine Engineering*, **3**, 1–6.

[80] Lee, J.T., Kim, M.W., Song, Y.S., Kang, T.J., and Youn, J.R. (2010). Mechanical Properties of Denim Fabric Reinforced Poly(Lactic Acid). *Fibers and Polymers*, **11**, 60–66.

[81] Hu, R. and Lim, J.K. (2007). Fabrication and Mechanical Properties of Completely Biodegradable Hemp Fibre Reinforced Polylactic Acid Composites. *Journal of Composite Materials*, **41**, 1655–1669.

[82] Ochi, S. (2006). Development of High Strength Biodegradable Composites Using Manila Hemp Fibre and Starch-Based Biodegradable Resin. *Compos Part A*, **37**, 1879–1883.

[83] Placketta, D., Andersen, T.L., Pedersenc, W.B., and Nielsenc, L. (2003). Biodegradable Composites Based on L-Polylactide and Jute Fibres. *Composites Science and Technology*, **63**, 1287–1296.

[84] Srebrenkoska, V., Gaceva, G.B., and Dimeski, D. (2009). Preparation and Recycling of Polymer Ecocomposites: Comparison of the Conventional Molding Techniques for Preparation of Polymer Eco-ComPosites. *Macedonian Journal of Chemistry* and *Chemical Engineering*, **28**, 99–109.

[85] Tao, Y., Yan, L., and Jie, R. (2009). Preparation and Properties of Short Natural Fibre Reinforced Poly(lactic acid) Composites. *Transactions of Nonferrous Metals Society of China*, **19**, s651–s655.

[86] Liu, L., Fishman, M.L., Hicks, K.B., and Liu, C.K. (2005). Biodegradable Composites from Sugar Beet Pulp and Poly(lactic acid). *Journal of Agricultural and Food Chemistry*, **53**, 9017–9022.

[87] Gregorova, A., Hrabalova, M., Kovalcik, R., and Wimmer, R. (2011). Surface Modification of Spruce Wood Flour and Effects on the Dynamic Fragility of PLA/Wood Composites. *Polymer Engineering & Science*, **51**, 143–150.

[88] Qin, L., Qiu, J., Liu, M., Ding, S., and Shao, L. (2011). Mechanical and Thermal Properties of Poly(lactic acid) Composites with Rice Straw Fibre Modified by Poly (butyl acrylate). *Chemical Engineering Journal*, **166**, 772–778.

[89] Islam, M.S., Pickering, K.L., and Foreman, N.J. (2010). Influence of Alkali Treatment on the Interfacial and Physico-Mechanical Properties of Industrial Hemp Fibre Reinforced Polylactic Acid Composites. *Composites Part A*, **41**, 596–603.

[90] Petinakis, E., Yu, L., Edward, G., Dean, K., and Liu, H. (2009). Effect of Matrix–Particle Interfacial Adhesion on the Mechanical Properties of Poly(lactic Acid)/Wood-Flour Micro-Composites. *Journal of Polymers and the Environment*, **17**, 83–94.

[91] Yu, T., Ren, J., Li, S., Yuan, H., and Li, Y. (2010). Effect of Fibre Surface-Treatments on the Properties of Poly(lactic acid)/Ramie Composites. *Composites Part A*, **41**, 499–505.

[92] Lee, S.H. and Wang, S. (2006). Biodegradable Polymers/Bamboo Fibre Biocomposite with Bio-Based Coupling Agent. *Composites Part A*, **37**, 80–91.

[93] Faruk, O. and Ain, M.S. (2013). Biofibres Reinforced Polymer Composites for Structural Applications. *Developments in Fibre-Reinforced Polymer (FRP) Composites for Civil Engineering*, Nasim Uddin, Editor, Woodhead Publishing, pp. 18–53.

[94] Ramamoorthy, S.K., Åkesson, D., and Rajan, R. (2018). Mechanical Performance of Biofibres and Their Corresponding Composites. *Mechanical and Physical Testing of Bio Composites, Fibre Reinforced Composites and Hybrid Composites*, Mohammad Jawaid, Mohamed Thariq & Naheed Saba, Editors, Woodhead Publishing, pp. 259–292.

[95] Huda, M.S., Drzal, L.T., Mohanty, A.K., and Misra, M. (2007). The Effect of Silane Treated and Untreated-Talc on the Mechanical and Physico-Mechanical Properties of Poly(lactic acid)/Newspaper Fibres/ Talc Hybrid Composites. *Composites Part B*, **38**, 367–379.

[96] Sawpan, M.A., Pickering, K.L., and Fernyhough, A. (2011). Improvement of Mechanical Performance of Industrial Hemp Fibre Reinforced Polylactide Biocomposites. *Elsevier, Composites Part A, Applied Science and Manufacturing*, **42**(3), 310–319.

[97] Totaro, G., Sisti, L., Fiorini, M., Lancellotti, I., and Andreola, F.N. (2019). Formulation of Green Particulate Composites from PLA and PBS Matrix and Wastes Deriving from the Coffee Production. *Journal of Polymer and the Environment*, **27**, 1488–1496.

[98] Papadopoulou, E.L., Paul, U.C., Tran, T.N., Suarato, G., and Ceseracciu, L. (2019). Sustainable Active Food Packaging from Poly(lactic acid) and Cocoa Bean Shells. *ACS Applied Materials and Interfaces*, **11**, 31317–31327.

[99] Mallegni, N., Phuong, T.V., Coltelli, M.B., Cinelli, P., and Lazzeri, A. (2018). Poly (lactic acid) (PLA) Based Tear Resistant and Biodegradable Flexible Films by Blown Film Extrusion. *Materials*, **11**, 148.

[100] Farah, S., Anderson, D.G., and Langer, R. (2016). Physical and Mechanical Properties of PLA, and Their Functions in Widespread Applications-A Comprehensive Review. *Advanced Drug Delivery Reviews*, **107**, 367–392.

[101] Packaging, A. Plastics Comparison Chart. Available online: http://www.alphap.com/bottle-basics/plasticscomparison-chart.php (accessed on 24 August, 2021).

[102] Omnexus. Glass Transition Temperature Values of Several Plastics. Available online://omnexus.specialchem.com/polymer-properties/properties/glass-transition temperature (accessed on 11 August, 2021).

[103] Zhao, X., Cornish, K., and Vodovotz, Y. (2020). Narrowing the Gap for Bioplastic Use in Food Packaging: An Update. *Environmental Science & Technology*, **54**, 4712–4732.

[104] Battegazzore, D., Bocchini, S., and Frache, A. (2011). Crystallization Kinetics of Poly(lactic acid)-Talc Composites. *Express Polymer Letters*, **5**(10), 849–858.

[105] Mngomezulu, M.E., John, M.J., and Jacobs, V. (2014). Review on Flammability of Biofibres and Bio Composites. *Carbohydrate Polymers*, **111**, 149–182.

[106] Aaliya, B., Sunooj, K.V., and Lackner, M. (2021). Biopolymer Composites: A Review. *International Journal of Biobased Plastics*, **3**(1), 40–84.

[107] Thakur, V.K., Thakur, M.K., and Gupta, R.K. (2014). Review: Raw Natural Fiber–Based Polymer Composites. *Carbohydrate Polymers*, **104**, 87.

[108] Stark, N.M., White, R.H., Mueller, S.A., and Osswald, T.A. (2010). Evaluation of Various Fire Retardants for Use in Wood Flour-Polyethylene Composites. *Elsevier, Polymer Degradation and Stability*, **95**, 1903–1910.

[109] Azwa, Z., Yousif, B., Manalo, A., and Karunasena, W. (2013). A Review on the Degradability of Polymeric Composites Based on Natural Fibres. *Elsevier, Materials and Design*, **47**, 424–442.

[110] Batista, K.C., Silva, D.A.K., Coelho, L.A.F., Pezzin, S., and Pezzin, A. (2010). Soil Biodegradation of PHBV/Peach Palm Particles Biocomposites. *Journal of Polymer and the Environment*, **18**, 346–354.

[111] Beninia, K., Voorwald, H.J.C., and Cioffi, M. (2011). Mechanical Properties of HIPS/Sugarcane Bagasse Fiber Composites after Accelerated Weathering. *Procedia Engineering*, **10**, 3246–3251.

[112] Dittenber, D.B. and GangaRao, H.V. (2012). Critical Review of Recent Publications on Use of Natural Composites in Infrastructure. *Composites Part A, Applied Science and Manufacturing*, **43**(8), 1419–1429.

[113] Dong, Y., Ghataura, A., Takagi, H., Haroosh, H.J., Nakagaito, A.N., and Lau, K.T. (2014). Polylactic Acid (PLA) Biocomposites Reinforced with Coir Fibres: Evaluation of Mechanical Performance and Multifunctional Properties. *Composites Part A: Applied Science and Manufacturing*, **63**, 76–84.

[114] Azammi, A.M.N., Ilyas, R.A., and Sapuan, S.M. (2020). Characterization Studies of Biopolymeric Matrix and Cellulose Fibres-Based Composites Related to Functionalized Fibre-Matrix Interface. *Interfaces in Particle and Fibre Reinforced Composites*, K.L. Goh, A. M.K., R.T. De Silva & S. Thomas, Editors, Woodhead Publishing, pp. 29–93.

[115] Getme, A.S. and Patel, B. (2020). A Review: Bio-Fibres as Reinforcement in Composites of Polylactic Acid (PLA). *Materials Today: Proceedings*, **26**, 2116–2122.

[116] Kabir, M.M., Wang, H., and Lau, K.T. (2012). Chemical Treatments on Plant-Based Natural Fibre Reinforced Polymer Composites: An Overview. *Composites Part B: Engineering*, **43**(7), 2883–2892.

[117] Vinod, A., Sanjay, M.R., and Suchart, S. (2020). Renewable and Sustainable Bio-Based Materials: An Assessment on Biofibres, Biofilms, Biopolymers, and Bio Composites. *Journal of Cleaner Production*, **258**, 1–27.

12 Polymeric Nanocomposites for Artificial Implants

Funsho Olaitan Kolawole
Department of Metallurgical and Materials Engineering, University of Sao Paulo, Sao Paulo, Brazil

Department of Materials and Metallurgical Engineering, Federal University Oye-Ekiti, Ekiti State, Nigeria

Shola Kolade Kolawole
National Agency for Science and Engineering Infrastructure, Abuja, Nigeria

Felix Adebayo Owa, Chioma Ifeyinwa Madueke, and Oluwole Daniel Adigun
Department of Materials and Metallurgical Engineering, Federal University Oye-Ekiti, Ekiti State, Nigeria

CONTENTS

12.1 Introduction...322
12.2 Requirements of Materials for Artificial Bone Implants.........................324
    12.2.1 Structural Properties of Materials for Artificial Bone Implants...325
    12.2.2 Mechanical Properties of Materials for Artificial Bone Implants...325
    12.2.3 Wear Resistance Properties of Materials for Artificial Bone Implants...326
    12.2.4 Biocompatibility and Cytotoxicity of Materials for Artificial Bone Implants...326
12.3 Polymethylmethacrylate-Based Composites Implants............................327
    12.3.1 Mechanical Properties of Polymethylmethacrylate....................327
    12.3.2 Composition of Polymethylmethacrylate....................................328
    12.3.3 Polymethylmethacrylate Storage...328
    12.3.4 Viscosity of Polymethylmethacrylate...328
    12.3.5 Deformation of Polymethylmethacrylate....................................328
    12.3.6 Thermal Properties of Polymethylmethacrylate..........................329

DOI: 10.1201/9781003170549-16
321

12.1 INTRODUCTION

The average age worldwide of people needing implants increased from 29.0 to 37.3 years of age from 1950 to 2000, respectively, and may still rise to 45.5 years of age by 2050 [1], due to a rise in life expectancy and longevity. A worldwide increase in the average age of the population of people needing implants has resulted in increasing the number of surgical operations involving prosthesis implantation, because the load-bearing joints become more susceptible to ailments as the human body ages. The functionality of the implants determines their surface that can be fabricated from biomedical material like apatite, silicone, or titanium [1]. In a number of cases, cochlear and artificial pacemaker implants may have electronic devices attached to them. Some implants maybe bioactive, with subcutaneous drug delivery devices in the form of drug-eluting stents or implantable pills [2]. The Egyptians, South Central Americans, and Chinese practised dental implants in the 16th and 17th centuries using stone, ivory, and gold [3]. Implant failure can occur; some implant failures include breast implants, artificial heart valves, hip replacement joints, and rupture of silicone. The implant's nature and positioning in the human body determines the extent of the effect of its failure. Therefore, failure of the heart valve threatens human life, while failure of a breast implant or hip joint failure threatens life less [1].

An ideal biomaterial implant should possess excellent biocompatibility, bioactivity, degradability, no toxicity, and high mechanical property. Synthetic biomaterials are an excellent choice, due to their high reproducibility, availability, and adjustability. Its mechanical properties, degradation rate, and composition can easily be modified. Not all synthetic materials can be used as implants due to some limitations such as structural differences from native tissues and organs, due to tissue remodeling, lack of biocompatibility, and sites for cell adhesion, which possibly can restrict their regeneration tendencies in vivo [4]. Synthetic biomaterials can be obtained from ceramics, biodegradable and nonbiodegradable polymers, and metals, which are commonly used for clinical positioning for plastic intraocular lenses, metal hip implants, and several others used as biomedical implants [4].

Biodegradable polymers are used as artificial implants due to their unique properties such crystallinity, degradation rate, polydispersity, molecular weight, and thermal transition [5]. They have the tendency to be manufactured in massive amounts, with a lengthy life span, and a comparatively low price [5]. Degradable synthetic polymers implants are resorbed by the body, which enables complete self-healing without the aid of foreign bodies.

Polymethylmethacrylate (PMMA) copolymer is an active ingredients of bone cement. A German chemist named Otto Rohm was the first to develop polymethylmethacrylate in 1901 [6]. PMMA ensures elasticity between prosthesis and bone, enhancing shock absorption, contact forces distribution, and rigid fixation [7]. Since the inception of PMMA application, it has been demonstrated to be a resourceful and sustainable biomaterial in oncologic surgery, spinal, and orthopaedic joint replacement. Polylactide is a semi-crystalline polymer synthesised during polymerisation or polycondensation. Its molecular weight range is from 180,000 to 530,000, melting point being set approximately at $174°C$, and glass transition at $57°C$. When used as an implant, PLA is degraded by hydrolysis, which maybe in form of poly-L-lactide (PLLA) and poly-D-lactide (PLDA). PLLA is a better candidate for implants because it is known for its slow degradation between 2 to 5 years, compared to PLDA, which loses its mechanical strength much faster [8].

Ultrahigh molecular weight polyethylene (UHMWPE) combines biocompatibility, high fracture toughness, and superior wear resistance in comparison to other polymers. Due to its low-friction surface properties, impact resistance, high rigidity, and wear resistance, it is mostly used as a matrix, making it a suitable artificial implant for hip joint and knee replacements. Its application is linked with its unique properties [9]. Carbon, kaolin, quartz, montmorillonite, aluminum oxide, and hydroxyapatite (HA) have been used as fillers in UHMWPE composite [9]. Hydroxyapatite as a reinforcement for polymer composites was introduced first in 1980 by Bonfield and some researchers [9].

12.2 REQUIREMENTS OF MATERIALS FOR ARTIFICIAL BONE IMPLANTS

Bone implants are commonly used as fillers and scaffold to enhance bone formation and promotion of healing wounds. Most grafts are biomaterials with no antigen-antibody reaction; they serve as a mineral reservoir, which induces newborn formation [10]. Missing bones are usually replaced through surgical operations by bone implants, which are materials from one's own body, which could be an artificial, synthetic, or natural substitute. Implantation is achievable due to the ability of regeneration of bone tissue, as long as space is left for growth to occur. Natural bone tends to replace the implant with time [11–13].

Bone implant materials are exposed in body fluids where interaction take place. This may be unfriendly and tremendously responsive and can restrict an excellent biomaterial's applications. This occurs because tissues in human body are very responsive to external materials, which can lead to stimulation of the body to be poisoned or rejection of implants. Due to this reason, it is necessary for a biomaterial to be biocompatible, meaning that the implant must possess the capacity to carry out a relevant host response in a specific application. A biocompatible biomaterial ought to be devoid of damaged cellular functions, persistent inflammation, toxicity, and carcinogenics. The implant's mechanical and physical properties are important to reach the body requirement. The ideal biomaterial should possess the following properties:

- Material chemical composition that is compactible to the human body to steer clear of unfavorable tissue reactions
- Exceptional degradation resistance
- Sustainable strength for cyclic loading of the joint
- Low modulus for minimal bone resorption
- High wear resistance to diminish debris generation

Implant requirements can be subdivided into compatibility, cytotoxicity, mechanical properties and wear resistance, and ease of fabrication.

Compatibility and Cytotoxicity
Artificial implant materials need to be compatible with body systems. Their interaction with tissue and change in mechanical, physical, and chemical properties causing degradtion should not release products that are harmful to the body.

Mechanical Properties
Elasticity, yield stress, ductility, toughnes, time-dependent deformation, creep, ultimate strength, fatigue strength, hardness, and wear resistance are part of the requirements before materials can be selected for any application. For artificial

implants involving bone replacement, compressive and flexural strength, modulus, and strain are the basic requirements. According to standards reported in many articles, a compressive strength of 70 MPa and flexural strength of 50 MPa have been set as benchmarks for any materials to be suitable as artificial implants for bone replacements. This implies that for emerging materials to be suitable as biomedical materials for bone replacement, they should have at least a compressive strength of 70 MPa and flexural strength of 50 MPa.

Fabrication
Beside the materials' property requirements, ease of fabrication is part of the requirement for selecting materials for biomedical applications. This covers consistency and conformity to all requirements, quality of raw materials, superior techniques to obtain excellent surface finish or texture, capability of material to get safe efficient sterilisation, and cost of product.

12.2.1 Structural Properties of Materials for Artificial Bone Implants

The morphology of most artificial bone implants has a greater role in determination of the effectiveness of mechanical property of a reinforced implant. Two-dimensional nanomaterials are superior reinforcing agents than one-dimensional or zero-dimensional nanomaterials. The two-dimensional nanomaterials are known for their large surface area combined with the increase in cross-linked density of polymeric composites, consequently leading to notable increase in various mechanical properties such as tensile strength, fracture toughness, creep strain, and compression strength [14]. Different carbon (MWCNTs, SWCNTs, MWGONRs, SWGONRs, and GONPs) and inorganic nanomaterials (MSNPs and WSNTs) as reinforcing agents for PPF polymer have wide applications in bone-tissue engineering. Inorganic nanomaterials are usually better implant materials than carbon nanomaterials [14].

12.2.2 Mechanical Properties of Materials for Artificial Bone Implants

Most implants are saddled with the responsibility of bearing loads like hips, knees, and jaws are made to pass through a combination of axial, which could be bending and torsional, compressive, and tensile forces during function and can be used to anticipate the lifetime survival rate. High stress would bring about implant failure at low-cycle and single-cycle, while low stresses would predominately cause fatigue failure; this occurs mostly when dealing with metallic implants [15]. Reduction of applied loads can lessen the probability of failures based on the biomechanical approach. Lack of bone healing is responsible for most orthopaedic implant fails from fatigue. Non-healing can be caused by one of several reasons: skeletal tissues, too much movement, blood supply shortage, infection, immune suppression, ailments, and stress shielding of implants. When an implant is exposed to stress above its tolerance limit, failure is bound to occur within the implant structure. Nevertheless, if exposed to lower stress levels, it can resist millions of cycles, even though this may lead to microscopic fracture zones contributing to catastrophic long-term failures caused by fatigue. This occurs due to a localised concentration of

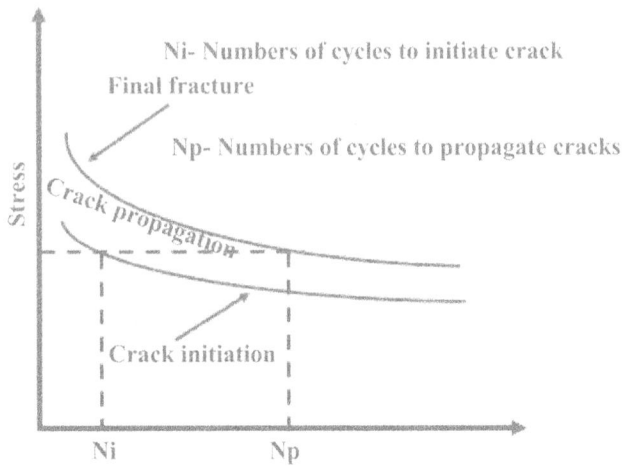

FIGURE 12.1   S-N Curve reveals an increase in fatigue life as stress decreases.

stress on specific spots of the implant before it spreads to other areas, initiating cracks that will in due course propagate all around the implant [15].

Tensile, compressive, and torsional loads act at the same time on the load-bearing implant during its function. Biomechanical studies are important and govern their mechanical properties for full understanding of failure criteria in implants. Applied stress against a number of cycles (S–N curve) (Figure 12.1) is basically used to characterise fatigue behaviour of biomaterials for failure.

12.2.3   WEAR RESISTANCE PROPERTIES OF MATERIALS FOR ARTIFICIAL BONE IMPLANTS

When dealing with artificial implants, wear is unavoidable, which mostly occurs with total or partial joint, total hip, and knee replacements. In dealing with metallic implants, wear debris can cause particulate disease, which is mostly encountered in clinicals and can result in carrying out a second surgery. Disastrous peri-implant bone loss due to metallic and polyethylene wear in knee implants in several instances has been reported. A diseased fracture caused by a far-away femur can be associated with disastrous polyethylene failure, and even a failure affiliated with a three-body wear of the polyethylene from detached CoCr beads. Wear behaviour has been already carried out on total joint replacements and other implants by several researchers [16].

12.2.4   BIOCOMPATIBILITY AND CYTOTOXICITY OF MATERIALS FOR ARTIFICIAL BONE IMPLANTS

Biomaterials may have some level of risk as far as human safety is concerned, and this may be dependent on material type and level of contact with the patient. International Organization of Standardization (ISO 10993-1/EN 30993-1) for biomaterials should be used to prevent toxicity. It is also important to use the recommended steps for biological analysis of prospective biomedical materials consisting of an *in vitro* assessment of cytotoxicity and genotoxicity [17,18]. Due to the low bio-degradation rates of hydroxyatatite (HA), the addition of beta-tricalcium phosphate aids generation of biphasic calcium phosphate (BCP) composite, which plays an important role during assisted bone regeneration [19]. Mitri et al. (2012) [19] assessed the cyto-compatibility of dense and porous HA and dense and porous BCP by three different cell viability parameters (crystal violet dye elution, XTT, neutral red assay) on human mesenchymal cells. There was not many differences observed between cell density (crystal violet dye elution) and mitochondrial activity (XTT), because dense materials induce a lower level of total viable cells by neutral red assay [19].

12.3 POLYMETHYLMETHACRYLATE-BASED COMPOSITES IMPLANTS

Polymethylmethacrylate (PMMA) is a synthetic polymer, which is known for its light weight and is economical. PMMA is not harmful and does notcontain any toxic subunits. Additionally, synthetic polymers like PMMA are easier to handle and process and are cheap. Generally, PMMA has been used where problems of craniofacial tissue defects occur; examples are skin and dentures [4–21].

The mechanical properties of PMMA are excellent, and due to this it has found wide applications in hip-joint transplantation due to its inert properties; the degradation of PMMA is slow. Therefore, manufacturing a composite of polymer blend with PMMA produces a polymer that will serve well as an implant (lenses, bone cement, drug delivery systems, bone substitutes, and dental roots). PMMA is used for bone tissue regeneration because it is a non-biodegradable polymer that has a permanent, and mechanically stable structures [4,20]. At high temperatures, the properties of PMMA are altered because the long-chain backbone divides up into several units and reacts with itself. In the presence of oxygen, PMMA experiences thermal degradation and thermal oxidative, along with photodegradation, oxidative degradation, and UV degradation. Ductility reduction, chalking, colour changes, and cracking occur due to thermal degradation, leading to change in the properties of PMMA. Depolymerisation of PMMA at 300°C–400°C generates a volatile monomer of methyl methacrylate (MMA) [4,20]. PMMA is used to remove scars and wrinkles from skin tissue indefinitely, having an ideal porosity, thermal and electrical passiveness, and low modulus of elasticity that matches that of human beings [4–21].

12.3.1 MECHANICAL PROPERTIES OF POLYMETHYLMETHACRYLATE

Polymethylmethacrylate might be strong but has a lightweight and compressive strength (85–110 MPa) and tensile strength (30–50 MPa) [6,22]. PMMA is known for its relatively high coefficient of thermal expansion and the temperature can reach 40°C–56°C during polymerisation in situ [6,7,23,24]. Therefore, shrinkage of approximately 67% can occur, causing tissue necrosis during the curing process in vivo. PMMA's mechanical properties is adjustable by changing the mixture ratios of monomer and precursor during polymerization. Tensile strength and fatigue strength of PMMA usually decreases when it absorbs water; at body temperature for several weeks, it can absorb water. Degradation of PMMA is insignificant in aqueous environments, even though it has a poor stability. PMMA possesses excellent optical properties in aqueous environments [6,23,25].

12.3.2 COMPOSITION OF POLYMETHYLMETHACRYLATE

The composition of polymethylmethacrylate is mainly of polymer powder and monomer liquid (ratio 2:1). The monomer is colourless and has an odour. MMA is 97%–99% liquid, 0.4–2.3% of the liquid weight is N, and N-Dimethyl-p-toluidine is for acceleration [22]. Early polymerisation can be hindered by 15–75 ppm hydroquinone (traces) which stabilises MMA [22]. Its variation in powder form in composition can cause variation in properties; 83%–99% of the powder is known to be microspheres of copolymer or ground PMMA [22], while the remaining components include zirconium dioxide (ZrO_2), barium sulphate ($BaSO_4$), or radio-opacifier [22]. Other variations of initiators include accelerator 2,5-dimethylhexane-2,5-hydroperoxide, tri-n-butyl borane, chlorophyll dye, ethanol, and ascorbic acid [26].

12.3.3 POLYMETHYLMETHACRYLATE STORAGE

The cementation structural properties of PMMA are affected by its storage. Lewis and Son (2015) [26] recommended the cement constituents temperature for storage as 21°C for surgical operations. Most often PMMA is packaged as a powder copolymer and liquid monomer component, which is mixed at the time of arthroplasty and undergoes polymerisation, leading to the formation of a viscous mouldable biomaterial that is placed into regions that require it to be fixed. Essentially, the cement curing time is to be made known to be able to manipulate and insert it before it becomes out of use [26].

12.3.4 VISCOSITY OF POLYMETHYLMETHACRYLATE

Sterilisation methods are the most common way in which bone cement's viscosity can be changed [23]. Shrinkage of cement by approximately 6%–7% in volume is caused by molecules of monomers converting into fewer polymer molecules that are long chains [7,23]. The predominate phase during the curing process PMMA is divided into low or high viscosity. Cements with longer waiting phases and shorter working phases possess low viscosity, while those with shorter waiting phases and longer working phases possess high viscosity. A viscosity increase in PMMA cement is because of a faster polymerisation process occurring, causing higher released heat and shorter setting time [7].

12.3.5 DEFORMATION OF POLYMETHYLMETHACRYLATE

A material's deformation under load is defined as creep; the viscoelastic property of bone cement determines if creep will occur or not. A material prone to creep at constant load will only deform based on the duration and intensity of load applied. An increase in creep can occur due to copolymer particles, lipids, and unreacted monomers with hydrophilicity. Creep reduces as polymerisation proceeds. Studies have showed creep with prosthetic subsidence; however, additional findings recommend the relaxation of cement stresses and decreased failure rate due to the creep's positive effects on the mantle–prosthesis interface [7].

12.3.6 THERMAL PROPERTIES OF POLYMETHYLMETHACRYLATE

An exothermic reaction usually occurs whenever powder and liquid monomer are blended, causing an exothermic reaction, reaching a peak temperature of approximately 113°C in the vertebral bodies' anterior cortex [7,23]. In addition, the temperature range used for in vivo studies is between 40°C–56°C and when a reaction takes place between PMMA's monomer and other monomers leads to enlarged polymer chain formation [7,23]. PMMA powder instigates polymerisation, thereby forming a feasible dough; for bone necrosis to take place, the use of high viscous PMMA cement has been proposed [7].

12.3.7 USE OF POLYMETHYLMETHACRYLATE COMPOSITES

Gap mismatches that occur between the bone–prosthesis interfaces can be easily filled with PMMA. The biomechanical characteristics of the PMMA causes the distribution of forces from implant to bone to be subjected to more physiologic transition. Body parts like total hip arthroplasty, bone cement, total knee arthroplasty, and total shoulder and total ankle arthroplasty have benefited from the application of PMMA [21].

12.4 POLYLACTIC ACID–BASED COMPOSITES FOR BONE IMPLANTS

Polylactic acid (PLA) is known as a linear aliphatic biopolymer, and is usually obtained from agricultural residues that are from renewable sources (e.g., sugarcane and corn). Stereo-chemically, PLA exists as poly (L-lactide) (PLLA), poly (D-lactide) (PDLA), and poly (DL-lactide) (PDLLA). PLA has gained wide attraction as a material for biomedical applications, leading to the replacement of regular petroleum-based polymers in industry [26,27]. Today, PLA is known for is vast field of applications in clinical practice, due to its biocompatibility and controllable degradation rates in vivo [28]. The elastic modulus of PLA is known for its similarity to human bone, making it the best possible matrix for bone scaffolds [5,8]. PLA is mainly applied in tissue engineering scaffolds, short-term and long-term implants, anchors, bone screws, spinal cages, prostheses, sutures, vascular grafts, and drug encapsulation and delivery. Biodegradability is a major advantage of PLA and has been widely studied by researchers and scientists.

However, the drawbacks of PLA are its low degradation rate and poor cell adhesion, caused by its biological inertness, inflammation, and hydrophobicity in vivo from acidic degradation products. Such drawbacks limit the use of PLA for bone regeneration treatments when considering specific interactions between implants and cells. Hydrophobicity is the major drawback of PLA, which halts water/living cells penetration, leading to necrosis, forms biofilms, and bacterial adhesion [5,8].

12.4.1 POLYLACTIDE-BASED COMPOSITES

Polylactic acid–based composites are commonly used for tissue engineering, bone regeneration, skin regeneration, blood vessels, and organs. A major way of solving these issues is to fabricate PLA-based composites for applications in biomedicine. Mixing PLA with additional materials will ensure an improvement in the biological and physical properties. Improving the PLA properties and reducing the negative effects in biomedical applications. Lately, nanoparticles incorporated in PLA has aided in the improvement for better performance. Nanocomposites are known for their unique properties, hence attracting much attention. Bio-nanocomposites are an interesting field of study based on nanotechnology, biology, and materials science [5,8]. Bio-nanocomposites are now a very relevant research field in controlled local drug delivery because of their bioavailability, biocompatibility, drug encapsulation, and release ability. Addition of nanoparticles (metallic nanofillers, cellulose nanocrystals (CNCs), and carbon-based nanoparticles) to PLA aids in producing high-performance PLA bio-nanocomposites. The presence of metallic nanoparticles/nanofillers (Pt, Au, Ag, and several others) increases PLA degradation because they produce high thermal conductivity to the PLA composites. Enhanced cell adhesion and surface roughness can be improved when metal nanoparticles dispersion in PLA is uniform. Mg alloy coated with films and further incorporated with ZnO nanoparticles recently was submerged in PLA. In vitro and in vivo experiments revealed ZnO nanoparticles enhanced the surface morphology and degradation rate of the composites, supporting cells, and inhibited bacterial growth [5,8].

Cellulose nanocrystals are nano-biomaterials derived from renewable sources (corn stalks and cotton seed hulls). CNCs can be potentially applied in packaging, sensors, pharmaceutical engineering, and biomedicine. CNCs has gain wide attention in field of nanocomposites due to their aspect ratio, high surface area, biocompatibility, biodegradability, renewability, optical properties, good mechanical performance, and low costs. It has been proven that PLA's thermal stability could be improved through the addition of microcrystalline cellulose fluids (MCCFs). PLA/MCCFs composite's tensile strengths and breaking elongation has been improved and known for super-hydrophilicity because its water contact angle is close to $0°$. The use of carbon-based nanomaterials and graphene derivatives has successfully strengthened the mechanical properties and thermal stability of PLA, as shown by previous studies [8].

12.4.2 POLYLACTIC ACID–BASED COMPOSITE FABRICATION TECHNIQUES

Polylactic acid–based composite scaffolds possessed with distinct porosity and morphologies are synthesised via phase separation, 3D printing, electrospinning, solvent casting, injection moulding, salt leaching, gas foaming, microsphere sintering, hydrogel, freeze-drying, and others. The application requirement for a scaffold will determine the kind of preparation methods to be used [5,8].

12.4.3 Mechanical Properties of Polylactic Acid–Based Composite

The mechanical of PLA composites depends on scaffold's porosity and micro-structure, determining their abilities for differentiation, proliferation, cell attach-ment, and migration. Notwithstanding, mechanical properties of scaffolds can be reduced due to an increased porosity. Mechanical behaviour of PLA composite is controlled by their pore shape, porosity, pore size, pore orientation, and material type [5,8]. A mimic biochemical process of scaffold can only be successful when it has similar micro-architecture and physicochemical properties with human bones. A scaffold having excellent porosities and mechanical properties can be used as a vessel to store the transplanted cells in a bioreactor. It is usually a serious setback for the development of biomimetic scaffolds with satisfactory porosities and me-chanical strength in the field of bone tissue engineering [5,8].

12.4.4 Thermal Properties of Polylactic Acid–Based Composite

Fiber's alignment can be used to strongly enhance PLA/CNCs' mechanical prop-erties and thermal stability. The composite's porous microstructure was responsible for the elongation at the breaking point. Increased crystallinity, efficient stress transfer, and ordered arranged fibers improve the performance of the PLA-based composite [29]. A synthesised derivative of triazine was introduced into CNCs and incorporated as reinforcements for PLA. Consequently, weakening the hydro-philicity and improving the thermal stability of CNCs happened due to the in-corporation of triazine. The interfacial interactions of modified PLA/CNCs composite were improved, resulting to an improvement in the mechanical properties and thermal stability of PLA/triazine modified CNCs [30]. Bio-based PLA/CNCs developed from foams using sucrose were prepared by casting and leaching. The derived foam had hydrophobicity and cellular morphology.

12.4.5 Rheological Properties of Polylactic Acid–Based Composite

The rheological properties like shear viscosity are key to a thermal process like in-jection moulding, sheet forming, fiber spinning, film blowing, and paper coating. PLA is a non-Newtonian pseudoplastic fluid under high shear conditions and rheological models has been used to study its rheological properties. Thermodynamics, which is the study of the transfer and conversion of energy, to have a better understanding of PLA's processability is necessary to study its thermal and rheological properties [5,8].

12.5 ULTRAHIGH MOLECULAR WEIGHT-BASED COMPOSITES FOR BONE IMPLANTS

Ultrahigh molecular weight polyethylene (UHMWPE) is a polymer with unique properties such as excellent chemical, physical, and mechanical properties [9,31,32]. Most importantly, it is known for its lubricity, chemical inertness, impact resistance, and abrasion resistance [9,31,32]. Due to these unique properties UHMWPE is sui-table for a wide range of implants; for example, it can be utilised as a bearing in artificial joints for orthopaedics. Tensile strength and elastic modulus are the most important properties to be considered before determining their possible use as bio-materials, most importantly orthopaedic implants. UHMWPE has a low coefficient of friction and low implant wear and patient risk compared to other polymers. Almost every year approximately 1.4 million joints implant surgical operations are carried out in the world. However, even with success achieved, UHMWPE have only a limited lifetime. The implant's survival of UHMWPE components is dependent on wear and damage [9,31]. UHMWPE's structure is formed by a C=C covalent bond, which has both amorphous and crystalline arrangements. The crystalline phase contains highly oriented folded chain lamellae, with an orthorhombic crystal structure [33]. The la-mellae are randomly oriented within the amorphous phase with molecules linking individual lamellae to one another are 10–50 nm and 10–50 m thick and long, re-spectively. Enhancing UHMWPE and their composites' degree of crystallinity is important, and an integral dose of gamma radiation and high pressure of crystal-lisation can be used to affect them [9,31–34].

A lower coefficient of friction and an increase in resistance to scratch at both micro- and nano-scales can be obtained by increasing the degree of crystallinity of UHMWPE. They noted that a stiffer, harder, and less ductile behaviours in UHMWPE can be achieved by a higher degree of crystallinity. Wear resistance can be affected by the lamellar structure when dealing with nano-scale and lower wear depth and width are achieved with an increase in crystallinity [34,35]. Generally, like many other polymers, UHMWPE also displays the same three stages of thermal transitions as its glass transition temperature (Tg); 120°C is different from its melting point (130°C–135°C). Mechanically, UHMWPE has tensile strength ranges from 39 to 48 MPa. It experiences no break when subjected to impact loading, implying that it has an excellent impact energy or toughness. Its tensile elongation falls between 350% and 525%. Its coefficient of friction is within the range of 0.05–0.22 and its respective thermal capacity and conductivity are 1.7–1.8 and 0.41, as reported in [34].

12.5.1 ULTRAHIGH MOLECULAR WEIGHT POLYETHYLENE/HYDROXYAPATITE COMPOSITES

Ultrahigh molecular weight polyethylene and hydroxyapatite (HA) can be used to manufacture UHMWPE/HA composites, which are very attractive in biomedical applications because of their combined improved tribological, mechanical, and biological properties. Characterisation of all materials is very important; their ap-plications are to have a better understanding of their chemical composition and morphology [9,31].

12.5.2 GRAPHENE DERIVATIVE–REINFORCED ULTRAHIGH MOLECULAR WEIGHT POLYETHYLENE COMPOSITES

CNT and graphene are also materials used as reinforcements for UHMWPE in biomedical applications. UHMWPE/CNT composite and UHMWPE/graphene composite have been studied by several researchers. UHMWPE/graphene composites are biocompatible, and reveal a 0.1% survival rate of osteoblast morphology and survival rate of graphene nanopelets-(GNP). UHMWPE/fluorinated graphene (FG) composite has good biocompatibility [9,36]. Currently, because of the good mechanical and tribological properties, polymers like UHMWPE are mostly chosen for the fabrication of HA-based composites as biomaterials, when compared to most non-biodegradable polymers [9,36].

12.5.3   PROCESSING OF ULTRAHIGH MOLECULAR WEIGHT POLYETHYLENE/ HYDROXYAPATITE COMPOSITES

Several processing methods can be used to fabrication of UHMWPE/HA composites. The most common methods used are sol-gel, twin-screw extrusion, hot pressing, planetary ball milling, single screw extrusion, melting blending technique, biomimetic process, rotatory ball milling, and compression molding. The usual twin screw extrusion of UHMWPE composites is difficult because of the ultrahigh viscosity of UHMWPE melt. Addition of solvent brings about a reduction in viscosity of the swelling UHMWPE [9,31].

12.5.4   MECHANICAL PROPERTIES OF ULTRAHIGH MOLECULAR WEIGHT POLYETHYLENE COMPOSITES

The wear rate is the primary factor considered for selecting materials to be used as bone implants. Polyethylene possesses attractive properties for partial and total replacement of bone. Most times, prosthesis revision directly links to material wear; reinforcing organic/inorganic fillers in UHMWPE improves the material. Different types of organic/inorganic fillers including those that can be synthesised from agricultural waste show a reduction in the wear rate. Although, a lot of them possess low bioactivity, the use of HA, which is a bio-ceramic, is good for this purpose, due to its ability to form strong bond together with bone tissue. Furthermore, improving resistance to bioactivity and wear is helpful in producing high-quality artificial joints [9]. Universal Micro-Tribometer in a ball-on-disc configuration that can be used to investigate wear behaviour and friction of UHMWPE/bovine bone hydroxyapatite composites with human plasma lubrication [9].

Generally, low resistance to creep can affect the survival and function of an implant, hindering its further application as a biomaterial. Micro/nanoparticles inorganic material incorporated in a polymer matrix improves creep resistance. A hydroxyapatite (BHA) micro-particle addition to a polymer matrix has aided to improve the creep and hardness of the composite [9].

12.5.5   BIOCOMPATIBILITY OF ULTRAHIGH MOLECULAR WEIGHT POLYETHYLENE COMPOSITE

Ultrahigh molecular weight polyethylene/hydroxyapatite composites are used for bone regeneration processes and are known for their biocompatibility, osteoconductivity, and osteoinductivity, and should further be able to regulate interactions with cells and integration together with host tissues. HA are a highly biocompatible, bioactive, and non-reactive both in their natural and synthetic form; hence they are widely used in biological applications. HA is a substitute material for increasing alveolar ridges, repairing bony defects, and enhancing guided tissue regeneration. They can serve as a biomaterial in implantology (orthopaedic, dental, and maxillofacial applications) due to their chemical and crystallographic properties, which possess similarities with normal bone's mineral content. HA can be used as a reinforcement for biomedical materials to form composite due to its unique properties. UHMWPE's properties provide adequate conditions, bonding between it and bone directly producing a suitable biomaterial for bone replacement. Additionally, UHMWPE is frequently applied as a negative control for direct contact cytotoxicity tests because of their low cytotoxicity; due to this, UHMWPE and HA have been widely used to form composites. Adding HA to UHMWPE increases to facilitate biological fixation and micro-surface mechanical properties between the prosthesis and the hard tissue [9].

12.6 REVIEW OF CLINICAL PRACTICE ON ARTIFICIAL BONE IMPLANTS

Millions of people have already benefited from biomedical bone implants, which have turned out to improve and enhance human lives. Till date, biomaterial applications for the use of bone replacements have been on the increase due to the continuous and increasing practical needs of medicine and health care practices; there are currently several types of bone implants for replacements in the human body. Polymeric materials such as fibers, films, and rods have an extensive application for bone implantation because they possess unique properties, making their fabrication easy. Polymeric implants are not cancerous, toxic, or allergic, and possess elasticity, durability, and good strength. Without doubt, biomaterials have a great impact on patient care and contemporary medicine practice and have saved and improved the quality of human lives, and with present-day revolutions in technology, tremendous achievements will continue to be made in the field [37].

12.6.1 CLINICAL FACTORS FOR SUCCESSFUL IMPLANTS

The health of an individual is very important when considering the success or failure of an implant. Patients affected by suppressed immune systems, such as those using steroids on a long-term basis and diabetic patients, are vulnerable to higher risks of infections compared to nonsteroid users. Failure is likely to occur for a patient with a suppressed immune system because they are more prone to attack by the presence of an implant and opportunistic infections. Furthermore, failure mechanisms of implants in the geriatric population mostly might come from infection and other comorbid factors due to a poorer health system [37,38].

To reduce surgical errors, proficiency of surgeons is a necessary factor for a successful implant surgery. Failure to follow preoperative, intraoperative, and postoperative protocols can lead to unsuccessful implant surgery. Misorientation of an implant and simple mistakes during surgical operations can affect the wear

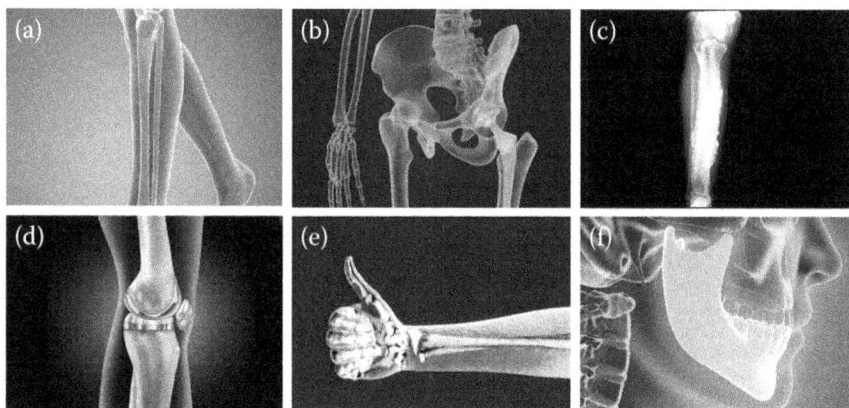

FIGURE 12.2 (a) Bio-friendly polymers and polymer nanocomposites implant for fibula and tibia (source: AGH University of Science and Technology), (b) 3D-printed implants (source: Shutterstock/Alex Mit), (c) bioactive glass implant from researchers in Ireland (source: National Institutes of Health), (d) plasma coating developed to decrease bone implant rejections (source: The University of Sydney), (e) drug-coated bone implants may help tackle postsurgery infections and bacteria (source: Indian Institute of Technology, Roorkee), (f) polymeric dental implant (source: Progressive Oral Surgery & Implantology of Long Island).

properties [37,38]. Figure 12.2 shows images of implants from different body parts that have undergone successful clinical operation.

12.6.1.1 Orthopaedic Implants

Orthopaedic implants aid to relieve problems affecting human bones and joints; they are commonly used in the treatment of osteoarthritis, bone fractures, spinal stenosis, scoliosis, and chronic pain and include a broad variety of rods, pins, plates, and screws that can be used to anchor fractured bones during the healing process. Metallic glasses based on Mg with Zn and Ca addition are prospective materials for biomedical implants [39].

A magnetic resonance imaging (MRI) machine is a very important piece of medical equipment for a patient having orthopaedic implants to study the musculoskeletal image in detail. Consequently, a lot of doubts have been observed due to an implant's migration and loosening; thermal damages are likely to occur when heating of the implant metal, affecting surrounding tissues; and distortion of the MRI scan that affects the imaging results. A study recently showed that the majority of the orthopaedic implants are inert to magnetic fields using a 1.0 Tesla MRI scanning machine. However, significant interaction between heel and fibular implants and MRI magnetic fields occurs at 7.0 Tesla [39].

12.6.1.2 Cosmetic Implants

Cosmetic implants are mostly prosthetics, which attempt to restore portions of the human body back to an acceptable aesthetic norm. They also serve as a follow-up after mastectomy, correcting disfiguration, and body modification (chin augmentation and buttock augmentation); and nose prosthesis, breast implant, injectable filler, and ocular prosthesis. Currently, the cosmetic procedures are any of the following: facial implants that involve inserting an implant on the bone or surgical exposure to deep tissue, pectoral implants, calf implants, breast implants (silane or silicone gel), and cheek implant.

Surgical operations may be used to put an implant on top of the cheekbones. The FDA has approved the use of dermal fillers or soft tissue fillers as a medical device implant to help create a fuller and smoother face appearance. Treatment of cheeks, nasolabial folds, lips, and increasing volume at the back of the hand is made possible. Reduction of lines, wrinkles, scars, and augment soft tissue contours can take place. Repetition of these treatments are important because of their degradable nature on the body. They are commercially available in gel form, such as hyaluronic acid, poly-L-lactic acid (PLLA), calcium hydroxyapetite, and polymethylmethacrylate beads (PMMA). They can be injected into the deeper layers of the skin and underlying soft tissue of the face [40,41].

12.6.2 Development of Implant Materials

An array of minimum bioreactive metals is usually used as implants, and implant materials are grouped depending on material used and their biological response generated when implanted in the human body [42].

12.6.2.1 Metallic Implants

Stainless steel was mainly used in earlier periods as a feasible implant material for arthroplastic surgery due to availability and processing ease. Alloying additions of Cr, Ni, and Mo addition to ferrous metals to produce alloys use for implants like 316L (ASTM F138) [43]. Primarily used in making temporary devices like fracture plates, hip nails, and screws but stainless high modulus was a deterrent. The need for alloys that were more biocompatible and corrosion and wear resistant began; cobalt-base alloys then became a suitable alloy for implants and prosthetic stems and load-bearing components were then fabricated from wrought alloys. CoCrMo alloys (ASTM F75 and F799) was still observed to possess inferior biocompatibility and higher modulus than what was desired for implant materials [43]. Titanium-based alloys began to become attractive in the early 1970s because of their lower modulus, excellent specific strength, higher corrosion resistance, and superior tissue compatibility [44]. Pure Ti (ASTM F67) became the primary candidate used due to its oxide having excellent osseointegration properties; this is because there was success in the bonding of human bone cells and growth on the TiO_2 layer. However, due to low strength, the implants were only used for some parts, like endosseous dental implants, hip cup shells, pacemaker cases, heart valve cages, and dental crown and bridges [45].

Generally, some metals and alloys are known for their biomechanical properties, making them suitable implant material that are easy to process with good finishing. A common sterilisation procedure can be used to sterilise metallic implants; as such they are easy to use. Due to advances in technology and little success using metals (stainless steel, gold, and cobalt-chromium), most of these metals and their alloys are now obsolete and are being replaced with new ones [42].

12.6.2.2   Polymeric Implants

Polymers are universally used materials for biomedical devices for dental, ortho-paedic, cardiovascular, and soft-tissue applications, including tissue engineering and drug delivery. They consist of macromolecules with a large number of repeat units of chains of atoms that are covalently bonded [46]. The polymers include natural rubber, cellulose, collagen, sutures, deoxyribonucleic acid, polypropylene (PP), polyethylene (PE), polyvinyl chloride (PVC), polyethylene terephthalate (PET), polycaprolactone (PCL), polyethylene glycol (PEG), polymethyl metha-crylate (PMMA), polytetrafluoroethylene (PTFE), and nylon [47].

Early use of methyl methacrylate resin as implants failed often [48]. However, in 1969, polymers were reported to be biologically tolerable substances by Hodosh [49]. Polymethacrylate tooth-replica implants brought about the concept for de-veloping polymer dental implant by Milton Hodosh. The polymethacrylate tooth-replica functions and appearance were restored [48]. Polymers were selected for the following reasons [49]: (i) polymer's physical characteristics could be altered based on their application because their composition can be easily changed, thereby forming porous or softer polymers; (ii) polymers can be reproduced due to easy manipulated; (iii) polymers do not generate microwaves or electrolytic current; (iv) polymers possess tissue attachment with fibrous connective; (v) polymer's micro-scopical evaluation is easy; and (vi) polymers are more esthetically pleasing. However, their disadvantages are: (i) poor mechanical properties, (ii) poor adhesion to living tissues, and (iii) adverse immunologic reactions.

12.6.2.3   Ceramics Implants

Ceramics used for implantation include glass ceramics and glasses are used for different biomedical applications, mostly in orthopaedic and dental prostheses. Bio-ceramic-tissue attachments can occur by morphological fixation; through physical attachment or fitting of inert ceramic to the tissue, or by biological fixation; when bone ingrowth and mechanical attachment occur in porous cera-mics or by bioactive fixation; through chemical bonding of bones with the dense, nonporous ceramic, or temporary attachment of resorbable ceramic that is finally replaced by bones [47]. Calcium phosphates are commonly used as bioactive ceramics and naturally occur in the human body and minerals. Ceramics found their application as surgical implant devices due to their good strength, inert behaviour, and physical properties like poor electrical conductivity and minimum thermal. High brittleness and poor ductile are responsible for the limitation of the use of ceramics [42].

12.6.3   Trend of Development in Clinical Implants

12.6.3.1   Ancient Era (AD 1000)

Implants were traced to ancient South American and Egyptian civilisations; an artificial tooth carved with dark stone was found in a skull from pre-Columbian era. Albucasis de Condue, an Arabian surgeon, is credited with a paper of transplants as a means of replacing missing teeth [50].

12.6.3.2 Foundational Period (1800–1910)

Endosseous oral implantology started in this era. Au was used in the shape of the tooth root by Maggiolo in 1809 and the use of teeth made of porcelain fitted with Pb-coated Pt posts were reported in 1887 by Harris. Zamenski, in 1890, reported the implantation of teeth made of porcelain, gutta-percha, and rubber and R.E. Payne placed Ag capsules in the tooth socket in 1898. Lambotte developed implants of Al, Ag, brass, red Cu, Mg, Au and soft steel plated with Au and Ni in 1900 [50].

12.6.3.3 Premodern Era (1901–1930)

R.E. Payne reported a technique of capsule implantation in 1901. A sholl-implanted porcelain tooth with a corrugated porcelain root was done in 1903. In 1913, the basket of iridium and 24-carat gold introduced into the alveoli; also introduced was the concept of submerged implant, the healing tissue, and dental implant immobility by Dr. E. J Greenfield [50].

12.6.3.4 Modern Era (1935–1978)

Ceramics, metal alloys, and synthetic polymers started replacing the naturally de-rived materials because of their better performance and more predictable results than the natural ones in this era. A vitallium screw within bone and immediately mounted with a porcelain crown was anchored by Strock. He was the first one to achieve an implant survival for more than one and half decades [50].

12.6.3.5 Implants in the 21st Century

The excellent biocompatibility of titanium and its alloys, due to the formation of a stable oxide layer on its surface, has led to it being successfully used as an implant material and the use of polymeric nanomaterials [50].

12.7 CONCLUSION AND SUMMARY

This chapter was focused on the use of polymeric nanocomposites for implants in different parts of the human body. This chapter discussed requirement for bioma-terials to be used as artificial implants in human beings, stressing the importance of any selected materials to be used as implants in the human body must be bio-compatible, with excellent mechanical properties and fabrication. Polymers such as polymethylmethacrylate (PMMA), polylactic acid (PLA), and ultrahigh molecular weight polyethylene (UHMWPE) have been used in the past as artificial implants in the body. However, to further improve these polymers, other biomaterials like me-tallic nanofillers, cellulose nanocrystals (CNCs), hydroxyapatite (HA), carbon na-notubes (CNT), graphene, fluorinated graphene (FG), high-density polyethylene (HDPE), multi-wall nanotube (MWNT), and bovine bone hydroxyapatite (BHA) produce nanocomposites for artificial implants. Most of these polymers mentioned can be obtained from agricultural wastes by processing them in polymeric nano-composites. The use of these agricultural wastes would reduce the amount of agri-cultural waste that is dumped in the environment and ensure cheaper artificial implants are fabricated. There was a brief review of the types of artificial implants and clinical practices, which are used to implant these replacements in the human body.

ACKNOWLEDGEMENTS

The authors are grateful to the Petroleum and Technology Development Fund (PTDF), Nigeria for funding this research work and University of Sao Paulo, Sao Paulo (USP), Brazil for their support for this research work.

REFERENCES

[1] Wong, J.Y., Bronzino, J.D., and Peterson, D.R., eds. (2012). *Biomaterials: Principles and Practices*. CRC Press, p. 281.

[2] Download Product Code Classification Files. FDA.org/medicaldevices. Food and Drug Administration. 4 November 2014.

[3] Misch, C.E. (1999). Contemporary Implant Dentistry. *Implant Dentistry*, **8**, 90.

[4] Manoukian, O.S., Sardashti, N., Stedman, T., Gailiunas, K., Ojha, A., Penalosa, A., Mancuso, C., Hobert, M., and Kumbar, S.G. (2019). Biomaterials for Tissue Engineering and Regenerative Medicine. *Encyclopedia of Biomedical Engineering*, **1**, 462–482. 10.1016/B978-0-12-801238-3.64098-9

[5] Liu, S., Qin, S., He, M., Zhou, D., Qin, Q., and Wang, H. (2020). Current Applications of Poly(Lactic Acid) Composites in Tissue Engineering and Drug Delivery. *Composites Part B*, **199**(108238), 1–25. 10.1016/j.compositesb.2020.108238

[6] Samavedia, S., Poindexterb, L.K., Dykec, M.V., and Goldstein, A.S. (2014). Synthetic Biomaterials for Regenerative Medicine Applications. Regenerative Medicine Applications in Organ Transplantation. *Principles of Regenerative Medicine and Cell, Tissue, and Organ Bioengineering*. 81–94.

[7] Webb, J.C.J. and Spencer, R.F. (2007). The Role of Polymethylmethacrylate Bone Cement in Modern Orthopedic Surgery. *Journal of Bone & Joint Surgery British*, **89**, 8517.

[8] Dedukh, N.V., Makarov, V.B., and Pavlov, A.D. (2019). Polylactide-Based Biomaterial and Its Use as Bone Implants (Analytical Literature Review). *Pain Joints Spine*, **9**(1), 28–35. doi: 10.22141/2224-1507.9.1.2019.163056

[9] Macuvele, D.L.P., Nones, J., Matsinhe, J.V., Lima, M.M., Soares, C., Fiori, M.A., and Riella, H.G. (2017). Advances in Ultrahigh Molecular Weight Polyethylene/ Hydroxyapatite Composites for Biomedical Applications: A Brief Review. *Materials Science and Engineering C*, **76**, 1248–1262. 10.1016/j.msec.2017.02.070

[10] Kumar, P., Vinitha, B., and Fathima, G. (2013). Bone Grafts in Dentistry. *Journal of Pharmacy & BioAllied Sciences*, **5**(1), S125–S127. Doi: 10.4103/0975-7406.113312

[11] Giannoudis, P.V., Dinopoulos, H., and Tsiridis, E. (2005). Bone Substitutes: An Update. *Injury*, **36**(3), S20–S27. 10.1016/j.injury.2005.07.029

[12] Laurencin, C., Khan, Y., and El-Amin, S.F. (2006). Bone Graft Substitutes. *Expert Rev Med Devices*, **3**(1), 49–57. 10.1586/17434440.3.1.49

[13] Titsinides, S., Agrogiannis, G., and Karatzas, T. (2019). Bone Grafting Materials in Dentoalveolar Reconstruction: A Comprehensive Review. *Japanese Dental Science Review* **55**, 26–32. 10.1016/j.jdsr.2018.09.003

[14] Lalwani, G., D'Agati, M., Farshid, B., and Sitharaman, B. (2016). Carbon and Inorganic Nanomaterial-Reinforced Polymeric Nanocomposites for Bone Tissue Engineering. *Nanocomposites for Musculoskeletal Tissue Regeneration*, Huinan Liu, Editor, Elsevier Ltd. Stony Brook University, Stony Brook, pp. 31–66. 10.1016/ B978-1-78242-452-9.00002-9

[15] Geesink, R.G., Groot, K., and Klein, C.P. (1988). Bonding of Bone to Apatite-Coated Implants. *The Journal of Bone and Joint Surgery, British*, **70-B**(1), 17–22. 10.1302/0301-620X.70B1.2828374

[16] Kashi, A. and Saha, S. (2010). Mechanisms of Failure of Medical Implants During Long-Term Use. *Biointegration of Medical Implant Materials*, Chandra Sharma, Editor, Elsevier, pp. 407–426.

[17] Saska, S., Mendes, L.S., Gaspar, A.M.M., and Capote, T.S.O. (2015). *Bone Substitute Materials in Implant Dentistry, Current Concepts in Dental Implantology*, IntechOpen, pp. 25–57. DOI: 10.5772/59487

[18] Jantova, S., Theiszova, M., Letasiova, S., Birosova, L., and Palou, T.M. (2008). In Vitro Effects of Fluorhydroxyapatite, Fluorapatite and Hydroxyapatite on Colony Formation, DNA Damage and Mutagenicity. *Mutation Research*, **652**, 139–144.

[19] Mitri, F., Alves, G., Fernandes, G., Konig, B., Rossi, A.J., and Granjeiro, J. (2012). Cytocompatibility of Porous Biphasic Calcium Phosphate Granules with Human Mesenchymal Cells by a Multiparametric Assay. *Artificial Organs*, **36**, 535–542.

[20] Lam, C.X.F., Hutmacher, D.W., Schantz, J.T., Woodruff, M.A., and Teoh, S.H. (2008a). Evaluation of Polycaprolactone Scaffold Degradation for 6 Months In Vitro and In Vivo. *Journal of Biomedical Materials Research*, **90A**(3), 906–919.

[21] Sayeed, Z., Padela, M.T., El-Othmani, M.M., and Saleh, K.J. (2017). Acrylic Bone Cements for Joint Replacement. *Biomedical Composites*, Luigi Ambrosio, Editor, Elsevier, 199–214. 10.1016/B978-0-08-100752-5.00009-3

[22] Jaeblon, T. (2010). Polymethylmethacrylate: Properties and Contemporary Uses in Orthopaedics. *Journal of American Academy of Orthopaedic Surgeons*, **18**, 297–305.

[23] Kuehn, K.D., Ege, W., and Gopp, U. (2005). Acrylic Bone Cements: Composition and Properties. *Orthopaedic Clinic North America*, **36**(1), 17–28. doi: 10.1016/j.ocl.2004.06.010

[24] Chehade, M., and Elder, M.J. (1997). Intraocular Lens Materials and Styles: A Review. *Australian and New Zealand Journal of Ophthalmology*, **25**(4), 255–263. doi: 10.1111/j.1442-9071.1997.tb01512.x

[25] Lewis, P.M., Ackland, H.M., Lowery, A.J., and Rosenfeld, J.V. (2015). Restoration of Vision in blind Individuals Using Bionic Devices: A Review with a Focus on Cortica Visual Prostheses. *Brain research*, **159**(5), 51–73.

[26] Zhang, Y., Xiong, Z., Ge, H., Ni, L., Zhang, T., Huo, S., Song, P., and Fang, Z. (2020). Core–Shell Bioderived Flame Retardants Based on Chitosan/Alginate Coated Ammonia Polyphosphate for Enhancing Flame Retardancy of Polylactic Acid. *ACS Sustainable Chemistry and Engineering*, **8**(16), 6402–6412. 10.1021/acssuschemeng.0c00634

[27] Xue, Y., Shen, M., Zheng, Y., Tao, W., Han, Y., Li, W., Song, P., and Wang, H. (2020). One-Pot Scalable Fabrication of an Oligomeric Phosphoramide towards High-Performance Flame Retardant Polylactic Acid with a Submicron-Grained Structure. *Composites Part B: Engineering*, **183**(107695), 1–12. 10.1016/j.compositesb.2019.107695.

[28] Ran, S., Fang, F., Guo, Z., Song, P., Cai, Y., Fang, Z., and Wang, H. (2019). Synthesis of Decorated Graphene with P, N-Containing Compounds and Its Flame Retardancy and Smoke Suppression Effects on Polylactic Acid. *Composites Part B: Engineering*, **170**, 41–50. 10.1016/j.compositesb.2019.04.037

[29] Diaz, E., Valle, M.B., Ribeiro, S., Lanceros-Mendez, S., and Barandiaran, J.M. (2018) Development of Magnetically Active Scaffolds for Bone Regeneration. *Nanomaterials*, **8**(678), 1–16. doi: 10.3390/nano8090678

[30] Wu, J., Chen, N., Bai, F., and Wang, Q. (2018). Preparation of Poly(Vinyl Alcohol)/Poly(Lactic Acid)/Hydroxyapatite Bioactive Nanocomposites for Fused Deposition Modeling. *Polymer Composite*, **39**, E508–E518. 10.1002/pc.24642

[31] Panin, S.V., Kornienko, L.A., Sonjaitham, N., Tchaikina, M.V., Sergeev, V.P., Ivanova, L.R., and Shilko, S.V. (2012). Wear-Resistant Ultrahigh-Molecular-WeiGht Polyethylene-Based Nano- and Microcomposites for Implants. *Journal of Nanobiotechnology*, **2012**, 1–12. doi: 10.1155/2012/729756

[32] Kurtz, S.M. (Ed.). (2004).*The UHMWPE Handbook: Ultra-High Molecular Weight Polyethylene in Total Joint Replacement*, Elsevier Academic Press.

[33] Park, J.B. and Lakes, R.S. (1992). *Biomaterials an Introduction*, Second ed., Plenum Press, pp. 394.

[34] Martínez-morlanes, M.J., Castell, P., Martínez-nogués, V., Martinez, M.T., Alonso, P.J., and Puértolas, J.A. (2011). Effects of Gamma-Irradiation on UHMWPE/ MWNT Nanocomposites. *Composites Science and Technology*, **71**(3), 282–288, 10.1016/j.compscitech.2010.11.013

[35] Bistolfi, A., Turell, M.B., Lee, Y.L., and Bellare, A. (2008). Tensile and Tribological Properties of High-Crystallinity Radiation Crosslinked UHMWPE. *Journal of Biomedical and Materials Research: Part B: Applied Biomaterials*, **90B**, 137–144, 10.1002/jbm.b.31265

[36] Zhang, Y., Tanner, K.E., Gurav, N., and Silvio, L.D. (2007). In Vitro Osteoblastic Response to 30 vol% Hydroxyapatite-Polyethylene Composite. *Journal of Biomedical and Materials Research: Part A*, **81**, 409–417. 10.1002/jbm.a.31078

[37] Jung, Y.H., Kim, J.U., Lee, J.S., Shin, J.H., Jung, W., Ok, J., and Kim, T. (2020). Injectable Biomedical Devices for Sensing and Stimulating Internal Body Organs. *Advanced Materials*, **32**(1907478), 1–25. DOI: 10.1002/adma.201907478

[38] Kashi, A. and Saha, S. (2020). Failure Mechanisms of Medical Implants and Their Effects on Outcomes. *Biointegration of Medical Implant Materials*, Chandra Sharma, Editor, pp. 407–426. 10.1016/B978-0-08-102680-9.00015-9

[39] Loeb, G.E. (2005). Galvani's Delayed Legacy: Neuromuscular Electrical Stimulation. *Expert Review of Medical Devices*, **2**(4), 379–381.

[40] Lipatov, V.A. (2021). Breast Implants in Cosmetic Surgery. *Journal of Medical Implants and Surgery*, **6**(3), 105.

[41] Lin, D.J., Wong, T.T., Ciavarra, G.A., and Kazam, J.K. (2017). Adventures and Misadventures in Plastic Surgery and Soft-Tissue Implants. *Radio Graphics*, **37**(7), 2145–2163. 10.1148/rg.2017170090

[42] Sykaras, N., Iacopino, A.M., Marker, V.A., Triplett, R.G., and Woody, R.D. (2000). Implant Materials, Designs, and Surface Topographies: Their Effect on Osseointegration. A Literature Review. *International Journal of Oral Maxillofacial Implants*, **15**, 675–690.

[43] Long, M.J. and Rack, H.J. (1998). Titanium Alloys in Total Joint Replacement—A Materials Science Perspective. *Biomaterials*, **19**, 1621–1639.

[44] Dowson, D. (1992). Friction and Wear of Medical Implants and Prosthetic Devices, Friction, Lubrication, and Wear Technology. *ASM Handbook, ASM International*, **18**, 656–664.

[45] Lee, C.M., Ho, W.F., Ju, C.P., and Chern Lin, J.H. (2002). Structure and Properties of Titanium-20Niobium-xIron Alloys, *Journal of Materials Science: Material Medical*, **13**, 695–700.

[46] Dee, K.C., Puleo, D.A., and Bizios, R. (2002). *An Introduction to Tissue-Biomaterial Interaction*, John Wiley and Sons.

[47] Ratner, B.D., Hoffman, A.S., Schoen, F.J., and Lemons, J.E. (2004). *Biomaterials Science: An Introduction to Materials in Medicine*, 2nd ed., Elsevier Academic Press.

[48] Waerhaug, J. and Zander, H.A. (1956). Implantation of Acrylic Roots in Tooth Sockets. *Oral Surgical, Oral Medicine & Oral Pathology*, **9**, 46–54.

[49] Hodosh, M., Povar, M., and Shklar, G. (1969). The Dental Polymer Implant Concept. *Journal of Prosthetic Dentist*, **22**, 371–380.

[50] Block, M.S., Kent, J.N., and Guerra, L.R. (1997). *Implants in Dentistry*, W.B. Saunders Company, Philadelphia, p. 4.

Index

For Product Safety Concerns and Information please contact our EU
representative GPSR@taylorandfrancis.com
Taylor & Francis Verlag GmbH, Kaufingerstraße 24, 80331 München, Germany

www.ingramcontent.com/pod-product-compliance
Lightning Source LLC
Chambersburg PA
CBHW060802220326
41598CB00022B/2520

9 780367 772703